Coping with Water Scarcity

Coping with Water Scarcity

Addressing the Challenges

By

Luis Santos Pereira

Agricultural Engineering Research Center, Institute of Agronomy,
Technical University of Lisbon, Portugal

Ian Cordery

School of Civil and Environmental Engineering, University
of New South Wales, Sydney, Australia

and

Iacovos Iacovides

I.A.CO Ltd., Environmental and Water Consultants, Nicosia, Cyprus

 Springer

Luis Santos Pereira
Universidade Técnica de Lisboa
Inst. Superior de Agronomia
Tapada da Ajuda
1349-017 Lisboa
Portugal
lspereira@isa.utl.pt

Ian Cordery
University of New South Wales
School of Civil &
Environmental Engineering
Sydney NSW 2052
Australia
i.cordery@unsw.edu.au

Iacovos Iacovides
I.A.CO Ltd.
Environmental & Water
Consultants
3 Stavrou Ave.
2035 Strovolos, Nicosia
Cyprus
iaco@cytanet.com.cy

This book is a new revised and extended edition based on the first edition of the document "Coping with Water Scarcity" by Luis Alberto Santos Pereira, Ian Cordery and Iacovos Iacovides published in 2002 via UNESCO, Paris as IHP-VI-Technical Documents in Hydrology- publication No. 58

Cover illustration: Dry tree in the savanna at sunset, photograph taken by Ana Vilar (olhares.com) in South Africa, 2007; and ceramic jars, photograph taken by Ana Vilar (olhares.com) in Mozambique, 2007

ISBN 978-1-4020-9578-8   e-ISBN 978-1-4020-9579-5

DOI 10.1007/978-1-4020-9579-5

Library of Congress Control Number: 2008942791

9 8 7 6 5 4 3 2 1

springer.com

Foreword

In the 21st century water scarcity is almost universal – something that is unfortunately either not recognised by or is concealed from the majority of the people. The world's population is growing steadily, the standard of living and the associated demand for water are both increasing, freshwater sources are becoming increasingly polluted and the climate is changing due to the enhanced greenhouse effect. As a result, the volumes of fresh water that have traditionally been available to many centres of industry, food production (irrigation) and urban population have been declining for some time. The quantity of naturally occurring, good quality fresh water available to each person throughout the world is decreasing daily.

The motive underlying the preparation of this monograph is an attempt to discuss all forms of water scarcity and to provide suggestions and guidelines for: (1) anticipating and preparing for water scarcity; (2) the way thinking about traditional practices in communities has to change, particularly among community leaders and politicians, to manage situations where water is becoming increasingly scarce; and (3) possible practical remedies for locations where water scarcity is already acute.

Thanks are due to UNESCO for authorizing the use of material that the authors produced in the context of the 6th International Hydrologic Program in the preparation of this book. We sincerely acknowledge the help provided by Ms. P. Paredes in editing the figures, tables and text. Also acknowledged is the support of Ms. V. Rodrigues and Ms. M. M. Gabriel. Thanks are also due to Dr. I. Bordi and Dr. E. Duarte for revising some sections and providing opportune advice. Finally, thanks are due to the publishers for their patience in allowing this book to see the light.

Luis S. Pereira, Ian Cordery and Iacovos Iacovides

Contents

Chapter 1
Introduction

Water scarcity is among the main problems to be faced by many societies and the World in the XXI century. Water scarcity is commonly defined as a situation when water availability in a country or in a region is below $1000\,m^3$/person/year. However, many regions in the World experience much more severe scarcity, living with less than $500\,m^3$/person/year, which could be considered severe water scarcity. The threshold of $2000\,m^3$/person/year is considered to indicate that a region is water stressed since under these conditions populations face very large problems when a drought occurs or when man-made water shortages are created. However, the concept of water availability based on indicators driven from the renewable water resources divided by the total population should be taken with great care. It is often the case that the renewable resource is augmented by desalination, non-renewable groundwater resources and wastewater re-use to compensate for their renewable water scarcity. Where there is little opportunity for irrigation, smaller per capita volumes may be adequate. In these cases a simple volume per person of renewable water may not be a good indicator of adequacy of supply.

Figure 1.1 shows the world water availability per capita and per country using the above mentioned indicator. It gives only a partial view of the problem because within the same country regions exist with high water scarcity together with others having abundant water resources. This is the case for most of South American countries mapped as regions of plenty.

Water scarcity causes enormous problems for the populations and societies. The available water is not sufficient for the production of food and for alleviating hunger and poverty in these regions, where quite often the population growth is larger than the capability for a sustainable use of the natural resources. The lack of water does not allow industrial, urban and tourism development to proceed without restrictions on water uses and allocation policies for other user sectors, particularly agriculture.

Natural fresh water bodies have limited capacity to respond to increased demands and to receive the pollutant charges of the effluents from expanding urban, industrial and agricultural uses. In regions of water scarcity the water resources are often already degraded, or subjected to processes of degradation in both quantity and quality, which adds to the shortage of water. Health problems are commonly associated with scarcity, not only because the deterioration of the groundwater and surface waters favours water borne diseases, but because poverty makes it difficult to

L.S. Pereira et al., *Coping with Water Scarcity*, DOI 10.1007/978-1-4020-9579-5_1,
© Springer Science+Business Media B.V. 2009

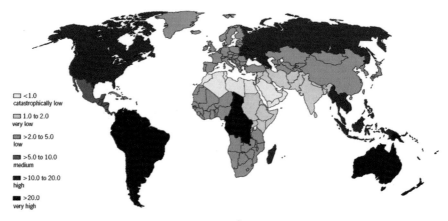

Fig. 1.1 World water availability per capita (1000 m^3/year) and per country (source: UNEP 2002a)

develop proper water distribution and sewerage systems. Water conflicts still arise in water stressed areas among local communities and between countries since sharing a very limited and essential resource is extremely difficult despite legal agreements. Poverty associated with water scarcity generates migratory fluxes of populations within countries or to other countries where people hope to have a better life, but where they may not be well received. Last, but not least, water for nature has become a low or very low priority in water stressed zones. Preserving natural ecosystems is often considered a superfluous use of water compared with other uses that directly relate to healthy human life, such as domestic and urban uses, or that may lead to the alleviation of poverty and hunger, such as uses in industry, energy and food production. However, the understanding that natural ecosystems, namely the respective genetic resources, are useful for society is growing, and an effort to protect reserve areas is already developing, even in water scarce regions.

Regions where water has been always scarce gave birth to civilisations that have been able to cope with water scarcity. These societies developed organisational and institutional solutions and water technologies and management skills within the local cultural environment that allowed for appropriate water for domestic use, food production and local industrial purposes. Lifestyle and development changes during the last decades created new needs for water, provided contradictory expectations on cultural and institutional issues, and led to very strong increases on the demand for water. The existing balances among demand and supply were broken and new equilibriums are required, mainly through the use of modern technologies and management tools which must be adapted to the local culture, environment, and institutions. Finding such new equilibriums is the challenge for the societies living in water stressed areas, and for professionals in a multitude of scientific and technological domains which impinge on the cultural, social and environmental facets of water resources. To explore some of the issues that may help in finding and implementing such equilibriums to cope with water scarcity is the objective of this book.

The book intends to call attention to problems and possible solutions to cope with water scarcity. It was first produced as a UNESCO IHP Technical Document to serve as a guide to support establishment of regional or local guidelines oriented to help in developing and implementing new conceptual and managerial ideas that may assist in coping with water scarcity (Pereira et al. 2002a). Meanwhile, other United Nations agencies also produced various publications relative to water scarcity such as FAO, UNDP and WHO. The basic idea behind the book is that water scarcity will continue to exist and, for many regions, unfortunately will continue to grow due to ever growing demands for water from increasing populations with rising standards of living and due to impacts of climate change. Human and societal skills will need to be developed to cope with water scarcity and to assist the local people to live in harmony with the environmental constraints, particularly those concerning water resources. This book does not produce an exhaustive review on every aspect covered but attempts to provide basic information to assist decision-makers, water managers, engineers, agronomists, economists, social scientists and other professionals to have coherent and hopefully harmonious consolidated views on the problems. Therefore, to help assist those who may need more detail, supporting references are given on most topics.

A range of concepts relative to water scarcity are first discussed. Natural and man-made water scarcity is first defined. Natural scarcity usually results from arid and semi-arid climates and drought, whereas man-made scarcity is associated with desertification and water management. This approach brings into focus the effects of water quality as well as water quantity on water scarcity. For example degradation of water quality makes unavailable a resource that otherwise could be available. The consideration of desertification shows how water scarcity can bring about the impoverishment of the land and other natural resources due to erosion, salinization and loss of fertility caused by human activities and climate variation in arid to sub-humid climates. This conceptual approach implies that measures and practices to cope with water scarcity need to be related to the causes, and therefore to the actual problems prevailing in any given region. An introductory discussion of the respective water management issues and implications for sustainable development are consequently presented.

The physical characteristics and processes leading to water scarcity complement the discussion of concepts. Climatic conditions dominant in water scarce regions are analysed, particularly rainfall variability and evaporation. The essential hydrologic characteristics are briefly reviewed, referring to the runoff regimes prevailing in water stressed regions, the processes affecting groundwater recharge and availability, and sediment loads and water quality, including those aspects leading to desertification and water-shortage. Special reference is made to droughts, particularly concerning definitions and the possibilities of forecasting. Also included are references to the need for data collection, data quality assessment and use of data handling and archiving techniques so that data are readily available for planning and management studies.

The impacts of droughts and desertification in many semi-arid and sub-humid areas are growing and the importance of respective processes is receiving increased

consideration. Therefore, an in-depth examination of drought concepts, identification indices, prediction, and forecasting is produced as well as revisiting monitoring
and risk management tools. Desertification concepts, indicators, and monitoring and
information tools are also dealt in the perspective of supporting the identification of
vulnerability domains and measures and practices that may combat desertification.

There is also an in-depth examination of conceptual thinking in coping with water
scarcity. This is oriented to provide a basis for innovation in examining the value
of water in conditions of water scarcity, to encourage different thinking in assessment of the social, environmental and economic values of water and to consequently
establish appropriate priorities for water allocation and use, taking such values into
account.

Ideas for maximising the availability of surface water for human use are proposed. These ideas are presented to support planning and management initiatives to
provide more water for domestic or industrial supply or for crop growth for production of food and fibre. These refer to both large and small scale projects, including
water harvesting. Particular attention is given to reservoir management. The need for
reservoirs in water scarce regions is well identified despite the controversy on the
environmental impacts of river dams in less stressed environments. Requirements
for water scarcity management, relative to operation of single and multiple reservoir
systems, including for groundwater recharge, as well as implications for design and
management of water resource systems are discussed. Attention is called to the control of water losses and non-beneficial uses of water, namely concerning the location
of losses, the reduction of evaporation, and the encouragement needed for reduction
of wastes. Water harvesting as a main tool for rainfall and runoff conservation in
arid and semiarid zones is reviewed under a variety of situations, such as rainwater
collection, and use of terracing, small dams and tanks, runoff enhancement, runoff
collection, flood spreading, and water holes and ponds. The environmental issues
relative to surface water storage and use are also discussed, embracing the protection of stored water, the control of sediments and water quality, the impacts on the
riparian ecosystems and combat of water borne diseases.

Principal aspects of groundwater use and recharge are reviewed in the perspective of achieving sustainable groundwater development, i.e. avoiding groundwater
degradation. The analysis concerns both major aquifers and well fields and minor
aquifers of local importance, such as in islands. It includes brief reviews on groundwater reservoir characteristics, their conditions for discharge, replenishment and
storage, issues for exploitation and management, as well as requirements for aquifer
monitoring and control and maintenance of wells, pumps and other facilities. Attention is paid to the effects and environmental impacts of aquifer over-exploitation
upon groundwater levels, energy costs for pumping, water quality deterioration
resulting from sea water intrusion, land subsidence, reduction of surface stream
base-flow, drying of wetlands and impacts on landscapes. Also attention is given
to artificial recharge needs and methods, and related problems and solutions, particularly environmental impacts. The need for conjunctive use of surface- and
groundwater is analysed.

Taking account of its importance in water scarce regions, the use of non-conventional water resources to complement or replace the use of usual sources of fresh water is discussed. The use of wastewater is considered, mainly for irrigation purposes, but other less stringent uses such as for aquaculture are mentioned. Attention is paid to the need to know the characteristics of wastewater and effluents, and particularly of the possible occurrence of toxic substances and pathogens in wastewaters.

The examination pays particular attention to the control of health risks for both consumers and workers, and so it includes an overview on wastewater treatment measures and practices used to minimise health hazards, and issues in monitoring and control for safe wastewater use. Another resource important for irrigation in water scarce regions is saline water. As for wastewaters, the need to know the characteristics and impacts of saline waters is stressed, together with a need to adopt appropriate criteria and standards for assessing the suitability of that water for irrigation. Related practices for crop irrigation management, including leaching requirements to control the impacts on soil salinity as well requirements for monitoring and evaluation are discussed. The use and treatment of desalinated water for non-agricultural uses are also analysed. Other non-conventional resources considered are fog water capture, water harvesting, groundwater harvesting, cloud seeding, and water transfers.

Water use concepts and performance that may be useful to analyse water conservation and saving are dealt with some detail. New indicators are proposed in view of consideration of water reuse and, in particular, to identify beneficial and non-beneficial water uses. An analysis of water productivity concepts useful not only in irrigation but for other uses is also included. These approaches permit to better establish the concept of efficient water use and to deal with water conservation and saving relative to the various water scarcity regimes.

Water conservation and water saving practices and management constitute a central, extended part of this book, consisting of technical guidelines oriented to practice. The importance of water conservation and water savings in coping with water scarcity is reviewed, considering particularly the scarcity conditions resulting from aridity, drought, desertification and man induced water shortage. The main concepts relative to water use, consumptive use and water losses are discussed prior to the introduction of the topics of water conservation, water saving practices and management issues. These refer to uses in urban areas, domestic water uses, landscape and recreational uses, industrial and energy uses, and rain-fed and irrigated agriculture. Since irrigated agriculture is the largest user of water, even in many water scarce regions, considerable space is devoted to water saving practices for irrigation. It is pointed out that there is much scope for saving of water in irrigation by good management both at the individual farm and irrigation district levels. A detailed discussion is presented covering opportunities for water saving at both levels. This is accompanied by complementary presentation of developments in irrigation technology and of the skills and knowledge needed to successfully operate and maintain the newer technologies. Essentially the focus is on demand reduction and water loss/waste control practices and tools. Complementarily, irrigation water supply management is discussed to emphasise measures that provide for operation and water allocation aimed at higher system performance and reduced impacts of water saving on users. An overview of water conservation and saving under severe water scarcity conditions, such as may occur during drought completes this technical subject.

Complementing the engineering aspects referred above, social, economic, cultural, legal and institutional constraints and issues are discussed with respect to water scarcity. These issues need to be considered for each local community, urban centre, rural areas, user group, and administrative, public and private organisation which are concerned with water supply.

The educational aspects constitute the last part of this guide. The main purpose of education is to change people's attitudes to water. Therefore the focus of the discussion is to consider how attitudes can be changed and how an educational program may be established. Education is meant in the widest sense, aimed at children and youths, women, with their role in the family and the community, farmers and industrial water users, managers, operational and maintenance personnel, educators, agronomists and engineers. The need for innovation in developing public awareness of water scarcity issues is also stressed.

Chapter 2
Water Scarcity Concepts

Abstract A range of concepts relative to water scarcity are discussed in this chapter. Natural scarcity resulting from arid and semi-arid climates and drought is one issue, and a second is man-made scarcity which is associated with desertification and water shortage due to poor management. Separation of these two causes brings into focus the effects of both water quality and water quantity on water scarcity. An introductory discussion of the respective water management issues and implications for sustainable development are consequently presented.

2.1 Concepts

2.1.1 Introduction

At the beginning of the XXI century, the sustainable use of water is not only a priority question for water scarce regions and for agriculture in particular, but for all sectors and regions. Imbalances between availability and demand, degradation of surface and groundwater quality, inter-sectoral competition, inter-regional and international conflicts, all bring water issues to the foreground.

Decades ago, water was viewed as a non-limited natural resource because it was renewed every year in the course of the seasons. Man progressively appropriated this resource and used it with few restrictions. Developments in controlling and diverting surface waters, exploring groundwater, and in using the resources for a variety of purposes have been undertaken without sufficient care being given to conserving the natural resource, avoiding wastes and misuse, and preserving the quality of the resource. Thus, nowadays, water is becoming scarce not only in arid and drought prone areas, but also in regions where rainfall is relatively abundant. Scarcity is now viewed under the perspective of the quantities available for economic and social uses, as well as in relation to water requirements for natural and man-made ecosystems. The concept of scarcity also embraces the quality of water because degraded water resources are unavailable or at best only marginally available for use in human and natural systems.

Figure 2.1 illustrates the causes for water scarcity, which may be natural, dominated by climate features, or man-made. Pollution and contamination degrade the

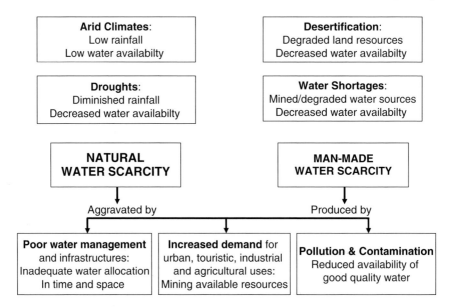

Fig. 2.1 Natural and man-made water scarcity

water quality and lead to water unavailability for many uses. Degradation of land alters hydrological processes that negatively influence water availability for humans and the environment. Demand may grow much in excess of availability. In other words, natural scarcity may be aggravated by human influences, such as population growth and poor water management. Man-made water scarcity is a consequence of these and other human activities.

Worldwide, agriculture is the sector which has the highest demand for water. As a result of its large water use irrigated agriculture is often considered the main cause for water scarcity. Irrigation is accused of misuse of water, of producing excessive water wastes and of degrading water quality. However, irrigated agriculture provides the livelihood of an enormous part of the world rural population and supplies a large portion of the world's food. At present, irrigated agriculture is largely restricted by the scarcity of water resources. Efforts from funding agencies and managers are continuously providing incentives to innovate and develop management practices to improve water management, to control the negative impacts of irrigation, to diversify water uses in irrigation projects, and to increase yields and farmers' incomes. Similarly, great progress in engineering and economic management are producing new considerations for water use and water quality control for non agricultural purposes, particularly for domestic consumption and sanitation.

The sustainable use of water implies resource conservation, environmental friendliness, technological appropriateness, economic viability, and social acceptability of development issues. The adoption of these sustainability facets is a priority for using water in every human, economic and social activity – human and domestic consumption, agriculture, industry, energy production, recreational and leisure uses – particularly in water scarce regions. The perception that water is increasingly scarce gives to water management a great relevance.

2.1.2 Causes of Water Scarcity

Water scarcity may result from a range of phenomena. These may be produced by natural causes, may be induced by human activities, or may result from the interaction of both, as indicated in Table 2.1.

Aridity is a natural permanent imbalance in the water availability consisting in low average annual precipitation, with high spatial and temporal variability, resulting in overall low moisture and low carrying capacity of the ecosystems.

Aridity may be defined through climatological indices such as the Thornthwaite moisture index, the Budyko radiation index of dryness, or the UNESCO precipitation/evapotranspiration index (Sanderson et al. 1990). Under aridity, extreme variations of temperatures occur, and the hydrologic regimes are characterised by large variations in discharges, flash floods and long periods with very low or zero flows.

The current increased water demand in arid, semi-arid and sub-humid climates exacerbates the natural low water availability and makes water scarcity visible to all. Related problems increase when poor water management such as uncontrolled pollution and contamination or inadequate water allocation policies lead to reduced availability and inequities in access to water.

Drought is a natural but temporary imbalance of water availability, consisting of a persistent lower-than-average precipitation, of uncertain frequency, duration and severity, of unpredictable or difficult to predict occurrence, resulting in diminished water resources availability, and reduced carrying capacity of the ecosystems. Precipitation in all regions is variable over time. Drought is the occurrence of a low precipitation part of this naturally variable time series.

Many other definitions of drought exist (Yevjevich et al. 1983, Wilhite and Glantz 1987, Tate and Gustard 2000). The U. S. Weather Bureau defined drought (Dracup et al. 1980) as a lack of rainfall so great and so long continued as to affect injuriously the plant and animal life of a place and to deplete water supplies both for domestic purposes and the operation of power plants, especially in those regions where rainfall is normally sufficient for such purposes. Generally, these definitions clearly state that drought is mainly due to the break down of the rainfall regime, which causes a series of consequences, including agricultural and hydrological hazards which result from the severity and duration of the lack of rainfall.

Drought impacts are higher when the demand is close to, or even higher than the long term average availability of water. Drought effects are especially severe when the water resource development is limited and there is not sufficient water storage

Table 2.1 Water scarcity regimens

Water scarcity regime	Nature produced	Man induced
Permanent	**Aridity**	**Desertification**
Temporary	**Drought**	**Water shortage**

capacity to have stored water from high flow periods to supplement the low flows during drought. Drought impacts are also greater when pollution and poor water management negatively impact the access to water sources.

It is important to recognise the less predictable characteristics of droughts, with respect to both their initiation and termination, as well as their severity. These characteristics make drought both a hazard and a disaster. Drought is a hazard because it is an accident of unpredictable occurrence, part of the naturally variable climate system and that it occurs with some known or recognised frequency. Drought can be a disaster because it corresponds to the failure of the precipitation regime, causing the disruption of the water supply to the natural and agricultural ecosystems as well as to other human activities.

Desertification is a man-induced permanent imbalance in the availability of water, which, combined with damaged soil, inappropriate land use and mining of groundwater, can result in increased flash flooding, loss of riparian ecosystems and a deterioration of the carrying capacity of the land system.

Soil erosion and salinity are commonly associated with desertification. Climate change also contributes to desertification, which occurs in arid, semi-arid and sub-humid climates. Drought can also strongly aggravate the process of desertification by temporarily increasing the human pressure on the diminished surface and groundwater resources. Different definitions are used for desertification, generally focusing on land degradation, as in the case of the definition proposed by the United Nations Convention to Combat Desertification: land degradation in arid, semi-arid and dry sub-humid zones resulting from various factors including climatic variations and human activities. In this definition land is understood as territory and is not restricted to agricultural land since desertification causes and impacts do not relate only to the agricultural activities but are much wider, affecting and affected by many human activities, nature and the overall living conditions of populations. However, these definitions need to be broadened in scope to focus attention on the water scarcity issues. When dealing with water scarcity situations, it seems more appropriate to define desertification in relation to the water imbalance produced by the misuse of water and soil resources, so calling attention to the fact that the misuse of water is clearly a cause of desertification.

Some definitions of desertification focus only on land degradation and do not refer to water. However, when dealing with water scarce situations it seems more appropriate to define desertification in relation to the water and ecological imbalances produced by the misuse of water and land resources, thus calling attention to the fact that desertification, including land degradation, is both a cause of, and may be caused by man-made water scarcity.

Water shortage is also a man-induced, sometimes temporary water imbalance including groundwater and surface waters over-exploitation resulting from attempts to use more than the natural supply, or from degraded water quality, which is often associated with disturbed land use and altered carrying capacity of the ecosystems. For example withdrawals may exceed groundwater recharge, surface reservoirs may be of inadequate capacity and land use may have changed, revising the local ecosystem and altering the infiltration and runoff characteristics. Degraded water quality is often associated with water shortages and exacerbates the effects of demand being near to or greater than the natural supply. There is no widely accepted definition for this water scarce regime and the term "water shortage" is often used synonymously with water scarcity. However it is important to recognise that water scarcity can result from human activity, either by over-use of the natural supply or by degradation of the water quality. This man-induced water scarcity is common in semi-arid and sub-humid regions where population and economic forces may make large demands on the local water resource, and where insufficient care is taken to protect the quality of the precious resource. In these circumstances it is also common for there to be inadequate attention given to provision for equity in allocation among competing users, including for environmental uses.

2.1.3 Aridity and Drought Water Scarcity

In terms of water management, regions having dry climates due to natural aridity are often not distinguished from drought prone areas. This rough approximation of definitions can be misleading in water management.

In arid regions rainfall is low all the year around and is particularly lacking during the dry season, which may last for several months. The natural vegetation and ecosystems may be of the desert or steppe type in arid zones, or of savannah and chaparral types in semiarid zones. Rain-fed agriculture is uncertain in arid climates, but is generally viable under semi-arid conditions, especially when the use of rainfall is maximised, as for crops that are suited to the season of occurrence of the rain (e.g. monsoon season crops in the north of the Indian sub-continent, or winter crops in Mediterranean climates). However, to achieve higher yields and to produce vegetables and food and fibre cash crops in these climatic conditions, irrigation is often required. Irrigation in arid and semi-arid zones has been practised for millennia, supporting ancient civilisations in very densely populated areas of the world.

Peoples and nations in arid areas developed appropriate skills to use and manage water in a sustainable way. These included measures to avoid waste of water, the adoption of technologies appropriate to the prevailing conditions, making successful use of water for agriculture and other productive activities, and the development of institutional and regulatory conditions that largely influenced the behaviour of rural and urban societies. These adaptations to live with limited water availability also included measures and practices to cope with extreme scarcity conditions when drought aggravated the limited water supplies. In the civilisations that survived the present conditions correspond to an integration of a strong cultural heritage with modern management, engineering, social and institutional practice to respond to the challenges of development. In part due to the rapidity of recent population explosions, this integration is not totally satisfactory due to the contradictory forces of progress, and thus innovative thinking is required to make development sustainable.

Water scarcity due to drought needs appropriate approaches. To successfully cope with drought there is a need to understand the characteristics and consequences of those phenomena which make water scarcity due to drought very different from that caused by aridity. Dealing with water scarcity situations resulting from aridity usually requires the establishment of engineering and management measures that produce conservation and perhaps the seasonal augmentation of the available resource. On the other hand, droughts require the development and implementation of preparedness and emergency measures, in other words, risk management strategies.

Differences in the perception of drought lead to the adoption of different definitions (Fig. 2.2), which do not have general acceptance, nor have worldwide applicability, as reviewed by Wilhite and Glantz (1987) and Tate and Gustard (2000).

As shown in Fig. 2.2, precipitation deficits are first detected and cause meteorological droughts. Next detected are soil water deficits due to lack of rainfall to refill the soil water storage, thus producing the so-called agricultural droughts. When precipitation deficits continue the river flow regimes are affected and a hydrological drought occurs. Continuing this situation surface water storage is heavily affected causing a water supply drought. Last to be perceived are the ground-water deficits usually associated with long periods of below average precipitation. The time to perceive those deficits relates to the hydrologic processes involved, which require different time durations, less to deplete soil water storage, much more to impact storage in large aquifer systems. Perception varies then with the nature of those processes in relation to environmental and physical conditions of the area, and with the nature of water use activities affected. A farmer quickly perceives the onset of a drought while an urban citizen may not perceive it until water is not available at home.

The controversy over perceptions of drought, and the consequent uncertainty in defining them and their characteristics, do not help decision and policy makers to

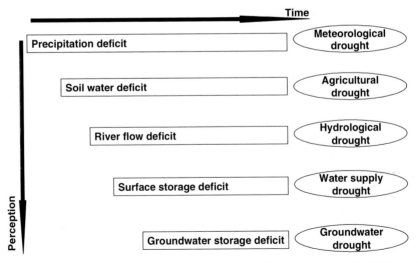

Fig. 2.2 Differences in perception of drought that lead to different concepts and definitions of drought

plan for droughts. Lack of clearly agreed definitions makes it difficult to implement preparedness measures, to apply timely mitigation measures when a drought occurs, or to adequately evaluate drought impacts.

Several authors point out that scientists, engineers, professionals, and decision-makers often do not agree on whether to regard drought as a hazard or as a disaster. As stated by Grigg and Vlachos (1990), this difference in perception is one of the central problems of water management for drought. As discussed above, drought is a hazard and a disaster. It is a hazard because it is a natural incident of unpredictable occurrence, and it is a disaster, because it corresponds to the failure of the precipitation regime to supply what is expected or taken for granted, causing disruption of water supply to the natural and agricultural ecosystems as well as to human and social activities.

The definition adopted herein (stated in Section 2.1.2) accommodates a variety of conditions and corresponds to the concepts of both hazard and disaster. That definition also allows a clear distinction between drought and the other water scarce regimes defined above.

2.2   Coping with Water Scarcity

2.2.1   Main Problems in Brief

Water availability in the different regions of the world is indicated in Table 2.2. Data shows that availability varies widely as a function of climate and population density. The lowest annual potential water availability per capita is for North Africa ($710\,m^3$/capita) where water scarcity is very high. Other regions affected by water scarcity are North China and Mongolia, Southern Asia and Western Asia (Near East). All these regions have an arid or semi-arid climate and high growth of population. All these regions had ancient civilizations that had developed over centuries.

In addition to these regions, potential water availability per capita is also low in Southern and Central Europe, mainly due to highly dense population, and the Southern European part of the former Soviet Union and Central Asia and Kazakh-stan, where climate aridity is dominant.

In most regions of the world, the annual withdrawal or use of water is a relatively small part (less than 20%) of the total annual internally renewable water resources. In water scarce regions, as is the case for the Middle East and North Africa regions, that share averages 73% of the total water resources (The World Bank 1992). This situation illustrates that not only the potential water availability in these regions is low but that sustaining an increased demand requires innovative approaches to cope with water scarcity. This also happens in other countries and sub-regions situated in regions where the overall potential water availability is higher, as it is the case for the densely populated Yellow River basin and North China Plain, the northeast of Brazil, western Peru and northern Mexico. An analysis at region or country level is not sufficient to detect the areas where water scarcity challenges are high but an in-depth analysis is required.

Table 2.2 Potential water availability by regions of the world (adapted from Shiklomanov and Rodda 2003)

	Area (10^6 km^2)	Population by 1994 (10^6 hab.)	Potential water availability (10^3 m^3/year)	
			per km^2	per capita
Europe	**10.46**	**684.7**	**277**	**4.24**
Northern	1.32	23.2	534	30.4
Central	1.86	293	333	2.12
Southern	1.79	188	335	3.19
North of Former Soviet Union	2.71	28.5	222	21.1
South of Former Soviet Union	2.78	152	181	3.32
North & Central America	**24.3**	**453**	**325**	**17.4**
Canada and Alaska	13.67	29	369	174
USA	7.84	261	234	7.03
Central America & Caribbean	2.74	163	406	6.82
Africa	**30.1**	**708**	**135**	**5.72**
Northern	8.78	157	12.6	0.71
Southern	5.11	83.5	86.5	5.29
East	5.17	193.5	147	3.94
West	6.96	211.3	158	5.22
Central	4.08	62.8	444	28.8
Asia	**43.5**	**3445**	**311**	**3.92**
North China & Mongolia	8.29	482	124	2.13
Southern	4.49	1214	476	1.76
Western	6.82	232	71.8	2.11
South East	6.95	1404	965	4.78
Central Asia & Kazakhstan	3.99	54	51.1	3.78
Siberia & Far East of Russia	12.76	42	252	76.6
Transcaucasia	0.19	16	390	4.63
South America	**17.9**	**314.5**	**672**	**38.3**
Northern	2.55	57.3	1310	58.3
Eastern	8.51	159.1	843	45.1
Western	2.33	48.6	738	35.4
Central	4.46	49.4	249	22.5
Australia & Oceania	**8.95**	**28.7**	**269**	**83.8**
Australia	7.68	17.9	45.8	19.7
Oceania	1.27	10.8	1614	190
The World	**135**	**5634**	**316**	**7.59**

In Middle East and North African regions near 53% of the per capita annual withdrawals for all uses, including irrigation, are below 1000 m^3/year and 18% are between 1000 and 2000 m^3/year. In Sub-Saharan countries, 24% of the population lives in areas where annual withdrawals are below 2000 m^3/person/year and 8% below 1000 m^3. The importance of water scarcity in these regions is obvious when it is noted that estimates for the average annual growth of the population in these regions are among the World's highest: 2.9% for 1990–2000, 2.3% for 2000–2035 for the Middle East and North Africa, and 3.0% and 2.4% for the same periods in Sub-Saharan Africa (Table 2.3). Forecasts for the next decades show that water scarcity or water stress may affect a very large number of countries (e.g. Engelman and LeRoy 1993).

Table 2.3 Population and average annual growth (adapted from: The World Bank 1992)

Country group	Population (millions)				Average annual growth (percent)		
	1980	1990	2000	2030	1980–1990	1990–2000	2000–2030
Low and middle income	3383	4146	4981	7441	2.0	1.9	1.4
Sub-Saharan Africa	366	495	668	1346	3.1	3.0	2.4
East-Asia & Pacific	1347	1577	1818	2378	1.6	1.4	0.9
South Asia	919	1148	1377	1978	2.2	1.8	1.1
Europe	182	200	217	258	1.0	0.8	0.6
Middle East & North Africa	189	256	341	674	3.1	2.9	2.3
Latin America & the Caribbean	352	433	516	731	2.1	1.8	1.2
High income	766	816	859	919	0.6	0.5	0.2
OECD members	733	777	814	863	0.6	0.5	0.2
World	4443	5284	6185	8869	1.7	1.6	1.2

Table 2.4 Dynamics of water use and consumption by continents (km^3/year) (adapted from: Shiklomanov and Rodda 2003)

	Assessment				Forecast	
	1900	1950	1980	1995	2010	2025
Water withdrawal						
Europe	37.5	136	449	455	535	559
N. America	69.6	287	676	686	744	786
Africa	40.7	55.8	166	219	275	337
Asia	414	843	1742	2231	2628	3254
S. America	15.1	49.3	117	167	213	260
Australia & Oceania	1.60	10.4	23.5	30.4	35.7	39.5
Total	579	1382	3175	3788	4431	5235
Water consumption						
Europe	13.8	50.5	177	189	234	256
N. America	29.2	104	221	237	255	269
Africa	27.5	37.8	124	160	191	220
Asia	249	540	1084	1381	1593	1876
S. America	10.8	31.7	66.7	89.4	106	120
Australia & Oceania	0.58	5.04	12.7	17.5	20.4	22.3
Total	331	768	1686	2074	2399	2764

Data on the dynamics of water use and consumption (Table 2.4 and Fig. 2.3) shows there has been enormous growth since 1900. During the 20th century, water withdrawals increased by more than 29 times in Australia and Oceania, 12 times in Europe, 10 times in North and South America and 5 times elsewhere. At the World scale withdrawals have grown by nearly 7 times. This expansion of water use has given rise to the idea that water is scarce, a concept that was hardly relevant 50 years ago. Moreover, looking to these numbers on a global scale, it becomes apparent that

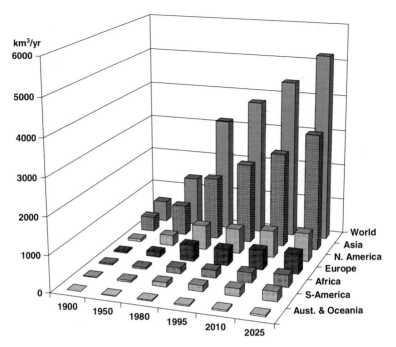

Fig. 2.3 Water withdrawals per continent from 1900 to 2025 (source: Shiklomanov and Rodda 2003)

water scarcity is already a worldwide problem. However, there are many regions of the globe where potential water availability is above $10000\,\text{m}^3$/capita/year and where withdrawals are below 20% of the internal renewable resource. In these regions the problem is just a question of improved water management (including water quality) and allocation while for water stressed regions the problem is to find innovative approaches to cope with water scarcity.

Agriculture is the sector with by far the highest demand for water (Table 2.5 and Fig. 2.4). Withdrawals for irrigation may increase by 6 times from 1900 to 2025, and related water consumption is foreseen to increase by 7 times, since technologies help to decrease the non-consumed fraction of water use. However, the demands for industrial and domestic water uses are increasing much faster than for irrigation: withdrawals for industry are foreseen to increase in the same period by nearly 30 times and those for domestic uses by more than 26 times.

Agriculture will continue to be the largest water user and mainly responsible for non-point pollution with nitrates and agro-chemicals despite recent technological efforts to decrease the demand for water and to control pollution. Industry, including mining, and urban water uses constitute the main source for point pollution and contamination, with very serious impacts on water quality and health when effluents are disposed to the water courses without appropriate treatment, so causing the above mentioned water shortage problems.

Table 2.5 Dynamics of water use and consumption (km³/year) in the World by sector of economic activity (adapted from Shiklomanov and Rodda 2003)

Sector	Assessment			Forecast		
	1900	1950	1980	1995	2010	2025
Population (million)		2542	4410	5735	7113	7877
Irrigated land area (Mha)	47.3	101	198	253	288	329
Water withdrawal						
Agricultural Use	513	1080	2112	2504	2817	3189
Industrial Use	21.5	86.7	219	344	472	607
Municipal Use	43.7	204	713	752	908	1170
Reservoirs	0.30	11.1	131	188	235	269
Total	579	1382	3175	3788	4431	5235
Water consumption						
Agricultural Use	321	722	1445	1753	1987	2252
Industrial Use	4.61	16.7	38.3	49.8	60.8	74.1
Municipal Use	4.81	19.1	70.9	82.6	117	169
Total	331	768	1686	2074	2399	2764

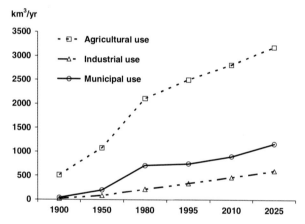

Fig. 2.4 Water withdrawals by economic sector from 1900 to 2025 (source: Shiklomanov and Rodda 2003)

Agriculture takes the highest share among water user sectors in Asia, Africa and South America, with generally more than 60% of water withdrawals at present (Table 2.6). The trend is however to decrease to values near to 40% except in Asia, where irrigation water withdrawals are foreseen to be above 70% in 2025. The share of water consumption is higher because the non-consumed fraction of water use is larger in industry and domestic uses. These continents, as referred above, include regions where water scarcity is more severe. Water withdrawals for domestic uses are the highest in Europe and North America, where they may reach more than 40% by 2025 (Table 2.6) due both to dominance of temperate and humid climates, and not having high demand for irrigation.

Table 2.6 The ratio of water withdrawal and consumption to total water withdrawal and consumption by sectors of economic activity (%) by continents (adapted from Shiklomanov and Rodda 2003)

Continent	1950			1995			2025		
	Agr	Ind	Dom	Agr	Ind	Dom	Agr	Ind	Dom
Water withdrawal									
Europe	32.2	25.4	41.2	37.4	14.7	44.8	37.2	14.0	45.8
N. America	53.5	7.9	36.0	43.5	10.7	41.5	41.4	12.3	41.3
Africa	90.5	7.0	2.6	63.0	8.1	4.4	53.1	18.0	6.0
Asia	93.4	2.4	4.2	80.0	6.9	9.9	72.0	9.5	15.2
S. America	82.4	9.5	7.9	58.6	17.2	15.4	44.2	22.7	23.8
Australia & Oceania	50.0	7.2	39.4	51.0	10.9	23.5	46.8	11.3	26.1
World	*78.1*	*6.3*	*14.8*	*66.1*	*9.1*	*19.9*	*60.9*	*11.6*	*22.3*
Water consumption									
Europe	67.7	12.6	15.6	71.4	5.6	15.3	66.8	4.3	22.3
N. America	83.5	4.7	3.6	75.1	5.0	7.2	72.4	6.0	7.5
Africa	97.9	1.6	0.5	63.8	1.5	0.8	60.5	3.4	1.3
Asia	98.0	0.7	1.1	91.0	1.5	2.3	88.4	1.8	4.1
S. America	95.0	2.5	1.9	76.4	4.0	3.2	67.4	4.7	8.3
Australia & Oceania	81.3	2.0	9.9	69.1	2.2	3.1	64.1	2.1	6.4
World	*94.0*	*2.2*	*2.5*	*84.5*	*2.4*	*4.0*	*81.5*	*2.7*	*6.1*

The general welfare and standard of living expectations which are developing in these regions will also increase the water demand (Fig. 2.5). Water withdrawals for industry are relevant in South America, largely related with the mining industry, but its share is foreseen to increase more in Africa. Differently, its share in Europe and North America is growing at a lesser rate because water recycling is being adopted in industrialized countries. Irrigation water forms the largest component of water use in all continents (Fig. 2.5).

The very high share of water demand and consumption for agriculture, particularly in water scarce regions, requires that innovative approaches be vigorously sought for water management in this sector. Irrigated areas have grown as much as 4–5 million hectares a year (Rangeley 1990). This corresponds to an average growth rate of almost 2% a year since 1960 (Waggoner 1994). However, this trend has declined to near 1% during the 90s (Jensen 1993) and less for the last decade. Forecasts in Table 2.6 show that this trend will continue. This growth is justified by the productivity of irrigated agriculture: irrigated areas represent about one-sixth of arable land but provide about 40% of the world's crop production; this share tends to increase as much as productivity in irrigated agriculture continues to increase.

Expansion of irrigated areas occurred when crop yields also increased dramatically due to a combination of factors: improved irrigation techniques, introduction of high yield varieties and higher rate use of fertilisers. In arid, semi-arid and sub-humid regions these improvements could only be successful under irrigation. However, the expansion of irrigation has been achieved with several negative environmental impacts, namely waterlogging and salt affected soils (Jensen 1993).

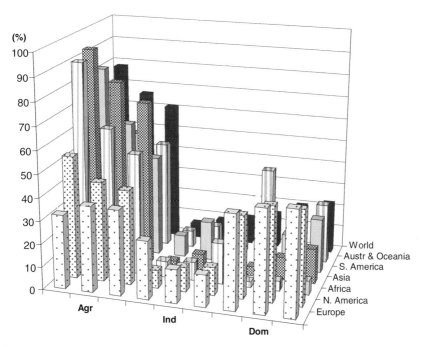

Fig. 2.5 Percent share of water use in agriculture, industry and domestic activities for 1950, 1995 and foreseen for 2025 (those years correspond to the 3 cells within Agr, Ind and Dom); (source: Shiklomanov and Rodda 2003)

Arid and semi-arid water stressed areas are particularly sensitive to these detrimental effects (Agnew and Anderson 1992). Controlling environmental effects and increasing the performance of irrigation are among future issues for water resources management (Hennessy 1993, Pereira et al. 1996, 2002b).

Water quality is an expanding problem. Surface waters are becoming more polluted (e.g. low dissolved oxygen and increased faecal coliforms) in developing countries, but this trend is being reversed in high income countries due to expanded adoption of treatment of effluents from urban and industrial uses, and water recycling in industry. Data on groundwater pollution and nitrate contamination is limited but shows that aquifer protection is a must. Problems resulting from increased salinization of aquifers are requiring increased attention in less developed regions in arid zones (Saad et al. 1995). Health problems are particularly acute as identified by the World Health Organization (WHO 2004).These problems are common in water scarce regions and bring into clear focus the need for availability of safe water and sanitation.

2.2.2 Water Management Issues

Policies and practices of water management under water scarcity must focus on specific objectives according to the causes of water scarcity. An integrated technical and scientific approach is essential to develop and implement the management practices appropriate to deal with water scarcity. Policies and practices must be based on the assumption that water will not become abundant, so that water management policies and practices have to be specific for water scarcity, i.e. for man to cope with the inherent difficulties.

To cope with water scarcity means to live in harmony with the environmental conditions specific to and dictated by limited available water resources. For millennia, civilisations developed in water scarce environments and the cultural skills that made it possible to live under such conditions are an essential heritage of those nations and peoples. Progress in XX century often questioned traditional know-how, which has in some situations been replaced by modern technologies and management rules directly imported from other areas having different physical and social environments. Water consumption and demand has increased every where for domestic and urban uses, for agriculture and irrigation, for industry and energy production, and for recreation and leisure. However, these increases became particularly evident in regions where water was already scarce or, at least, not abundant. Therefore, man-made dry regimes are now adding to the natural water scarcity conditions, in many cases aggravating the existing situation.

Solving water management problems that are faced in water scarce regions calls for innovative approaches to cope with the water scarcity. Innovation includes the adaptation of traditional know-how to the current day challenges, the adaptation of the externally available technologies to the prevailing physical and social conditions, the creation of new and well adapted technologies and management approaches, and in some regions cessation of irrigation. Innovation must be used to assist man to cope with the environmental constraints and engineering and managerial solutions must be found that are specific to the existing causes for water scarcity: nature produced aridity and drought, and man induced desertification and water shortage.

2.2.2.1 Aridity

Coping with water scarcity due to aridity conditions requires understanding of the main problems, the main objectives and defining issues for water management in agreement with dominant physical and social conditions (Fig. 2.6).

Main factors to be considered to cope with water scarcity in arid regions are indicated in Fig. 2.6. Factors such as population growth above the capacity of support of the available water and natural resources are beyond water management solutions. However it is known that policies supporting a control on population growth are required to decrease the pressure on water and natural resources. This is also associated with the need to eradicate poverty and improve education, and in many situations these cannot be achieved without policies and measures that deal with water and health issues and management. These issues relate with the need for appropriately valuing the water as an economic, social, cultural and environmental good, including for nature conservation.

It is important to note that water issues are often tied with land issues in arid zones since land productivity and combating soil salinization largely depend upon

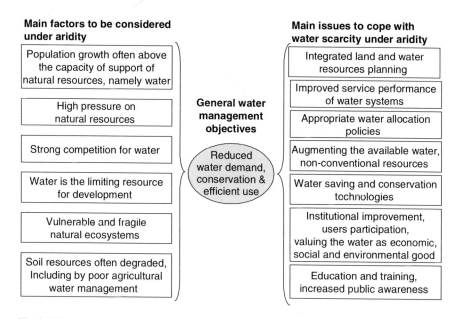

Fig. 2.6 General framework for coping with water scarcity due to aridity: main factors, objectives and issues

appropriate water use. As mentioned above, the empirical know-how cumulated through generations that provided for long term use of land and water has been questioned in recent decades when imported technologies were progressively adopted. However, problems referring to land quality preservation and to the conservation of water resources are far from being solved. As in the past, innovative thinking and innovative institutions are required together with public participation, education and training and increased public awareness on water, soil and ecological conservation.

2.2.2.2 Drought

Water management under drought requires measures and policies which are common with aridity, such as those to avoid water wastes, to reduce demand and to make water use more efficient. However other possible measures are peculiar to drought conditions and may be beyond the scope of the more common water management issues and policies. The peculiar characteristics of droughts may require use of specific measures (Fig. 2.7). The complexity and hazardousness of droughts make their management particularly difficult and challenging. Adopting risk management is a must, together with clearly defined preparedness measures. Since droughts have pervasive long term effects and their severity may be very high, they also require appropriate evaluation of impacts.

The effectiveness of related risk management depends upon the drought monitoring facilities, drought prediction capabilities, and means to provide information to users, as well as on the awareness of populations of drought water scarcity.

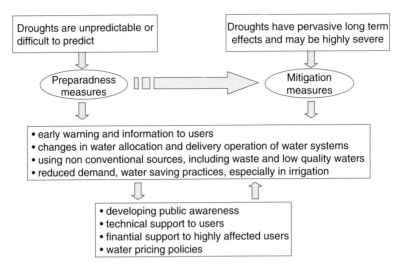

Fig. 2.7 General measures to cope with droughts

Supporting but modifying a local behaviour pattern is important, since all water use patterns/traditions are different and because the local water source access is often unique. Of course, in large towns, the objective may be to support the usual behaviour of the population, but with reduced water use, based upon long held traditions and culture.

2.2.2.3 Desertification and Water Shortage

Desertification and water shortage are man induced and are associated to problems such as soil erosion, land degradation mainly through salinization, over exploitation of soil and water resources, and water quality degradation as discussed in Chapter 4. As a result there is a need for policies and measures to be oriented to solve the existing problems and to prevent new occurrences of these problems. Therefore, combating desertification and water shortage needs to involve:

- Re-establishing the environmental balance in the use of natural resources,
- Enforcement of integrated land and water planning and management policies,
- Restoring the soil quality, including the adoption of soil and water conservation measures,
- Defining new policies for water allocation, favouring reduced demand and water conservation practices,
- Controlling ground-water abstractions and developing measures for recharging the aquifers,
- Minimising water wastage,
- Combating soil and water salinization,
- Controlling water withdrawals,
- Adopting policies and practices for water quality management, and

- Implementing policies of economic incentives to users, including graduated water pricing and penalties for misuse and abuse of natural resources.

2.2.3 Implications of Sustainable Development

Coping with water scarcity requires that measures and policies of water management be in line with the challenging and widely accepted concept of sustainable development. This usually requires new approaches towards development and water and soil resources management. New perspectives are required for integrated management of the soil and water. This is not only a question of implementing new technologies and management approaches for allocating and controlling the water and land uses. Rather there is the need for also considering the driving forces governing the pressures on the resources themselves, population growth, the behaviour of the users, and the diverse human and social objectives. Despite the enormous progress in development of technological and managerial tools, which are becoming available to improve management, there are still many gaps in knowledge and in development of skills for transferring scientific and technological knowledge into practice.

The concept of sustainable development is supported by a large number of definitions. The WCED (1987) introduced the concept as the "development which meets the needs of the present without compromising the ability of future generations to meet their own needs". A more specific definition has been adopted by the WCED (1987): "sustainable development is a process of change in which the exploitation of resources, the direction of investments, the orientation of technological development, and institutional change are all in harmony and enhance both current and future potential to meet human needs and aspirations".

This concept of sustainability relates to the human legacies that future generations will receive from the present. The legacy should enhance future prospects and opportunities and not restrict them in any way. As pointed out earlier, this is of great relevance for regions of water scarcity, where the cultural heritage relative to water has been threatened by population growth and the adoption of imported technological and management tools.

The FAO (1990) revised concepts proposed by many authors and formulated its own definition focusing on agriculture, forestry and fisheries: "sustainable development is the management and conservation of the natural resource base and the orientation of technological and institutional change in such a manner as to ensure the attainment and continued satisfaction of human needs for the present and future generations. Such sustainable development (in the agriculture, forestry and fisheries sectors) conserves land, water, plant and animal genetic resources, is environmentally non-degrading, technically appropriate, economically viable and socially acceptable". This definition fully agrees with that of WCED (1987), uses the same conceptual components and introduces biodiversity implications.

Analysing definitions of sustainable agriculture, the NRC (1991), states that "virtually all of which incorporate the following characteristics: long-term maintenance of natural resources and agricultural productivity, minimal adverse environmental impacts, adequate economic return to farmers, optimal crop production with minimised chemical inputs, satisfaction of human needs for food and income, and provision for the social needs of farm families and communities".

Adopting the concept of sustainability, water management to cope with water scarcity has to:

- Be based on the knowledge of processes which can lead to resource degradation and to the conservation of natural resources;
- Include resource allocation to non productive uses such as natural ecosystems;
- Consider technological development not only for responding to production objectives but to control resource degradation and environmental impacts;
- Value the non-productive uses of the land and water;
- Prioritise processes which help reverse degradation;
- Include institutional solutions which support the enforcement of policies and rules and socially acceptable decisions and measures for land and water management, and
- Rely on clear objectives focusing not only on the natural resources by themselves and the economics of the returns from their use, but also embracing human needs and aspirations.

The sustainable development items enumerated early in this chapter were presented on the basis of the issues referred to above. Population to be served and sustained, resource conservation and environmental friendliness are the main sustainability issues in relation to water and soil. Appropriate technologies constitute an essential challenge to achieve policies and practices well adapted for problem solving. The economic viability of measures to be implemented is a necessary condition for their adoption and a reason for provision of financial incentives. The need for appropriate institutional developments is considered as a pre-condition for implementation of innovative practices, whether of a technological, managerial, economic or social nature. Finally, the social acceptability of measures to cope with water scarcity is particularly dependent on the development of awareness, within the local population, of water scarcity problems and the associated issues.

One potentially beneficial approach is to consider the water not only as a natural renewable resource but also as a social, environmental and economic "good". Since scarcity favours assigning a high value to a good, valuing the water only as an economic, marketable good may provide only limited assistance in promoting sustainability. This is because water acts not only as the basis for production, but it also supports other natural resources and plays a major role in cultural and social contexts, particularly when scarcity gives it a special value. A coupled environmental, economic, and social approach is therefore required in valuing water in water scarce environments.

Chapter 3
Physical Characteristics and Processes Leading to Water Scarcity

Abstract The physical characteristics and processes leading to water scarcity complement the discussion of concepts presented in the preceding chapter. Climatic conditions dominant in water scarce regions are analysed, particularly rainfall variability and evaporation. Other contributors to water scarcity such as the essential hydrologic characteristics of the region, the processes affecting groundwater recharge and availability, and sediment loads and water quality, including those aspects leading to desertification and water-shortage are also discussed. The importance of data collection and data quality assessment for planning and management studies are outlined.

3.1 Introduction

The water availability of a region depends primarily on its climate and then on the topography and geology. Its sufficiency depends on the demand placed upon it.

The climate is largely dependent on the geographical position of the region. The climatic factors connected to the availability of water are the rainfall and its variability of occurrence in time and space, the humidity, the temperature and wind, all of which affect the rate of evaporation and plant transpiration. The topography is important for water availability since it controls the way rainfall is discharged both in terms of quantity as well as rate, the development of lakes, marshlands and provision of opportunity for surface water to infiltrate. Geology affects the topography and controls the availability of suitable underlying rocks that form aquifers to which water can infiltrate and be available for exploitation.

Water availability, sufficiency or scarcity, are relative terms that greatly depend on the demand placed upon the existing water resources.

Unfavourable climatic conditions coupled with unfavourable topography, geology, demand for irrigation or water supplies for concentrated populations and industry in excess of the available supply are processes leading to water scarcity (Fig. 3.1).

In semi-arid, arid and dry sub-humid regions affected by water scarcity, the processes leading to the water scarcity have specific characteristics, quite different from those of humid or temperate areas. It is important to underline that these

L.S. Pereira et al., *Coping with Water Scarcity*, DOI 10.1007/978-1-4020-9579-5_3,
© Springer Science+Business Media B.V. 2009

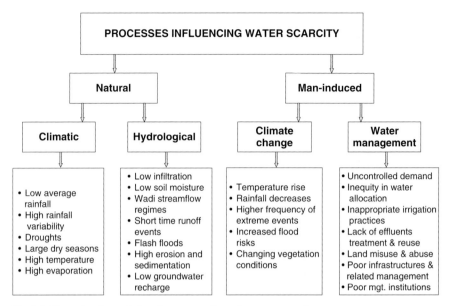

Fig. 3.1 Natural and man-induced processes favouring water scarcity

characteristics that act strongly upon the availability of water and its management. Most of these areas that are likely to be affected by water scarcity have some similar factors that make up the identity of their ecosystems and particularly the functioning of the water cycle. However, all water scarce regions have particular characteristics that make them to be different, sometimes unique. Common features are to be found in the climate, the rainfall regime, the conditions of surface runoff and soil infiltration, and in the replenishment regime of deep and surface aquifers.

Of equal importance to the above, are also some non-physical processes that may lead to water scarcity such as population growth, mismanagement of resources and climate change (Fig. 3.1). Population growth is normally associated with the development of metropolitan areas that provide increased opportunities for employment, and with tourist areas, which especially develop at the seacoasts. Population growth exerts considerable stress on the available water resources and on the water infrastructure, which normally does not keep pace with the demands made on the water resources. As a result the water scarcity problems are accentuated and coping with them becomes even more pressing and difficult.

Quite often water scarcity is exacerbated as a result of mismanagement of the available limited water resources (Fig. 3.1). Careful, prudent and imaginative management is especially called for in areas of limited water resources to enable the demand to be met in the best possible way.

Climate change is nowadays commonly considered as a cause of water scarcity (Fig. 3.1). However, there is frequent confusion between climate change and climate variability. Climate change is a long-term change and recently is associated with global warming due to natural factors and human activities. Differently, climate variability is part of the Earth's climate system and occurs at various temporal and spatial scales. That natural variability related to very large cycles in rainfall regimes may partly explain why in the long term North Africa regions that were granaries during the Roman Empire are now semi-arid and have no capabilities to reliably produce cereals without irrigation; however this is also climate change. Shorter cycles, of 2–3 decades, may refer to the variability in occurrence of severe droughts in various regions (Moreira et al. 2006, 2007). However, without enough long time series of precipitation it is difficult to definitely conclude whether this is variability or climate change and therefore whether the related impacts on water availability are permanent or temporary. Much shorter scales of climate variability, often called inter-annual variability, are currently observed as related to the occurrence of droughts, which are a feature of climate. Even smaller time scales, referring to intra-annual variability, largely affect flood regimes, typically in arid climates. To summarise, climate variability affects the water resources systems in terms of floods, dry spells, droughts, and waterborne diseases. As referred in a recent United Nations report (UNDP 2006) it is not just the extremes of climate variability that are of concern to the water sector but the increasing and extreme variability in the hydrological cycle and climate systems which are impacting several countries' water resources.

The United Nations Framework Convention on Climate Change (UNFCCC) defines climate change as: a change of climate which is attributed directly or indirectly to human activity that alters the composition of the global atmosphere and which is in addition to natural climate variability observed over comparable time periods. The UNFCCC uses the term Climate Change to mean only those changes that are brought about by human activities. Climate change, which may result from an increase of temperature due to changes in the atmosphere caused by human activities, affects the water availability and the demand for water. The drier the climate, the more sensitive is the local hydrology. Relatively small changes in temperature and precipitation may cause relatively large changes in runoff. Arid and semi-arid regions are particularly sensitive to changes that are likely to reduce rainfall, increase temperature and the evaporative demand of the atmosphere. Drought frequency in some areas is likely to increase the effects of such change. Climate change models (CCM) suggest that the frequency and severity of droughts in some areas could increase, together with the frequency of dry spells and the evaporative demand (US National Assessment 2000). However, the uncertainties of model predictions are high and are greatly affected by the environmental conditions under which the temperature observations were obtained. Despite uncertainties, it is very likely that climate change will produce an increase in demand for water for agricultural production and cause disturbances in the demand for water for natural ecosystems. These disturbances will affect the vulnerability of those natural systems, and produce imbalances in supply, mainly in water scarce regions (Kaiser and Drennen 1993, Rosenzweig and Hillel 1998).

Another process that affects water scarcity is the misuse and abuse of land and water resources which leads to desertification. As analysed in Chapter 2, imbalances in water availability are associated with the degradation of land resources in arid, semi-arid, and dry sub-humid areas. Desertification is caused primarily by human activities and is influenced by climatic variations. Desertification reduces the land's resilience to natural climate variability. Soil, vegetation, freshwater supplies, and other dry-land resources tend to be resilient, so they can eventually recover from climatic disturbances, such as drought, and from non-severe human-induced impacts, such as overgrazing. When land is degraded, however, this resilience is greatly weakened. This has both physical and socio-economic consequences.

Some of the physical characteristics of the processes which cause scarcity of water are briefly discussed in the following sections.

3.2 Climatic Conditions

3.2.1 General Aspects

The climatic conditions commonly present in regions of water scarcity are associated with low and irregular rainfall, high temperature and high evaporation (Fig. 3.1). It is also common that meteorological and rainfall measurement systems are rarely well developed in areas affected by water scarcity. At best, in these arid areas the main climatic parameters are monitored at synoptic stations which have a very low spatial density. Poor levels of observation worsen the problem since it leads to inadequate management of the mediocre quantities of water available.

Figure 3.2 presents the delimitation of arid and semi-arid regions of the world as defined by the Map of the World Distribution of Arid Zones (UNESCO 1979). This delineation is primarily based on a bio-climatic aridity index, the P/ETP ratio (where P is the mean value of annual precipitation, and ETP is the mean annual potential evapotranspiration). The three zones are the *hyper-arid* zone (P/ETP<0.03), the *arid* zone (0.03<P/ETP<0.20) and the *semi-arid* zone (0.20<P/ETP<0.50). In addition to these criteria, temperature is taken into account based on the mean temperature

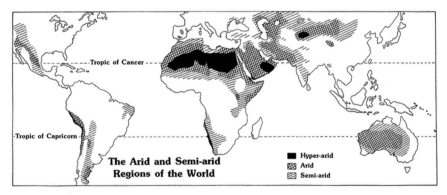

Fig. 3.2 The arid and semi-arid regions of the world (Hufschmidt and Kindler 1991)

of the coldest and the hottest month of the year. Consideration is also given to the rainfall regimes (dry summers, dry winters) and to the position of the rainfall period in relation to seasonal temperatures.

In arid climates annual precipitation ranges 0–200 mm and in semi-arid climates it ranges 200–500 mm. In the first, most rainfall (70% or more) is used as evaporation and evapotranspiration, less than 30% becomes runoff, and groundwater recharge is generally negligible. In semiarid climates the rainfall converted into runoff is about the same, groundwater recharge increases to near 20% and real evaporation and evapotranspiration then average around 50% (UNESCO 2006). In Australia, even in the semi arid zones the runoff is only 5%; there may be 10–50 rainfall events in the year but only one or two of these will produce any runoff, usually in very small amounts. Only once in every two or three years will significant flows occur, but even these will be of a very localised nature. Under these conditions, the evapotranspiration component reaches 95% and runoff decreases to near 5%. Differently, in sub humid temperate climates, where annual precipitation ranges from 500 to 1500 mm, the partition into evaporation, groundwater and runoff is much better balanced, meaning much more water availability from both streamflow and groundwater. Using the concepts of green and blue water relative to the parts of rainfall that are respectively used directly by the natural and man-made ecosystems and those that are mobilized for human uses, the differences among climates regarding water availability become very clear: in arid climates the blue water fraction is much smaller than in semi-arid zones and is extremely small compared to sub-humid temperate climates.

3.2.2 Rainfall Variability in Time and Space

Rainfall quantities and the pattern of their occurrence vary considerably in different climate regimes. In general the lower the annual rainfall amounts the greater their variability from one year to the next. In semi-arid and arid regions the rainfall is irregular and unreliable. In these regions, the rainfall variability and spatio-temporal differences are very pronounced. Many hydrological and hydrometeorological studies give evidence of such behaviour (e.g. Agnew and Anderson 1992, Lin 1999, Kalma and Franks 2000, Wheater and Al-Weshah 2002). Most of the world's arid and semi-arid regions have climatic regimes in which precipitation is characterized by some or all of the following:

- One, rarely two, short rainy seasons, followed by long and often hot dry periods;
- Short rainy periods, rarely more than two days, unevenly scattered throughout the season;
- Violent showers, having high rainfall intensity and large spatial variability over a small area, even at a scale of ten square kilometres, and
- Irregular inter-annual rainfall depths and great local differences that often render the usual statistical tools in climatology ill-adapted, e.g. asymmetric or multi-modal frequency/probability histograms.

Lack of rainfall has varying significance in the different climatic regimes in the world. In some of the arid zones, there may be several years in which no measurable precipitation occurs and the flora and fauna are adapted to these normally desiccating conditions. In other arid areas, where very little rainfall occurs, the deficiency of the rainfall below the normal results in serious water shortages requiring a number of measures to be taken.

Low annual amounts of rainfall in a region may result from its geographical location relative to the general circulation of the atmosphere, its location on the lee-side of a mountain range, or absence of a topographic high that would favour the formation of precipitation from clouds passing over it. This has a bearing on the surface runoff that is generated, and on the quantities of water that may be recharging the ground water systems. At the same time, the occurrence of rainfall in short and infrequent outbursts creates flash floods, which do not provide sufficient opportunity for water to infiltrate to ground water systems.

It is important to have good rainfall measurements for a sufficiently long duration and of adequate spatial distribution for the appropriate assessment, planning and management of the water resources at a regional level. The need for observed rainfall time-series, both for analysis of water availability and forecasts in the long and medium term, is the requirement most often expressed by development planners and decision-makers. This requirement is currently not usually satisfied in regions suffering from water scarcity. It is therefore necessary to strengthen observation networks for planning and operational purposes and to allow research on rainfall variability and spatial differences.

The rainfall measurement in a given location provides point information usually only valid for the area attributed to that rain gauge. A large number of such observation points would be needed for arid and semi-arid regions to provide meaningful information due to the exhibited peculiarities of spatial variation of rainfall in these areas (Cordery 2003). The knowledge of climate and precipitation is inevitably based on the development and maintenance of such an observing network system in the long term.

A drought is usually considered to be a period in which the rainfall consistently falls short of the climatically expected amount, such that the natural vegetation is affected, and availability of water for other uses is severely limited. A drought in low rainfall regions could be devastating and disastrous since the already limited existing quantities of water in such an area could not suffice to cope with the situation. Thus drought has widely different connotations according to location and likely consequences as outlined in Chapter 4.

The intervening periods between rainfall events – dry spells, not to be confused with droughts – in regions where aridity prevails are normally quite long and, with the usually high existing potential evaporation, create soil moisture deficiencies, which have to be recovered before runoff will be initiated or there will be infiltration to the aquifers. Thus, in these areas, the effect of a rainfall event on runoff generation and aquifer recharge is reduced compared to the effects of similar events in temperate regions, where rainfall occurs more regularly.

3.2.3 Evaporation

Evaporation and evapotranspiration are important in assessing water availability and they must be considered in water resource planning and management. Water evaporates from soil water and groundwater, vegetation canopies, natural and man made lakes and ponds, streams, and canals and wet surfaces. Evaporation and evapotranspiration are the object of many papers and books, mainly relative to observation and computational procedures (e.g. Burman and Pochop 1994, Allen et al. 1996, 1998), including the use of satellite information (e.g. Bastiaanssen and Harshadeep 2005, Courault et al. 2005, Allen et al. 2007b). The discussion here focuses on the particular aspects of evaporation that are of vital concern in arid and drought prone regions and that affect water availability and demand in these regions.

Evaporation affects the quantity of effective rainfall, the yield of river basins, the storage in surface reservoirs, the yield of aquifers and, especially the consumptive use of water by crops and natural vegetation. Evaporation is necessarily dependent on the water availability at the evaporating surface. Evapotranspiration from natural or agricultural vegetation depends upon the availability of water in the soil and the capability of plants to extract this water when retained at high soil water tension. Thus, evapotranspiration is controlled by soil moisture content and the capacity of the plants to extract water from the soil and to transpire, which is conditioned by the climatic demand of the atmosphere. A large amount of the water that is precipitated in a region is returned to the atmosphere as vapour through evaporation from wet surfaces and through transpiration from the vegetation. The water scarcity experienced in regions with low rainfall and long dry summers is exacerbated by high evaporation and transpiration demand.

More important climatic factors affecting evaporation and evapotranspiration are:

- Solar radiation, being the main source of energy affecting evaporation. Solar radiation varies with latitude and season, being generally higher for low latitudes, where its seasonal variation is smaller than at higher latitudes. The net available radiation also depends on the reflectivity of the surface (albedo), the rate of long wave radiation from the earth's surface and the transmission characteristics of the atmosphere.
- Temperature of both the air and the evaporative surface, which is also dependent on the major energy source, the solar radiation. As a consequence of the energy balance, the surface temperature increases when more solar energy is available and a larger fraction of that energy becomes sensible heat. Conversion of energy to sensible heat is enhanced when only a small fraction of the available energy is used for evaporation. The surface temperature decreases when the evaporation rate increases, and higher surface temperature occurs when the evaporative surfaces become dry.
- Vapour pressure deficit (VPD) of the air, which corresponds to the amount of water vapour that can be absorbed by the air before it becomes saturated, has a large controlling effect on the evaporation. More evaporation can be expected in

inland areas where the air tends to be drier, than in coastal regions where the air is damp due to the proximity to the sea.

- Wind speed above the evaporating surface. Wind represents a major driving force to replace the air layer adjacent to the surface, which has been wetted by the evaporated water vapour, with dry air having a higher VPD. As water evaporates, the air above the evaporating surface gradually becomes more humid until it becomes saturated and can hold no more vapour. If the wet air above the evaporating surface is replaced with drier air the evaporation may proceed. Evaporation is greater in exposed areas that enjoy plenty of air movement than in sheltered localities where air tends to stagnate.

Other factors have a generally broader scale influence on evaporation. The prevailing weather pattern also affects evaporation. The nature of the evaporating surface, which may modify the wind pattern, may also influence evaporation. Over a rough irregular surface, friction reduces wind speed, but has a tendency to cause turbulence so that, with an induced vertical component in the wind, evaporation may be enhanced. This influence is often associated with the effects of topography on the nature of evaporative surfaces, the energy balance and the weather variables that determine evaporation.

Evaporation is difficult to control. Development of ground water reservoirs and artificial ground water recharge can reduce evaporation losses. At the same time, proper design of dams located in topography that would store water with minimal free surface area would be advisable. However, such favourable locations are few. In arid zones, not only are evaporation losses important themselves, but high evaporation tends to encourage salinity problems. In such regions, water and soil salinity tend to be high due to the concentration of salts in the soil and aquifers following continuous high rates of evaporation. The resulting saline marshlands, salt-lakes, sabkhas, saline soils and brackish and saline waters, add to the problems of regions with water scarcity conditions.

3.3 Hydrologic Characteristics

3.3.1 Runoff Regime

The topography, the vegetation cover, the soils and the geology of an area, together with the climatic factors, affect the hydrologic characteristics of that region. These factors control the rainfall-runoff relations as well as the groundwater recharge and storage, so affecting the general water availability in the region. Rainfall-runoff relationships and runoff regimes are dealt with in a large number of Hydrology books and papers (e.g. Maidment 1993, Pilgrim and Cordery 1993, Wootton et al. 1996, Wheater and Al-Weshah 2002). However, relatively few concern arid and dry environments. The subject is discussed in this Section aiming at providing an overall view of the main problems encountered in water scarce regions.

Stream-flow is the sum of surface runoff, subsurface flow, and ground water flow that reaches the streams. Runoff consists of the quantity of rainfall water left after interception by vegetation, surface water retention and infiltration. The flow regime of a river is the direct consequence of the climatic factors influencing the catchment stream flow, and can be estimated from knowledge of the climate of a region.

The generation of runoff, among other factors, depends on the initial soil moisture conditions in the catchment, i.e. the soil moisture deficit, which represents the difference between the soil water content at saturation and that when rainfall occurs. Dry spells in arid areas are quite frequent and extensive and the soil moisture decreases approximately logarithmically during these periods of no rain. The soil is often dry when rain starts. A crust can form on the dry soil and subsequently infiltration tends to be small, favouring runoff formation and low refilling of the root zone and reduced groundwater recharge. Flash floods are then generated and vegetation growth is not encouraged. Most watercourses contain water only briefly and are known by their Arab designation as *wadis*.

Interception, the amount of water that can be caught on vegetation and returned as vapour to the atmosphere, is generally small in arid zones. This is because the vegetation cover is sparse compared to that of temperate humid climates and because rainfall in arid regions is usually of high intensity and short duration. Small interception favours rapid runoff formation, which is typical of flash floods. This ephemeral and sudden runoff regime makes capture and use of surface waters more difficult and produces particularly vulnerable riparian ecosystems.

In most arid and semi-arid climate areas the spatial cover of rainfall events is variable, resulting in contributions of runoff only from part of a catchment. Thunderstorms, which are common in these regions, typically have these high intensities and spatial variability. The infiltration capacity of the soils, which are mostly unprotected by the sparse vegetation, is exceeded quite easily and the excess rain is then available to flow as surface runoff.

The river flows are determined by the local runoff conditions which are essentially controlled by the soil cover and land use, the geology and state of the land surfaces, the topography and physiography of the basin, and by the time and space variability of the rainfall. More generally:

- The streamflow is basically constituted of surface runoff. Base flow generally occurs only in large drainage basins. The complete depletion of flows in dry seasons is the general rule, even in very large basins, in arid and semi-arid regions. In arid areas streamflow usually persist only for a few days after the rainstorm;
- Flows in semi-arid and arid areas are rapid and often violent;
- Erosion and solid matter transport are always important in small drainage basins because of the aggressiveness of rains and the fragility of the soils which are poorly protected by vegetation;
- The time variability of streamflow, at all time scales, is more important from a water resources point of view than the irregularity of the rainfall.

The irregularity of runoff and the low flows that are often experienced require careful analysis before attempts are made to utilize the water for development.

Long-term data are required for any meaningful analysis of the flow regime in areas of such high variability of flow.

The hydrological regime of large rivers in arid zones is seldom well known. On many occasions it is assumed that preliminary studies for the building of large dams can provide such evaluations. This has led to failures and mismanagement of water resources. Matters are worse in the case of smaller catchments. The sustained water availability and the effects of long periods of drought cannot be assessed on the basis of short-term measurements or on the basis of only a few measuring points.

The evaluation of surface water resources can be divided into (1) the measurement of flows in the drainage network, (2) knowledge of the hydrological regime of large basins and (3) the measurement and estimation of flows on small areas that can provide runoff to the main stream. In effect the study should encompass the whole hydrological balance, i.e. partitioning the rainwater among the different components of the water balance (runoff, subsurface and base-flow, soil water storage, evaporation and evapotranspiration, groundwater recharge), and considering the expected modification of the water quality, according to processes likely to be dominant in the catchment hydro-system. Studies of the water balance at different time and spatial scales are essential to allow understanding of schemes which are aimed at gaining control of the usable water in small or large catchments.

3.3.2 Groundwater

3.3.2.1 Infiltration and Soil Water

In regions of low and occasional rainfall, infiltration is relatively low since the soil has to become saturated before allowing deeper penetration of water. After large rainfall events, the drainage of water from the saturated upper layers of soil occurs by gravity. Drainage of the excess water continues until only the water retained by soil particles in the capillary pores and inside soil aggregates remains. This is water held in the so called *soil matrix*. The amount of water retained in the soil after free drainage corresponds to the soil water content at field capacity. If there is no rain to saturate the soil again, the soil water is gradually depleted by the vegetation and by evaporation through the soil surface. A soil water deficit is created again, which corresponds to the amount of water required to restore the soil water to field capacity.

When the water content of a soil is below field capacity and surplus rainfall collects on the surface, the water penetrates the soil at a rate depending on the existing soil moisture content, the soil characteristics, and other factors such as salinity, surface conditions and soil cover. As the rainfall supply continues, the rate of infiltration decreases as the soil becomes wetter and less able to take up water.

Infiltration from rainfall into the soil depends on a large number of factors such as the initial soil water content, the intensity of rainfall and the textural and structural characteristics of the soil. Soil surface conditions change with time after rain starts because of the impact of raindrops. These may re-locate soil particles and cause a destruction of soil aggregates, depending on the energy of the rain drops, the intensity of the precipitation, the nature of colloids assuring the liaison between soil particles, the percentage and type of clay minerals, and the organic matter in the soil. The cover of the soil surface by vegetation, mulches or litter protect the soil from erosion and are highly favourable for maintaining high infiltration rates. The nature and degree of coverage by vegetation and the dispersive characteristics of the soil surface materials also affect infiltration. When the soil aggregates are not stable, the soil is uncovered, the raindrops are large and the rainfall intensity is high, the aggregates tend to be destroyed, the soil particles become dispersed, the soil pores become sealed, and an extensive soil crust is formed. This results in a sharp decrease in infiltration rates. When the same soil is well protected by dense vegetation, litter or mulch, the macropores remain open at the surface and water penetrates into the soil at much higher rates.

The unsaturated soil zone may have the capacity to store a large amount of the rain. Unless the soil reaches a high degree of saturation, the water will just be held there until it is evaporated directly or extracted by vegetation. Rain falling on coarse soils will rapidly infiltrate and even heavy rains may not generate runoff. On the other hand clayey soils will resist infiltration and even very light rains will result in the generation of runoff.

The cycle of water in the soil, where infiltration into the vadose zone is followed by evapotranspiration, favours development of hardpans and impervious layers. These are often caused by the deposition of salts and calcium carbonate brought to upper levels by the upward fluxes of water due to high surface evaporative demand. These hardpans impede infiltration, acting as hydraulic barriers even if the underlying ground has sufficient storage capacity for more infiltration. This is quite common in arid and semi-arid regions. Hardpans can also be formed by continuous tillage to fixed depths in agricultural soils, the so-called plow-pans.

The flashy nature of runoff, as discussed earlier, restricts infiltration since there is not enough time for water penetration. Furthermore, the shortness and irregularity of rainy periods allows only scanty and occasional deep infiltration. In areas with water scarcity, the study of moisture content of soils generally indicates very low variation in the water stored in the unsaturated zone when a yearly time scale is adopted. Temporary storage during the rainy periods rarely exceeds the depth of the plant roots, and as a result deep drainage to the groundwater is quite meager. Such conditions are conducive to a strong and rapid return of water to the atmosphere by evapotranspiration.

3.3.2.2 Recharge of Groundwater

In regions with annual rainfall less than 250 mm, the recharge of aquifers is very small for all except the most permeable soils. The amount of recharge depends on the permeability and retention capacity of the soil, and the distribution of rainfall in relation to the evaporative demand. A more detailed analysis of groundwater exploitation and recharge is presented in Chapter 7 and available in literature (e.g. Simmers 1997, Gale 2005, Giordano and Villholth 2007).

The retention capacity of the soil controls the quantity of soil water that is held in storage. This quantity of water is depleted during dry periods, and the difference between it and that needed to saturate the soil is the amount of rainfall required to satisfy the soil-moisture deficiency before recharge can take place. If rain occurs in a number of events during the wet season and dry weather intervenes between them, then a substantial part of the total rainfall would be used to refill the depleted soil moisture storage, so leaving only a very small quantity, if any, for ground water recharge. This generally, is not the case for sands and sand dunes and alluvial stream channels which have a low specific retention of less than 5%. These allow penetration of rain deep enough and beyond the zone of seasonal drying so that it becomes recharge to the aquifer.

The channel beds are normally the areas through which most of the recharge takes place. Recharge also occurs in flood plains at zones where there is pronounced presence of coarse alluvial sediments, at zones where the outcrop is weathered, and at fractured exposures of bedrock. Runoff in arid regions is often heavily laden with sediments. If the water is not too turbid and if the channel bottom is permeable enough, water will flow into the gravel and sand of the channel bottom. Part of this water will return to the atmosphere by evaporation and part will go into recharge of the underlying aquifer. Most of the recharge tends to take place at points where the channel is constricted rather than where the flow spreads out over a large area. This is because at narrow points of the channel the sediments are more permeable, being of larger size due to the velocity potential of flow.

In arid regions the usually great depth to the water table in the upland areas and the very small hydraulic gradient suggest there will be only relatively small amounts of recharge in these regions.

3.3.2.3 Groundwater

Most of the distinctive hydrogeologic features of arid regions are related to the quantity and quality of available ground water. In dry regions, groundwater is a very important source of freshwater for domestic, agricultural, and industrial use, and it may be the only source of water supply over large parts of the year.

Groundwater resources play a significant role in the hydrologic cycle and the water balance of a region since they act as a buffer, providing significant amounts of water during years of low precipitation or storing a significant part of runoff on the occasion of high precipitation. In view of meagre recharge in areas of low rainfall, aquifers are destined for ultimate depletion unless prudent and very conservative development is carried out. Such development should not seek to extract more than the long-term amounts of recharge.

Due to the great quest for more water in these regions, aquifers are in many cases overexploited, and suffer much degradation, such as lowering of the water level, salinization and intrusion of marine water, and mineral and organic pollution. This unbalanced situation is typical of man induced water shortage and desertification. The very strong demand exerted on the most easily accessible aquifers brings about salt accretion to groundwater that becomes a serious problem near the coast or in the vicinity of aquifers with low quality water. This is more pronounced in areas of increasing water scarcity where water from different aquifers of varying quality is imported and blended with surface waters to maintain existing land uses and to meet other demands.

The exploitation of aquifers in regions of low natural recharge and scarce surface water is to a large extent controlled by the low rainfall and the relatively small amounts of runoff, which is often associated with high erosion when the vegetative cover is poor and flash floods are the rule.

Extensive aquifers found in arid regions are often the result of past climates which were considerably more moist than the present conditions. Other aquifers at the fringe of high mountain ranges may owe their replenishment to higher rainfall in the uplands.

Closed basins filled with fine sediments of recent origin, are often highly saline especially in areas where water remains at the surface for some time before it evaporates. Aquifers with good quality water are only found along the margins of such basins or in deeper aquifers not affected by present arid conditions. The most significant aquifers are formed by river deposits that are usually a mixture of poorly sorted, and of generally low permeability, sediments. The most permeable zones are found only where the runoff persists long enough to sort and deposit only the coarser material in the streambed. These conditions can normally be expected only in former narrow valleys and channels and they are normally of limited extent.

Furthermore, due to the high erosion experienced in arid regions the thickness of alluvial material is limited and the bedrock in many cases is exposed or quite near to the surface. On the other hand, widespread consolidated and semi-consolidated aquifers are sometimes found in regions of water scarcity with considerable quantities of ground water in storage which was recharged in past geologic times under different climatic conditions. The mining of these aquifers also entails increasingly heavy constraints, even if the estimated storage is huge compared with the annual consumption. Their development and use must be based on sound hydrologic studies (cf. Vrba and Lipponen 2007) since overexploitation will inevitably result in their depletion and extinction.

3.3.3 Sediments

In arid and semi-arid regions, several environmental conditions favour water and wind erosion. On the one hand, soil erodibility tends to be high because lack of organic matter and the presence of salinity lead to low stability of the soil aggregates. On the other hand, rainfall and wind erosion are quite strong. Rainfall events are often very intense and wind storms also have a high potential to detach the soil particles and transport them large distances. Erosion is aggravated by the reduced protection offered by vegetation, which is mostly sparse and offers little soil coverage, and by the lack or insufficiency of conservation measures in agricultural lands. Then, detached soil, of silt and clay size particles, is easily transported to the ephemeral streams by overland flow, thus providing a ready supply of sediments to the streamflow.

In low rainfall regions, runoff can be generated from even small rainfall events falling on rock outcrops without vegetation or on hard clay surfaces. However, only in the very low intensity rain events is the runoff likely to be free of suspended sediments. More commonly, in low rainfall regions, depending on the soil types, the relative lack of vegetation and the sudden bursts of short but intensive storms cause the runoff to erode large amounts of soil resulting in mudflows and highly turbid water.

In addition to the intensity and conditions of a rain event, the amount of erosion depends considerably on the type and state of the land surface. The extent of plant cover, roughness and micro-relief of the surface, porosity, texture, structure, and salinity of the soil and the soil moisture content are some of the basic controlling factors. In low rainfall areas, the lack of vegetative cover, the moisture deficiency and surface crusting are critical. Problems are worse when the topography comprises steep, long slopes, and the lithologic features encourage sheet erosion and the transport of detached particles by overland flow. Gullies are very often formed under these conditions. Therefore, in water scarce areas, surface flows are very often associated with large sediment charges and the deposition of fine sediments in flat areas and stream beds make the infiltration, that would provide for groundwater recharge, more difficult.

Erosion and salinity are among the main problems affecting soils in water scarce regions. This physical degradation, together with the soil moisture stress and loss of fertility associated with the soil losses, is among the most important causes of desertification. Other causes are the unbalanced land and water uses, which are very often much above the potential capability of those natural resources. When too much pressure is placed on their exploitation, their resilience is not sufficient for recovery and unbalances become evident and, in many cases, quite permanent.

3.3.4 Water Quality

In humid and sub-humid zones the blending of large quantities of surface runoff tends to maintain reasonable water quality standards. These large quantities of water are absent from arid and semi-arid climates. The result is that the reduced flows are laden with increased amounts of salts leached from the soil profiles and the groundwater, which is then enhanced by the very high evaporation. Furthermore, fertilizers and pesticides from agricultural activities that find their way to the groundwater and streamflow appear in increased concentrations in those water bodies due to the low recharge and the low discharges in the streams. Effluents from urban and industrial areas significantly add to the problem, mainly affecting the quality of streamflows. The fact that floods are sudden and are highly charged with sediments worsens the water quality problems.

The harsh evaporative environment in areas of low rainfall adds to the amount of salts present in surface waters. The high temperature and the addition of salts and nutrients to surface water bodies easily create conditions of eutrophication, which largely affect water quality and the riparian ecosystems.

Groundwater quality in regions of water scarcity is often very low. The contribution of ground water to surface runoff often increases the salts present in the surface flows. Water quality degradation, which affects both surface and groundwater, often results from man made water shortages and is a cause of desertification.

When high quality water becomes contaminated with salts, charged with untreated municipal and industrial effluents containing toxic substances and heavy metals, that water becomes unavailable and there is pressure to explore and exploit other sources of water. Consequently, water becomes scarce or the degree of water scarcity is increased. If the degradation of water quality can be reversed after appropriate management, which is more likely to be possible with surface water, the water shortage problem may be overcome, or at least reduced. If reversal of that degradation requires long term measures, such as are more likely for groundwater, then a desertification process may have been installed. Recovering the quality of groundwaters to an exploitable condition in areas of very limited recharge and high demand pressure may be a task for several generations or may be impossible. Considerable attention must therefore be placed on avoidance of man-induced water scarcity, which can mainly be combated through protection, preservation and good management of the resources.

3.4 Climate Change and its Impacts on Water Scarcity

Climate change drives much of the change evident in natural hydrological cycles, which is one of the greatest environmental, social and economic threats facing the planet. Recent warming of the climate system, irrespective of the causes is indisputable, and is now evident from observations of increases in global average air and ocean temperatures, widespread melting of snow and ice, and rising global mean sea level (Fig. 3.3).

The Intergovernmental Panel on Climate Change (IPCC 2007) concludes that observational evidence from all continents and most oceans shows that many natural systems are being affected by regional climate changes, particularly temperature increases. Other effects of regional climate changes on natural and human environments are emerging, although many are difficult to discern due to adaptation and non-climatic drivers.

Anticipated impacts of climate change on fresh water resources and their management are reported to be as follows (IPCC 2007):

- By mid-century, annual average river runoff and water availability are projected to increase by 10–40% at high latitudes and in some wet tropical areas, and decrease by 10–30% over some dry regions at mid-latitudes and in the dry tropics, some of which are presently water stressed areas. In some places and in particular seasons, changes differ from these annual figures.
- Drought-affected or water stressed areas will likely increase in extent.

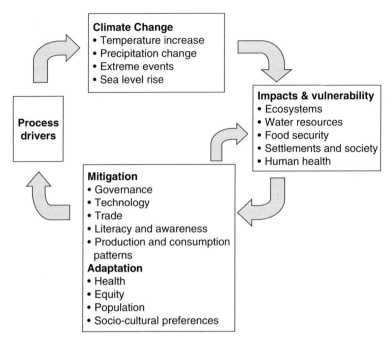

Fig. 3.3 Climate change, impacts, and mitigation and adaptation responses (modified from IPCC 2007)

- Heavy precipitation events are very likely to increase in frequency and intensity, and thus to augment flood risks.
- In the course of the century, water supplies stored in glaciers and snow cover are projected to decline, reducing water availability in regions supplied by melt-water from major mountain ranges, where more than one-sixth of the world population currently lives.

Impacts of climate change are of diverse nature (Fig. 3.3). They refer to losses in biodiversity due to changes in environmental conditions affecting the ecosystems. The present boundaries of natural ecosystems may change due to modifications in climate regimes; actual crop patterns may have to be modified due to changes in environmental conditions influencing the crop cycles, development and production. Rainfed crops are therefore more vulnerable than irrigated ones due to changes in precipitation, infiltration, evapotranspiration and soil moisture regimes. Food security is therefore threatened in more vulnerable regions and countries of the world. Changes in rainfall regimes will induce changes in streamflow regimes and lower baseflow is expected. Moreover, the water quality regimes will also change and contamination impacts may be larger, affecting public health. The latter may also be impacted due to increase of frequency and severity of heat waves and wildfires. Overall, the water availability is expected to decrease thus enhancing competition among users and making it more difficult to satisfy the increased urban water demand for residents and tourism.

Mitigation and adaptation measures are required and are already being partly implemented to reduce both the process drivers and the impacts and vulnerability to climate change. However, due to the wide nature of these measures (Fig. 3.4) large efforts are required from politicians and decision makers to create a sustainable process of mitigation and adaptation. Advice and guidance is also needed from researchers in a large variety of disciplines to develop technological, economic, managerial, cultural, educational and social measures appropriate to each country and environment to efficiently reduce vulnerability to, and expected impacts of climate change.

It is important to recognize climate change as a process driving exacerbated water scarcity and threatening development in developing countries. Unfortunately, many other processes and driving forces are contributing to degradation of Earth's environment and peoples welfare, including devastating wars. Nowadays, it is possible to identify within regions several situations that are expected to arise due to climate change (Box 3.1).

Box 3.1 Expected impacts of climate change by continents or regions (source information: IPCC 2007)

Africa: By 2020, between 75 and 250 million people are projected to be exposed to an increase of water stress due to climate change. If coupled with increased demand, this will adversely affect livelihoods and exacerbate water-related problems. Agricultural production, including access to food, in many African countries and regions is projected to be severely compromised by climate change and variability. The area suitable for agriculture, the length of growing seasons and yield potential, particularly along the margins of semi-arid and arid areas, are expected to decrease. This would further adversely affect food security and exacerbate malnutrition in the continent. In some countries, yields from rain-fed agriculture could be reduced by up to 50% by 2020.

Asia: Freshwater availability in Central, South, East and Southeast Asia, particularly in large river basins is projected to decrease due to climate change which, along with population growth and increasing demand arising from higher standards of living, could adversely affect more than a billion people by the 2050s.

Australia and New Zealand: As a result of reduced precipitation and increased evaporation, water security problems are projected to intensify by 2030 in southern and eastern Australia and, in New Zealand, in Northland and some eastern regions.

Southern Europe: Climate change is projected to worsen conditions (high temperatures and drought) in a region already vulnerable to climate variability, and to reduce water availability, hydropower potential, summer tourism, and in general, crop productivity. It is also projected to increase health risks due to heat waves and the frequency of wildfires.

Mediterranean Region: Future climate change could critically undermine efforts for sustainable development in the Mediterranean region. In particular, climate change may add to existing problems of desertification, water scarcity and food production, while also introducing new threats to human health, ecosystems and national economies of countries. The most serious impacts are likely to be felt in North African and eastern Mediterranean countries.

Latin America: In drier areas, climate change is expected to lead to salinization and desertification of agricultural land. Productivity of some important crops are projected to decrease and livestock productivity to decline, with adverse consequences for food security. Differently, in temperate zones soybean yields are projected to increase.

Water scarce regions are highly vulnerable to climate change impacts. Coping with water scarcity involves therefore requires that such impacts be recognized and appropriate mitigation and adaptation measures be developed and implemented to effectively cope with it (Fig. 3.4). In general with the increase in temperature it is likely that there will be an increase of potential evapotranspiration and therefore a higher vegetation and crops demand for water as well as impacts due to heat waves.

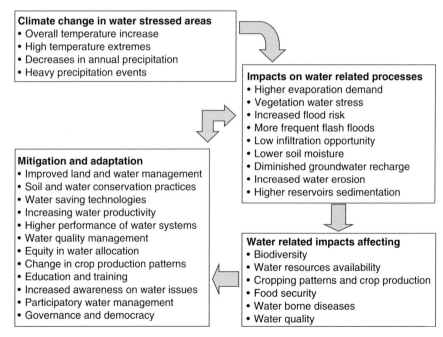

Fig. 3.4 Climate change impacts on water scarcity and related mitigation and adaptation measures

Changes in frequency of storm events are more likely to occur and the decrease in annual precipitation may strongly impact the soils and rivers hydrological behaviour due to lower infiltration and quick formation of runoff forming. Then less water will be stored in the soil and less will percolate to the groundwater.

A combination of impacts of temperature increases and lower soil moisture availability will cause stress on vegetation and rainfed production will loose viability in many areas, thus increasing the demand for irrigation and encouraging the need to change crop patterns. Land vulnerability to erosion is expected to increase when the erodibility from precipitation and runoff will increase. Also, land vulnerability to salinization may increase due to higher soil water evaporation and less leaching by the infiltrated water.

Expected changes in the hydrologic regimes refer to the likely increase in frequency of floods, which tend to have a smaller time of concentration and carry more sediments. This will affect the life and management of surface reservoirs. Expected decreases in infiltration to recharge the groundwater also diminish river baseflows. Water quality problems increase due to the reduction of flows and higher water temperatures, particularly where there is less flow to dilute contaminants introduced from natural and human sources. Low flows in many areas will lead to increases in salinity levels that may affect downstream water users. Overall, water availability will be diminished but the demand for water is likely to increase. Thus, there is a pressing need to better understand and cope with the issue of climate change, especially in regions facing water scarcity. The impact of climate change on the hydrological cycle and its extremes is very important in order to improve the management of both natural and man-made resources.

To overcome these problems, management strategies, adaptation measures, services and tools will need to be developed for coping with droughts and water scarcity as depicted in Fig. 3.4. These include technical measures and practices as well as policies of social and political nature. Among the latter are measures which are aimed at increasing awareness of water issues among water users and the general population, improving institutional arrangements for water management, and in particular increasing users participation in governance with some recognition of the importance of winning the support of all involved.

3.5 Meteorological and Hydrological Data Collection and Handling

The meteorological and hydrological data in an area must be sufficient, reliable and accessible to be of maximum use and benefit for the community. The hydrological systems are not understood well enough for reliable theoretical models of local conditions to be developed. In order to carry out water resources planning and design it is vital that some local data be available. For maximum usefulness there needs to be a combination of field data measurements and storage and retrieval systems together with the application of advanced processing techniques. Data acquisition and their management should be planned in such a way as to suit the local conditions, the available staff and the financial capability. This could be developed gradually and prioritized so as to solve the most immediate problems.

Quality control and well-developed operating systems, including field check observations, should form components of the primary data-processing system. These primary data should be purpose oriented and follow a plan for meeting defined objectives. Among others, one may distinguish objectives of different types: meteorological, in relation to the synoptic weather stations and to detect trends, such as caused by climate change; hydrological and water resources planning and management, relative to hydrological and water quality monitoring; agricultural, concerning agrometeorological networks for provision of irrigation advice and pest and disease information; drought management, referring to drought watch systems (Cordery 2003).

Data processing includes checking for completeness, long-term consistency, stationarity of the measured variables and statistical analysis of the data (Haan 1977, Conover 1980, Kottegoda 1980, Helsel and Hirsch 1992). The latter includes fitting of frequency distributions to the data and using the data in parametric models and multivariate time-series analysis. The validity of derived relationships should be tested on independent data. The degree of both detail and precision of the analysis should be consistent with the quality and sampling frequency of the available data and with the accuracy required by the application of the analysis. Using good data is always more important than using abundant data.

Time-series play a crucial role in water resources evaluations. Stochastic time series models are fitted to the corresponding input variables such as river flows, rainfall, evapotranspiration, and temperature. These models in turn can be used for the simulation of the operation of the system. Mean values and variances often give a good characterization of the phenomena under study; however, these values do not reflect the internal properties of the investigated time series and further statistical analysis is required. It is most important to note that non stationarity of the data (which may be caused by climate change, among other phenomena) may invalidate use of stochastic models.

The homogeneity of hydrological data is an important requirement for a valid statistical application. A detailed analysis of the data is the most effective method of evaluating data homogeneity and possible non stationarity. The methods of analysis are usually based on plotting different types of variables against time or data collected in other locations in the same environmental area, or by relating them to other variables to discover causes of a disturbance of homogeneity. Anthropogenic disturbances and changes of climate can only be detected using high quality data (e.g. Refsgaard et al. 1989, Kundzewicz and Robson 2004).

The WMO Guide to Hydrological Practices (WMO 1994) and numerous other textbooks and reports provide methods and techniques for processing and numerous tests for examining the normality and homogeneity of hydrological data (see Bibliography). The integrity and quality of data cannot be over-emphasized and one needs to ensure their quality before applying them for any hydrologic analysis.

Of particular interest are the frequency, severity, duration and spatial extent of droughts that seriously worsen the water scarcity conditions. This subject is treated with detail in Section 4.1.

The value of data in water resources management, especially under water scarcity conditions is immense. Nonetheless, water resources data collection programs are very vulnerable to cuts in government expenditures since they are not seen to have an immediate impact upon the community. Any adverse effects arising from a curtailment of these programs may not become evident until many years in the future. Because costs of data acquisition and primary analysis are very high, some governments only provide data to potential users for a fee. Then, to save costs, users often restrict to a minimum, often below what may be reasonable, the amount of data they use, so degrading the quality of the outcomes of any studies and projects based on the data. On the other hand, the increasing demand on limited water resources and increasing environmental degradation due to human activities, has ensured that the demand for reliable water resources data is increasing (Cordery and Cloke 2005). At the same time it is becoming evident that there is an increasing scrutiny of the reliability of data and that there are expectations from users for higher quality data.

The above clearly point to the need for a well thought out and technologically advanced information system for water resources management (Iacovides 2001). Such a system would define the type and extent of data required and should be such as to provide all the information necessary for integrated water resources management and for coping with water scarcity. To this end it should include a database system, and use of models and geographic information systems to facilitate decision support systems in water resources management.

Chapter 4
Droughts and Desertification

Abstract An in-depth examination of drought concepts, drought indices and their definitions, prediction, and forecasting is given. Monitoring of drought and drought risk management tools are discussed. Desertification concepts, indicators, monitoring and information tools are presented from the perspective of supporting the identification of vulnerability to desertification together with measures and practices that may be used to combat desertification.

4.1 Droughts

4.1.1 Definitions

As noted in Chapter 2, drought is considered to be a temporary natural water scarcity condition which is a feature of the climate. It may occur in all climatic zones and it differs from aridity which defines low precipitation regions and is a permanent feature of climate. The word drought should be used to describe the normal year-to-year dry season, which is part of the climatic regime. Rather drought is defined as a natural but temporary imbalance of water availability, consisting of a persistent lower-than-average precipitation, of uncertain frequency, duration and severity, whose occurrence is difficult to predict. Drought usually causes diminished water resources availability, and reduced carrying capacity of the ecosystems below average, or less than usual availability of water. The term is rarely applied to short periods, say of weeks or one or two months duration. The term drought is usually applied, as in this book, to periods of several months, up to much longer periods. However, where much longer periods of below average water availability occur it is possible that such an event is indicative of climate change.

Drought is usually considered to be part of the natural variability of the local climate. It should be considered relative to some long-term average conditions often perceived as "normal". Thus, drier than usual periods must be expected as part of the natural sequence of events. However, if a drier than usual period continues for many years it is unlikely this is part of the normal continuum of events. For example in the Sahel the observed rainfall in the early 1960s was well below average (Nicholson 1993). When these below average rainfall conditions continued for over

L.S. Pereira et al., *Coping with Water Scarcity*, DOI 10.1007/978-1-4020-9579-5_4,
© Springer Science+Business Media B.V. 2009

30 years it appeared that a climate change had occurred. Only in the 1990s has the rainfall been similar to that which was observed in the first half of the 20th century. Does the return to higher rainfalls suggest the 30 years or so of low rainfalls were indicative of a long period of persistently below average rains which are part of the normal variability? No categorical answer can be given at this time. It is known that long-term persistence occurs in rainfall and streamflow data but this is the longest significant deviation from usual conditions that has been properly documented. Long periods of above or below average rainfall conditions occur in ancient literature and oral history but this appears to be the longest period of an actually recorded local extreme. Similarly the south west of Australia has experienced a 30% reduction in annual rainfall from the 1960s to 1990, as reported by Nicholls and Lavery (1992). Since 1990 the rainfalls have been even lower, with very low availability of water resources. Widespread opinion is that the climate of this region has changed, or is in the process of changing, possibly as a result of global warming. Climate change as usually defined would be expected to produce such changes but are these observations indicative of very long term persistence, and therefore drought, or of climate change, with the new climate regime being indicated by the recent observations?

As also pointed out in Chapter 2, many definitions exist. These definitions have in common that drought is mainly due to the break down of the rainfall regime, which causes a series of consequences, including agricultural and hydrological hazards that result from the severity and duration of the lack of rainfall. Different durations of drought events lead to differences in perception of the phenomenon and of its impacts, and thus also to differences in conceptual approaches and measures to cope with droughts as summarized in Fig. 2.2. In other words, manifestations of droughts are various and require further discussion as described below.

4.1.2 Manifestations of Drought and Drought Impacts

As several authors have suggested (e.g. Heathcote 1969, Cordery and Curtis 1985) various sectors of the population define drought differently – as has already been noted in Chapter 2 – and so some care is needed to explain what is meant. For example, below average rainfall may be considered to constitute drought. However, if the below average conditions occur at a time of year when there is usually little or no streamflow the water resources may not be affected. Alternatively it is possible that the (small) rain that does fall may occur just at the right time to sustain crop growth, with no apparent drought for agriculture. Similarly water resources drought, i.e. sustained below average river flow or below average recharge of aquifers, can occur when rainfall is average or above average. If the rain occurs as a series of very small but frequent events, there may be no surplus of water over what is needed to supply the root zone of the vegetation, but the total rain may be quite large. In Colorado in 1977 there was a serious drought which led to restrictions of supply in the city of Denver, and no water in the irrigation supply reservoirs to the north of Denver. However the corn crop that year, even from the irrigation farms, to which no water was delivered from the reservoirs, was excellent. Rain occurred during the corn-growing season at just the right timing to provide optimum crop development, but was insufficient to cause runoff into any of the local reservoirs. Hence there is probably a need to distinguish between "meteorological drought", "agricultural drought", "water resources drought" and perhaps "economic drought" in which the water resources are insufficient to sustain local industries and income producing recreational activities.

In addition to differences in perception of droughts, it is also necessary to consider differences in impacts of droughts. A summary of drought impacts is given in Fig. 4.1. These impacts vary with the climate of the affected region and with the severity and duration of drought, as well as with the development of the society and the carrying capacity of the ecosystems.

Rainfed agriculture may be impacted by a relatively short dry spell if it occurs during a critical crop growth period, which may lead affected populations to identify such a dry spell as a manifestation of drought. However, rainfed agriculture is more likely to be impacted by a long duration drought even when its severity is not extreme. Irrigated crops are not affected by a short dry spell but are usually susceptible to the effects of a long duration severe drought that also affects streamflow and artificial storage lakes. Pasture lands and forests are generally not impacted by a dry spell but in a long lasting drought water availability to the vegetation root zone can be very much affected. Grasslands and cereal crops can usually only survive a couple of months drought, but very long lasting severe drought is usually needed to have a serious effect on native forests. This is because the native forest has adapted to the local climate. However plantation forestry may be severely impacted by drought if the predominant tree species are not native to the region.

Impacts on economic activities and urban water systems relate with the availability of infrastructures that provide for the respective supply. Large dams or rich aquifers can often provide for sustained water supply activities and may not be or are just barely affected by drought. However, in arid climates these favourable conditions rarely occur. Then also tourism may be affected. In humid and sub-humid

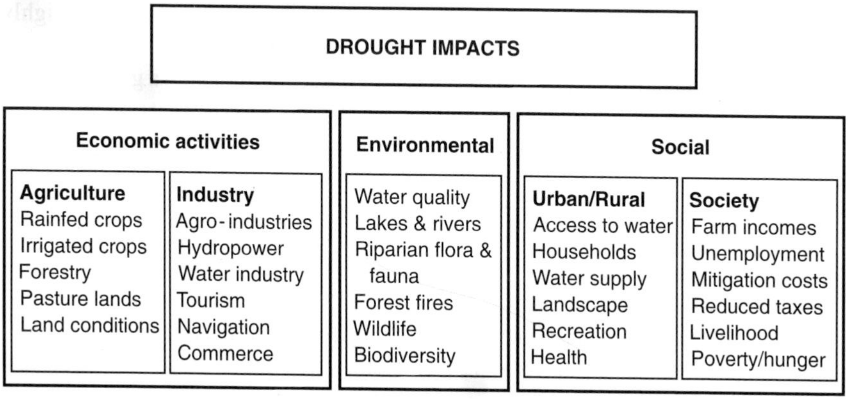

DROUGHT IMPACTS				
Economic activities		**Environmental**	**Social**	
Agriculture	**Industry**	Water quality	**Urban/Rural**	**Society**
Rainfed crops	Agro-industries	Lakes & rivers	Access to water	Farm incomes
Irrigated crops	Hydropower	Riparian flora &	Households	Unemployment
Forestry	Water industry	fauna	Water supply	Mitigation costs
Pasture lands	Tourism	Forest fires	Landscape	Reduced taxes
Land conditions	Navigation	Wildlife	Recreation	Livelihood
	Commerce	Biodiversity	Health	Poverty/hunger

Fig. 4.1 Enterprises and activities on which drought can have significant impacts

climates drought impacts are less important but hydropower generation or river navigation may be impacted. Rural areas may be highly affected when local water supply fails and household water has to be transported from distant water sources, with that distance and the associated costs increasing as the drought continues.

Social impacts relate to diminished incomes from the affected economic activities, which may lead to unemployment of workers and to loss of family incomes of farmers and rural populations. For prolonged and severe droughts, in addition to lack of incomes food also becomes scarce or inexistent, i.e., poverty and hunger are often associated with drought in low income regions. Then, populations may have to migrate to areas where a minimum of subsistence may be available. Such extreme, starvation conditions occur when the society is largely affected by poverty and where food reserves are not possible to be built at local level as was often practiced in the past. When the State centralizes food trade and assumes, in theory, the control over and responsibility for those reserves problems may become worse as observed during the long droughts in Sahel and the African Horn. Very probably drought manifested earlier at local level and with local authorities than it was perceived by the central powers. Then starvation could have been combated if local institutions had developed reserves which could then be distributed by organisations within the respective communities. This means that societal impacts of droughts not only depend upon the climate dryness, the severity and duration of drought, but also from the institutional and political structures that may support preparedness and mitigation.

The environmental impacts of drought relate in general to water quality because the concentration of salts and contaminants increase in rivers and lakes, including the pathogens. If under normal conditions water quality management is strictly enforced the resulting problems are reduced. But when poor water quality already affects the water courses and water bodies, these conditions aggravate during droughts and affect aquatic life, fishes die, the riparian ecosystems degrade and biodiversity is affected; moreover, health impacts on populations become of extreme importance (as discussed in Chapter 6 relative to surface waters and Chapter 7 for ground waters). In arid to sub-humid climates another environmental consequence is the increase of forest fires because vegetation dries out and becomes highly susceptible to fire.

Understanding which manifestations of droughts are dominant in a given location or region and which are the main impacts of droughts is relevant to the definition and implementation of preparedness and mitigation measures that are the most appropriate to cope with droughts at that location.

4.1.3 Drought Indices

The characterization of a drought using indices is controversial and often contradictory. Some authors, decision-makers and policy-makers prefer to adopt operational definitions that distinguish between meteorological, agricultural, and hydrological droughts. These usually focus on the indicator variable of prime interest, which could be the precipitation (meteorological drought), soil moisture (agriculture drought), streamflow discharges (hydrological drought) or groundwater levels (groundwater drought). Alternative definitions result from the complexity of the processes that control the temporal and spatial distribution of rainfall via the various paths within the large-scale hydrological cycle and the global circulation of the atmosphere. In certain regions, where water supplies mainly depend upon river diversions, when dealing with a regional drought it may be necessary to consider not only precipitation but also streamflow. However, streamflow is a dependent variable, controlled by the current and antecedent precipitation. In many cases, where the river discharges are regulated by dams and other hydraulic structures, the drought definition may need to be more a reflection of the river management decisions than of the natural supply. Thus, the use of precipitation or streamflow data may not be sufficient to characterize droughts at the local scale, but the definition may need to reflect the supply conditions at the river basin scale. As worldwide water consumption increases, basin scale considerations become more important, particularly as manifested by increasing tensions between States drawing water from multi-country rivers.

The interdependence between climatic, hydrologic, geologic, geomorphic, ecological and societal variables makes it difficult to adopt a definition and a related numerical index that fully describe the drought phenomena and the respective impacts. Meteorologists and hydrologists have developed indices, which depend on hydrometeorological parameters or, less often, rely on probability. Several drought studies (e.g. Yevjevich et al. 1983, Wilhite et al. 1987, Vogt and Somma 2000, Rossi et al. 2003, 2007a) give examples of these indices.

Over a long period a number of indicators or indices have been developed which purport to show that a drought is in progress or has occurred. In general these are relatively simple arithmetic descriptors of the rainfall or streamflows. For example an arbitrary definition of drought could be that there is a sequence of 3 months in which the total precipitation is less than half the long term average precipitation for those 3 months. Another possibility is that the flow in the river has been continuously well below average for several months. Such definitions and indices were in common use in the first half of the 20th century. During the 1950s and 1960s effort was made to develop more sophisticated indices which took account of the recorded variability of precipitation and surface water flows.

Precipitation based indices are those more often used. An easy-to-compute index is the percent of normal precipitation, but it has limitations in defining the appropriate threshold for drought alert. There are also difficulties in its use when the precipitation distribution is skewed rather than normal. Using deciles of the available time series overcomes some of these limitations. An example is the indicator developed by the Australian Bureau of Meteorology (www.bom.gov.au) aimed at evaluating the current precipitation against the historical data for the same location and calendar period. If the current period precipitation falls into the lowest 10% of all similar periods then a state of drought is assumed. This is a totally objective approach but it does not take any account of the timing of precipitation within the period or of soil water storage, and therefore at times it can be misleading (Heathcote 1969, Cordery and Curtis 1985).

Other precipitation indices have been developed. The theory of runs was often used after the 70s at local and regional scales, and is based on the choice of a critical threshold level (Guerrero-Salazar and Yevjevich 1975, Santos 1983, Rossi et al. 2003, Paulo and Pereira 2006). A run is a succession of the same kind of observations preceded and succeeded by one or more observations of different kind. A negative run occurs when the observed values x_t are consecutively smaller than the critical threshold y_c. Negative runs in rainfall time series are related to drought and the difference between y_c and x_t is referred as a deficit. A run can be characterized by its length (L), its cumulated deficit (D), and its intensity (I) given by the ratio D/L, as described in Fig. 4.2.

The theory of runs may be applied at various time scales according to the way monthly rainfall data are aggregated but these time scales do not reflect different lags in the response to precipitation anomalies. In addition, the theory of runs does not provide an objective classification of severity and the selection of an appropriate critical threshold is not objective.

The Palmer Drought Severity Index (PDSI) was developed by Palmer (1965) as a meteorological index to identify and assess the severity of a drought event. However, from the above discussion, its inclusion of an attempt to account for soil water means it is more than an indicator of meteorological drought. It is perhaps defining agricultural drought. The purpose of the index was to "measure the departure of the moisture supply". The PDSI is derived from the soil water balance, usually on a monthly basis. The input data consists on historical records of precipitation and temperature, in addition to information on water holding capacity of the soil. The soil is conceptually divided in two layers and the soil water balance is performed for both layers following various assumptions. From these computations a monthly moisture anomaly index is calculated which is then utilized to derive the PDSI using a backtracking calculation procedure. A regional calibration is required. It

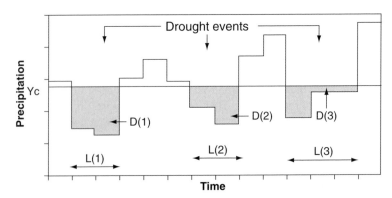

Fig. 4.2 Characteristics of local drought events with the theory of runs or when a precipitation threshold is used (from Paulo et al. 2003)

Table 4.1 Classification of PDSI values and drought categories

PDSI values	Palmer drought categories
0.49 to −0.49	Normal
−0.50 to −0.99	Incipient drought
−1.00 to −1.99	Mild Drought
−2.00 to −2.99	Moderate drought
−3.00 to −3.99	Severe drought
≤ −4.00	Extreme drought

was verified that the PDSI is sensitive to the evapotranspiration equation used and to the soil water characteristics adopted. In addition, the calculation procedure is not simple and the water balance is somewhat rough. Its calculation is well described by Alley (1984). However, despite its insufficiencies and related difficulties in use, which limit its widespread application, the PDSI is a robust drought indicator that incorporates some "memory" of the precipitation that occurred in the preceding months (Paulo and Pereira 2006). The index categorizes the severity of droughts as indicated in Table 4.1.

Various changes in PDSI calculations have been developed, including aimed at using this index for agricultural purposes. Pereira et al. (2007b) modified the water balance computation and adopted an estimate of the real evapotranspiration of an olive crop to produce a modified PDSI for Mediterranean environments. Thus, it could also be used as an agricultural drought stress index. An important modification aimed at assessing long term moisture anomalies that affect streamflow, ground-water and water storage is the Palmer Hydrologic Drought Index (PHDI) proposed by Karl (1986) and Karl et al. (1987) and widely used in USA.

The SPI, the standard precipitation index (McKee et al. 1993, 1995) is presently the most used drought index. Multiple time scales, usually from 3 to 24 months, are used. Shorter or longer time scales may reflect different lags in the response to precipitation anomalies. The SPI is an index based on the probability distribution of precipitation cumulated over the selected time scale. This index depends on the distribution function, on the sample used to estimate the parameters of the distribution, and on the method of estimation. The two-parameter gamma distribution function is often used but statistical goodness of fit tests are required before its adoption. The entire period of records is generally used to estimate the parameters of the gamma distribution function, which is then transformed, using equal probability, into a normal distribution. The computation of the SPI index is given in detail by Edwards (2000).

The initiation of a drought event is identified by a backtracking procedure: the drought event is confirmed only when in a series of continuously negative values of SPI the value −1 or less is reached. A drought ends when the SPI becomes positive. Each drought event is characterized by its:

- *lead-time*, which is the number of months within a drought event before SPI ≈ −1 is reached;
- *duration*, defined by the time between its beginning and end;
- *severity*, given by the SPI value for each month (Table 4.2);

Table 4.2 Classification of SPI values and drought categories

SPI values	Drought category
0 to −0.99	Mild drought
−1.00 to −1.49	Moderate drought
−1.50 to −1.99	Severe drought
≤ −2.00	Extreme drought

- *magnitude*, calculated by the sum of the monthly SPI values from the initiation to the end of each drought event;
- *intensity*, ratio between the magnitude and the duration of the event.

Various studies compared the described indicators for various regions and climates. Paulo and Pereira (2006) compared the theory of runs, the PDSI and the SPI and concluded that the consistency of results achieved with the PDSI and SPI (see example in Fig. 4.3), showed them to be superior. The preference is often given to the SPI results because of its easy of calculation and its lesser data requirements. Guttman (1998) compared the frequency of extreme drought months identified by the PDSI and the SPI and concluded that the PDSI overestimates that frequency.

Other indicators and indices are also used, e.g. the Surface Water Supply Index (SWSI), that was improved by Garen (1993), and represents water supply conditions relative to a given hydrological area, including regions influenced by snowpack. Other indices recently used or developed are reported by Vogt and Somma (2000), Rossi et al. (2003, 2007a) and Wilhite (2005a).

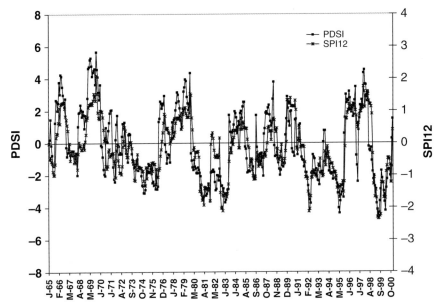

Fig. 4.3 Comparing the drought indices PDSI and the 12-month SPI for Elvas, Portugal, 1965–2000. Values above the zero line refer to wet periods and those below to dry ones

In addition to local indices, it is of interest to adopt regional indices, which relate to the area affected by drought in a given region and to the related intensity or severity (Rossi et al. 2003). The theory of runs has been the object of various developments with this perspective (e.g. Santos 1983). A simplified procedure has been adopted for computing a regional SPI (SPI$_{reg}$) for each month, which is the weighted average of the local SPI values using as weights the areas of influence of each rainfall station within the region (Paulo and Pereira 2006). It requires the selection of an areal threshold and allows the characterization of regional droughts through several parameters: the regional drought severity, the regional drought duration, and the monthly regional coverage under each drought category. An example of results is provided in Fig. 4.4.

As against adopting regional indices, an alternative is to identify regions where drought may have a similar behaviour, i.e., to identify homogeneous regions from the perspective of drought management and planning (e.g. Bonaccorso et al. 2003, Vicente Serrano et al. 2004, Bordi et al. 2006, Raziei et al. 2008).

In recent days popular media (web) have taken to publishing indicators such as the Southern Oscillation Index (SOI) and sea surface temperatures (Nino3, [NOAA, http://www.cdc.noaa.gov/]) as indices of drought or of the likelihood of drought occurring. The Southern Oscillation Index (SOI) is calculated from the monthly fluctuations in the air pressure difference between Tahiti and Darwin. Sustained negative values of the SOI often indicate El Niño episodes and positive values are associated with La Niña episodes. Nino3 is another indicator of the El Niño Southern Oscillation (ENSO), though it is actually a measure of sea surface temperature at a particular tropical location in the Pacific Ocean. These indices may be linked to drought conditions, but unfortunately their relation to drought in any particular region is far from the direct one suggested by the popular media which are prone to quote them. Using these indices requires further interpretation and analysis by specialists and care in adopting these kind of indices for popular uses.

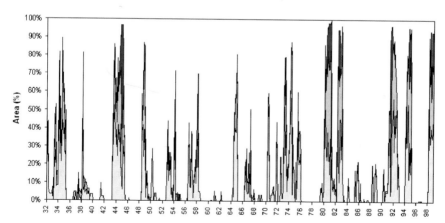

Fig. 4.4 Area covered (%) by droughts and respective severity computed with the regional SPI, Alentejo region, Southern Portugal, 1932–2000. (moderate drought ☐, severe drought ▨, and extremely severe drought ■,)

With some work (adaptation) at the local level, several of these indicators could be useful in providing qualitative suggestion of drought severity that could be of some assistance to water managers. However it needs to be stressed that none of these provide indications of the likelihood of the approach of a drought, they are more indicators of the severity of the drought that is in progress.

Indices must not be used in isolation. Indices of different phenomema – rainfall, surface water, groundwater, soil moisture – provide pieces of information that is indicative only, and that can be complementary to each other. Since there are no universal indices it is good to use several different ones for any decision making.

4.1.4 Drought Forecasting and Prediction

Usually, we tend to focus on drought when it is occurring and to react when crises strike. It would be far preferable to take a proactive approach to dealing with drought, anticipating the occurrence of the natural phenomenon and planning measures for minimizing its negative effects. Thus, around the world there has been considerable effort to develop practical methods of forecasting precipitation some months or even years ahead. It is clear that if reliable forecasts could be made there would be opportunities for development of strategies to reduce the devastating effects of drought. For example if a drought is about to begin crops demanding less water could be sown, and water supplies could be managed to conserve supplies before the drought starts, to ensure vital life-support-supplies could be sustained through the drought.

It is important to recognize the low predictability of droughts, which make drought both a hazard and a disaster. A recent statement on droughts by the American Meteorological Society (AMS 2004) gives particular attention to the disaster nature of droughts, their impacts and needs for prediction and warning. The hazard and disaster nature of droughts makes it important to develop prediction tools, including probabilistic ones, which may support early warning for timely implementation of preparedness and mitigation measures (Wilhite et al. 2000).

Droughts have a slow initiation and they are usually only recognized when the drought is already established. Forecasting of when a drought is likely to begin or to come to an end is extremely difficult (NDMC 2006). An adequate lead-time–the period between the release of the prediction and the actual onset of the predicted drought hazard – is often more important than the accuracy of the prediction. The lead-time makes it possible for decision and policy makers to implement policies and measures to mitigate the effects of drought in a timely manner (Nichols et al. 2005). Developing prediction and early warning tools appropriate to the climatic and agricultural conditions prevailing in different drought prone areas is a current research challenge.

Predictions may refer to simple relationships between precipitation and surface temperature anomalies that relate to seasonal rainfall (Cordery 1999). Recent developments in drought forecasting at large and regional scales using the SPI are reported by Bordi et al. (2004), Bordi and Sutera (2007) and Cacciamani et al. (2007), respectively. At present, more powerful tools explore teleconnections, mainly in relation to the El Niño–Southern Oscillation (ENSO) phenomenon (Cordery and McCall 2000, Tadesse et al. 2005, Kim et al. 2006), which is a measure of Pacific Ocean anomalies, but whose influences on atmospheric circulation patterns are apparent in regions very far from the Pacific Ocean, and the North Atlantic oscillation (Wedgbrow et al. 2002). However, atmospheric circulation patterns governing wet and dry rainfall regimes are quite complex, which makes it difficult to explore global circulation for prediction of droughts in various regions such as for Portugal (Trigo et al. 2004, Santos et al. 2005) despite the fact that conditions determining drought events are known (Santos et al. 2006).

Research is progressing on prediction of initiation of a drought, how drought develops once installed, and when it is terminating. Creating a lead-time prediction, even with a short time scale, improves the usefulness of drought monitoring and related information (Hayes et al. 2005). Adopting the SPI drought severity classes, a drought is said to be initiating when the near normal drought class tends to be maintained or is worsening, i.e. when monthly predictions of drought class transitions indicate a probable increase in severity. A drought is dissipating when transitions from severe/extreme drought classes to moderate and near normal classes occur. Predictions may refer to climate forecasts useful for improving the warning lead-time of droughts using drought indices (Steinemann 2006). A stochastic approach was recently developed by Cancelliere et al. (2007) to forecast monthly SPI for various time scales. Present trends also include neural networks and stochastic models applied to time series of precipitation or streamflow (Mishra and Desai 2006, Thyer et al. 2006). The stochastic properties of the SPI time series were explored for predicting short-term drought class transitions using Markov chain modelling (Paulo and Pereira 2008) and loglinear models were also successfully used with this purpose (Moreira et al. 2008). Large scale atmospheric circulation in terms of low frequency components (time scales of 1–3 months) tend to cluster around certain weather regimes whose resident and recurrence times, as well as the transition probabilities, seem to be well characterized by hidden Markov chains, as observed for the Euro-Atlantic region (Kondrashov et al. 2004, Deloncle et al. 2007). These approaches have to be combined, i.e. information from predictions of stochastic nature and weather regimes probabilistically linked to local climate conditions of precipitation and temperature should improve short-term predictions and hopefully extend them to seasonal predictions.

Some successes have been achieved with longer range forecasting (Hastenrath and Greischar 1993) in limited areas. Very promising results (McCabe and Legates 1995, Ruiz et al. 2006) have been achieved which suggest it should soon be possible to forecast drought up to 2 years ahead. The basis of these promising developments has been the demonstration of widespread influence on local precipitation of global ocean surface temperatures and large scale variations of these within the major oceans (e.g. the El-Niño Southern Oscillation in the tropical Pacific Ocean, the North Atlantic Oscillation in mid latitudes and variations in general ocean heat storage). Adopting a similar procedure, good streamflow forecasting has been shown to be possible in Australia (Ruiz et al. 2007).

4.1.5 Drought Monitoring

There have been proposals for maintaining a register of droughts around the world, but such a data base has not yet eventuated. Part of the problem has to do with maintenance of interest. When a drought is in progress many of those affected by the drought show concern. However maintaining interest in drought after the event has ended is very difficult – even for the traditionally scientific academic researchers, since it has commonly proved to be almost impossible to sustain funding of drought research when no drought is occurring.

Drought monitoring and, in particular, Drought Watch Systems (Rossi 2003) are often discussed as being a good idea but action is rarely taken. Attempts have been made at University of Nebraska (NDMC 2006) but even there sustaining interest between droughts has proved difficult. It seems the only means of sustaining such interest in most countries is the strong, altruistic, (perhaps indeed almost obsessive) commitment of one or two individuals.

Monitoring means keeping an ongoing watch and attempting to alert interested parties when a drought appears to be developing. Maintaining a historical record is part of drought monitoring, but if there is no intention of indicating and publicising the imminence of drought any monitoring scheme would seem to have little practical value.

Defining the beginning of a drought, or the likelihood of a drought in the near future has proved to be very difficult. There are proposals of forecasting schemes, which are discussed in Section 4.1.4, but even with these the entry into drought conditions is very uncertain, and so pronouncements of the beginning of a drought are usually stated with some probability of being correct, since such pronouncements can never be made with certainty. It is a simple matter to recognise drought conditions with hindsight, but the natural climatic variability means that early recognition of drought is uncertain, and difficult.

As mentioned above, for drought monitoring and warning, meteorologists and hydrologists have developed indices, which depend on hydro-meteorological parameters or rely on probabilities of drought occurrence Adopting different indices at the regional level with support of mapping probably has the best potential. The U.S. Drought Monitor (http://drought.unl.edu/dm) is an excellent example. It provides a variety of maps and adopts specific categories of drought severity from combining and interpreting the information made available. Similar services are provided by other organizations such as the Australian Bureau of Meteorology (www.bom.gov.au). The important feature of these web pages is that they produce a vast amount of information for users which cover not only the possible occurrence of droughts but also measures to cope with water scarcity and additional relevant scientific information. Figure 4.5 shows a map of rainfall deficiency adopting the deciles procedure referred to above in Section 4.1.3.

Drought monitoring may use a variety of variables relative to weather, surface hydrology, reservoirs, groundwater, water quality, soil moisture and vegetation, observation tools land or satellite based, and modelling tools for computing drought indices, exploring teleconnections, developing predictions and to produce public

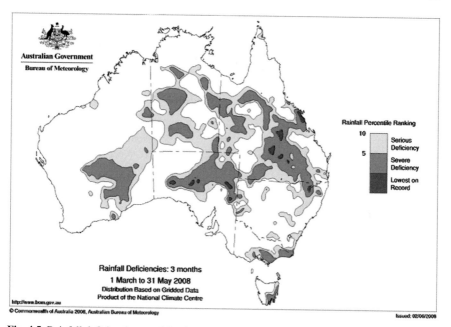

Fig. 4.5 Rainfall deficiencies over March – May 2008 in Australia (www.bom.gov.au)

information. As pointed out above, their usefulness depends upon the capabilities for providing early warning and creating interactions with water managers, decision-makers and policy-makers. As an example, in agriculture, development of data bases information and communication technologies (ICT) and communications pathways between farmers and grower associations and even extension services can greatly assist in timely adoption of non-routine measures to cope with drought (Fig. 4.6).

The controversy over perceptions of drought, and the consequent difficulties in defining them and their characteristics, do not help decision and policy makers to plan for droughts. Lack of clearly agreed definitions make it difficult to implement preparedness measures, to apply timely mitigation measures when a drought occurs, or to adequately evaluate drought impacts. Therefore, despite the logic behind monitoring as exemplified in Fig. 4.6, drought monitoring and warning are often not applied even when drought variables are observed and indices are computed.

Another cause of difficulties is of institutional nature. Observations of drought are often performed by different organizations and drought indices are differently computed and not analyzed in common by those institutions. Then, even if predictions or forecasts may be available it is extremely difficult to adopt them operationally. The result is that warning provided by different organizations for the same country or region lose credibility since they are released independently and for the promotion or prestige of the institution itself. In addition, specialized information that requires specialised knowledge and interpretation of different disciplines and disseminated through specific institutions, mainly universities and research centres is not adopted. It is therefore required that appropriate cooperative

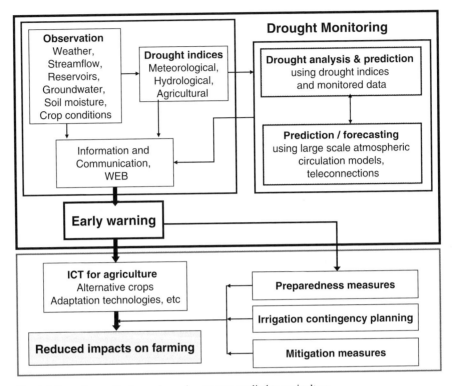

Fig. 4.6 Drought monitoring and warning system applied to agriculture

task forces be created combining institutions specifically to foster public interest benefit and service. However, as discussed above, maintaining such bodies through times of no drought has proved universally difficult, if not impossible – something to do with human nature!

4.1.6 Drought Risk Management and Communication

During a drought it is a relatively simple matter to enlist the support of water users for care in the use of available supplies. However, if reliable forecasts or predictions of drought could be provided there would be large opportunities for water managers and water users to adjust their controls and activities to conserve water before the start of the drought and to increase water availability during the drought. Unfortunately confident forecasts of drought are still only a hope – primarily among researchers.

These short time drought predictions are important for warning farmers about the probable initiation or establishment of a drought, about its continuation or its probable termination in a few months. This information may help them to make decisions to cope with that predicted situation. Short time drought predictions may also be used to alert water managers and decision or policy makers about the need to enforce appropriate preparedness measures before a drought is effectively installed, or to prepare for a post-drought period. The Australian experience constitutes a good example of moving from drought disaster mitigation into risk management (Wilhite 2005b).

To cope with droughts requires preparatory measures, contingency plans that support the timely implementation of mitigation measures and that forecast impacts which are likely to be experienced once the drought becomes established and evolves. This implies risk-based drought policies and effective monitoring and early warning systems. However this is only possible for a society that has strong institutions and where public participation forces policy-makers to adopt drought risk policies and make the society resilient to drought (Wilhite and Buchanan-Smith 2005). When technological and political capabilities are lacking and public participation is poor, the society is vulnerable to the full effects of drought (Fig. 4.7). Poverty, conflicts among users and overexploitation of resources will result while appropriate awareness will not improve but complacency will develop and there will be a misplaced expectation in accepting that next time it will be better. This type of situation produces a downward spiral to disaster.

Recognition that drought is inevitable and that its frequency of occurrence and even its intensity can be known provides opportunity to prepare for most of its effects. Drought is not an unexpected catastrophic occurrence. Careful study of available climatic data (at least precipitation, and streamflow if it is available) shows that in the normal variability of climate about long term mean values, there are

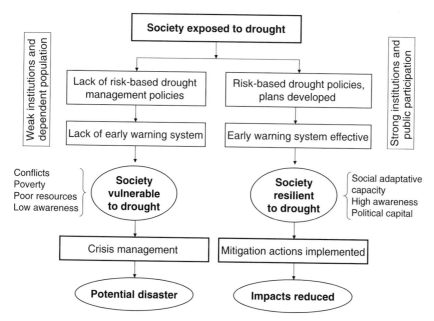

Fig. 4.7 Drought risk management to change a society from drought vulnerable to drought resilient

some very wet periods and some very dry periods (droughts). The people (all water users) need to be continually reminded of these facts. Education programs (as discussed in Chapter 12) are needed to make sure no-one is surprised when the next drought occurs. Depending on how drought is defined, abnormally dry periods can be expected to occur about 10% of the time. Therefore there is a need to store up essential life-support resources (water and food) during the average and wet periods so that when a drought inevitably occurs the population may be supplied the resources needed to survive.

Many populations, and their governments, seem to function as though drought cannot occur, or will never occur again as it did in the past. They live from day-to-day, consuming all of their life-support resources as they go. It is essential, even for poor people, to be storing resources for the next drought. In most instances successful storage and preservation of food for long periods is difficult if not impossible. Similarly storage of water is also very difficult unless there is a groundwater aquifer or a large, deep surface water reservoir. Storage of water in large reservoirs must usually be a government activity, and preserving the stored water to make sure it is available during the next drought needs very strong political will. There is always the desire to gain maximum economic returns from use of stored water and not save it "in case" a drought should occur some day.

For water resources systems, both for agricultural and non-agricultural uses, drought contingency plans are required. To appropriately build them, it is necessary to identify the mitigation measures and, during a drought, to evaluate their effectiveness. Ranking mitigation measures is therefore essential as proposed by Rossi et al. (2007b) and summarized in Fig. 4.8.

Since food and water can only be stored for a limited time, an alternative drought preparedness strategy may be to save money in a financial institution. If drought prevents production of a harvestable crop about one year in five then about one quarter of each years' crop needs to be sold and the proceeds deposited in a bank account to provide for drought survival. Unfortunately "no-drought today" produces complacency and the attitude "I will put something aside for drought next year". That may be too late. The next drought may be next year. Around the world, as mentioned earlier, it is almost impossible to invoke drought preparedness programs except when a drought is actually occurring. So there needs to be continual government and NGO publicity and educational programs to raise public awareness of the inevitability of the occurrence of drought. Educational strategies are discussed in Chapter 12.

In drought resilient societies, where drought risk management could be adopted, where awareness is appropriate for timely implementing of mitigation measures, there is the need for a proactive approach. This consists of planning during the no-drought period, implementation during drought and monitoring and evaluation during and after the drought event. Figure 4.9 synthesizes this approach and shows that implementation of measures relates to the monitoring and warning systems adopted. Contingency planning is initiated when a moderate drought is established, i.e., when risk for water scarcity is detected. More stringent mitigation measures as part of the contingency plan are adopted when severe or very severe drought

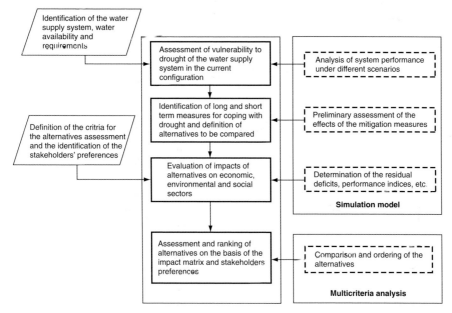

Fig. 4.8 Ranking mitigation measures to be applied in water resources systems (Rossi et al. 2007b)

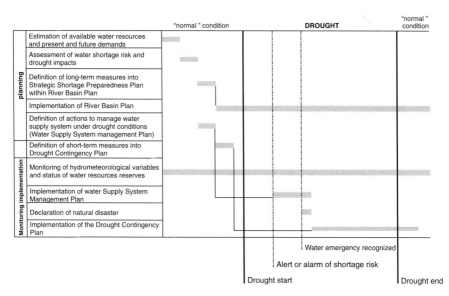

Fig. 4.9 Main steps of a proactive drought contingency plan applied to a water system (Rossi et al. 2007b)

initiates. Of course, severity conditions relate not only with weather parameters but with those parameters that identify the vulnerability of the water system under analysis.

4.2 Desertification

4.2.1 Concepts and Definitions

Desertification is a man-made water scarcity regimen as discussed in Chapter 2. It is defined here as a permanent imbalance in the availability of water combined with damaged soil, inappropriate land use and mining of groundwater. It produces increased flash flooding, loss of riparian ecosystems and a deterioration of the carrying capacity of the land system. It usually occurs in arid, semi-arid and sub-humid climates. Climate change contributes to desertification and drought can also strongly aggravate the process of desertification by temporarily increasing the human pressure on the diminished surface water and groundwater resources.

The United Nations Convention to Combat Desertification (UNCCD) adopts the following definition: "land degradation in arid, semi-arid and dry sub-humid areas resulting from various factors, including climatic variations and human activities" (http://www.unccd.int/). Combating desertification under the same Convention usually addresses activities of integrated development of land in arid, semi-arid, and dry sub-arid areas for sustainable development, aiming at (i) prevention and or reduction of land degradation; (ii) rehabilitation of partly degraded land; and (iii) reclamation of desertified land. The term "land degradation" usually refers to the reduction or loss of the biological or economic productivity and complexity of rainfed cropland, irrigated cropland, or range, pasture, forest, and woodlands resulting from inappropriate land uses or from a process or combination of processes, including processes arising from human activities and habitation patterns. Related concepts and definitions do not refer to water except in its relation to climate. However these concepts are not broad enough to cover the causes and impacts of desertification and essentially relate to physical factors and impacts, such as land relief, vegetation removal and soil degradation.

Different definitions are used for desertification, generally focusing on agricultural land degradation. Numerous papers about desertification have been produced which present a vast panoply of causes and concepts (e.g. in Balabanis et al. 1999, Enne et al. 2000, Geeson et al. 2002, Kepner et al. 2006). Research papers often focus on only a part of the land degradation problem and related responses. Also often, the concept of land is replaced by that of soil, which is very restrictive in scope. Nevertheless, there is a trend to better integrate the different factors that characterize the land, as in the case of studies relative to data, indicators and measures to prevent or to combat desertification (e.g. Enne and Zucca 2000, Enne et al. 2000, 2001, 2004, Briassoulis et al. 2003, Kepner et al. 2006). However, the water factor is generally given far too little consideration.

More integrative approaches, where the concept of land is broadened to approach the idea of landscape or territory and to include soil and water resources, land surface and vegetation or crops (UNCED 1992), are more often present when national perspectives are discussed (e.g. Roxo and Mourão 1998, Martin de Santa Olalla 2001, Louro 2004). The social and economic context favoring or being affected by desertification also has more relevance when national perspectives are considered, including regional development (Correia 2004). This situation identifies a possible gap between research studies and national identification of vulnerable areas and response measures. Research often focuses on the processes in great detail, and mainly the measurable physical ones, while approaches aiming to tackle identification of problems and issues in a national perspective tend to produce a holistic view with a broader, more interdisciplinary perspective.

The causes of desertification and its impacts do not relate only to the agricultural activities but have much wider effects, particularly on the overall living conditions of populations. Thus, land should be understood as territory and not be restricted to agricultural land, and concepts need to be broadened in scope to focus attention on the water scarcity issues. In this perspective, without omitting soil resources, it is more appropriate to define desertification in relation to the water and ecological imbalances produced by the misuse of water and land resources. Attention is then drawn to the fact that desertification, including land degradation, is both a cause and an effect of man-made water scarcity. In this perspective, it is also necessary to highlight the often adverse consequences of an economic development that disregards the effects of misuse and mining of natural resources and the damages inflicted to the quality and availability of natural assets. Social, economic and institutional issues deserve special attention since they play a key role in the evolution of the desertification processes and the effects of actions taken to combat desertification.

Broadening the concepts of desertification corresponds to consider not only the bio-physical aspects of the processes but also the impacts and role of the society in favouring or combating the desertification as expressed in Fig. 4.10.

Land degradation is a complex process produced and reproduced through interactions between biophysical and human – social factors. The evolution of the definition of the desertification process adopted herein evidently focuses on human

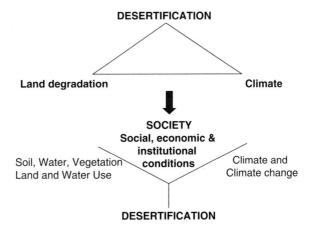

Fig. 4.10 Broadening the concept of desertification to include societal causes and impacts (from Pereira et al. 2006)

activities and social, economic and institutional processes as the principal causes, aggravated by natural climatic conditions. A territory where climate is marked by water scarcity, natural resources are poor and fragile, the population is poor and aged, the economy is based solely on agricultural activities, local institutions are weak, public participation is not effective and governmental policies do not appropriately focus on the development of rural areas, is highly vulnerable to desertification. A broad concept is therefore required, and a large space scale needs to be adopted, which is the approach taken in the definition proposed above.

4.2.2 Processes and Indicators

Common manifestations and symptoms of desertification and land degradation consist of: (i) reduction of crop yields in rainfed or irrigated farmland; (ii) gradual loss of soil productivity; (iii) reduction and loss of perennial woody biomass; (iv) reduced availability of water; (v) soil losses by water and wind erosion; (vi) cropland chemical degradation, mainly due to salinization, (vii) water quality degradation, (viii) increased flash flooding; and (ix) sedimentation of water bodies. Relative to agricultural land, the dimensions of land degradation are enormous (Table 4.3). The size of degraded areas justifies the attention given worldwide to the processes leading to land degradation and, in many regions, to desertification. Fortunately, desertification has a much smaller magnitude but figures indicate that many agricultural areas in dry regions are vulnerable to desertification.

The manifestations of desertification impact the life of communities living in rural and urban affected areas due to deterioration of life support systems. Impacts are generally associated with water scarcity due to climate aridity or dryness and poor water management. In addition, drought acts as a trigger by exacerbating the natural and man-made water scarcity and the pressure on the natural resources. These aspects identify a vicious circle of bio-physical degradation processes where climate and related water scarcity play a main role (Fig. 4.11).

The availability of water resources both in quantity and quality establishes a link between climatic driving forces and land conditions. On the one hand, the climatic

Table 4.3 Agricultural arid lands affected by land degradation in million hectares (adapted from Enne and Zucca 2000)

Continent	Irrigated land		Rainfed cropland			Rangeland			Total		
	Total (Mha)	Degraded (Mha) (%)	Total (Mha)	Degraded (Mha) (%)		Total (Mha)	Degraded (Mha) (%)		Total (Mha)	Degraded (Mha) (%)	
Africa	10.4	1.9 18	78.8	48.8	62	1342	995 74		1432	1046 73	
Asia	92.0	31.8 35	218.1	122.2	56	1571	1188 76		1881	1342 71	
Australia	1.8	0.2 13	42.1	14.3	34	657	361 55		701	376 54	
Europe	11.9	1.9 16	22.1	11.8	54	111	81 72		145	94 65	
N. America	20.8	5.8 28	74.1	11.6	16	483	411 85		578	429 74	
S. America	8.4	1.4 17	21.3	6.6	31	391	298 76		421	306 73	
Total	145.5	43.1 30	457.7	215.6	47	4555	3334 73		5158	3593 69	

Fig. 4.11 The circle of
bio-physical land degradation
characterising desertification

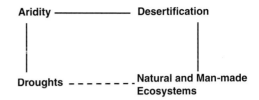

conditions determine the amount and time and space variability of precipitation, as well as the regime and quantity of evapotranspiration (ET) from natural and man-made ecosystems. Therefore, the precipitation and ET regimes influence and characterize land conditions. On the other hand, the land condition referring to infiltration and soil water retention and vegetation cover, influences or determines the partition of the precipitation into the various components of the hydrologic cycle, i.e., interception, direct evaporation, surface runoff, and infiltrated water, which partitions again into vegetation ET, subsurface flows and groundwater flows. Therefore, there is a close link between water resources availability and land condition to determine water balance at the local scale. By influencing the land condition, man influences the hydrologic cycle and the water balance, and thus the water scarcity at both local and basin scales. It is well recognized that water resources management, at both the local and at a large scale, influences desertification (WMO 1996, Enne et al. 2004, Pereira et al. 2005).

Differences in conceptual approaches and the difficulty in assuming the UNCCD definition operationally, i.e. considering the integration of the factors and processes affecting desertification, may lead to poor understanding of the causes of desertification and, consequently, of the vulnerability of the land or the territory to this degradation. A wider definition as adopted herein allows consideration that desertification not only affects the soil and water resources but the whole territory with consequences for all activities, not only for agriculture. Desertification impacts the urban, rural and agricultural societies where the degradation of natural resources and the pressure on the use of land and water creates adverse living conditions for the population of the region.

Public opinion commonly associates the term desertification with loss and ageing of population, lack of opportunities for development and lack of jobs for the younger generation. These opinions accurately express the dynamics in the territories affected by land degradation. In developed countries, such as those of the northern Mediterranean, the consequences of desertification in terms of reducing the capability of land to produce food are less significant than in less developed and poorer regions. However, changes in land use that produce a decrease in agricultural activities have enormous impacts and favour migration to urban areas, leading to land abandonment and ageing of remaining populations. These impacts are very large when the population is highly affected by poverty, illiteracy, health problems and lack of policies to improve their living conditions. Large, global scale approaches are advocated by many (e.g. Reynolds and Stafford Smith 2002), including consideration that combating desertification is a regional security issue

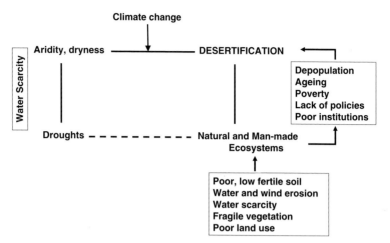

Fig. 4.12 The circle of desertification processes with identification of the role of social, economic and institutional factors (adapted from Pereira et al. 2006)

(Kepner et al. 2006). A holistic approach allows easier understanding of the processes involved in creating the circle of desertification (Fig. 4.11), considering the social, economic and institutional conditions, amplified in Fig. 4.12. Desertification is a serious environmental problem, but its social dimension is probably of greater importance.

Many indicators have been developed and are used to identify desertification problems. The aridity index (ratio of annual precipitation to potential ET, P/ET_o) is commonly used and the value $P/ET_o = 0.75$ is generally adopted as the upper climatic threshold, the ratio above which desertification is unlikely. However it should not be computed from average values of P and ET but from their probabilities of being exceeded, let say 20 or 10% of the years, to take into consideration the impacts of droughts, which are not considered when just using the mean or the median values. Rainfall variability and seasonal distribution is essential. In many regions indicators of wind erosion, windstorm frequency and intensity are also necessary.

More common indicators to describe the bio-physical land degradation processes are of an agricultural and environmental nature, mainly relating to soil, vegetation and land use. Small and large space scales are commonly used. However, indicators need to be selected according to the regional land-use conditions to identify the most important impacts that occur locally and not just to characterize the farmland, rangeland and forest lands. Indicators for water resources conditions – relating to surface water regimes and storage, groundwater and water quality – are also required but are difficult to adopt. Indicators are generally used to identify the vulnerability of a territory to desertification, which allow specification and implementation of preventive measures. Fighting against desertification when it is already well advanced is difficult because the processes tend to increase in intensity as the scourge progresses; thus, the most useful indicators are those which can indicate

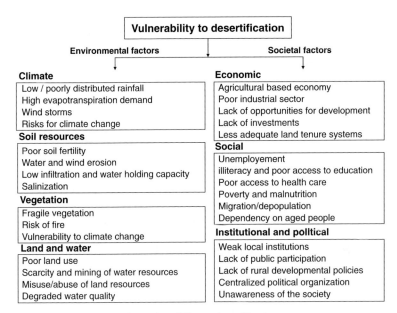

Fig. 4.13 Main factors determining vulnerability to desertification

the potential risk of desertification before it is too late for remedial action. Possible impacts of desertification are summarized in Fig. 4.13 and indicators should specifically refer to one or more of these impacts. Good examples of indicators are those proposed by Enne and Zucca (2000) and those reported by Abraham et al. (2003). Combining environmental and societal factors is an absolute necessity.

The indicators of desertification more commonly used are bio-physical/environmental and socio-economic. However, in addition to these, indicators referring to institutional problems, policies and political issues are needed. It is well known that innovation in land and water use, adaptation technologies, investments that create new jobs and better incomes, or fairness in the use of natural resources cannot be developed if local institutions are weak, public participation is not guaranteed, rural development policies are not effective and the political power is centralized. In addition to the local populations, the society as a whole needs to be aware of problems and to have a willingness to support related solutions, which usually require large public investments. Understanding the limitations due to institutional and political constraints is therefore required to identify the vulnerability to desertification and the limitations in finding solutions of a social, economic, bio-physical and environmental nature. Moreover, there is a need to promote democratic governance of natural resources.

The causes of desertification as embodied in the Convention for Combating Desertification refer to complex interactions among physical, biological, political, social, cultural and economic factors that may contribute to sustainable economic growth, social development and poverty eradication. Desertification affects sustainable development through interrelationships with important social problems such as poverty, poor health and nutrition, lack of food security, migration of populations, and demographic dynamics. Obviously, the great complexity of the system of interactions renders the task of identifying unequivocal and efficient indicators an arduous and challenging one.

4.2.3 Monitoring and Information

The UNCCD commits affected countries to engage in rigorous monitoring of both the vulnerability to desertification and the results delivered by National Action Programmes (NAPs). It also stimulates cooperation among countries in Regional Action Programmes (RAPs). MEDRAP, for the North Mediterranean region (Enne et al. 2004) is a good example. Considerable improvements in the efficiency of such programmes can be achieved by carrying out continuous surveillance and assessing the results of desertification control measures, as well as by providing public information and feeding the outcomes into NAPs and "on the ground" projects.

Three types of monitoring may be distinguished:

- Desertification monitoring, which consists of a systematic surveillance of land degradation. Indicators that capture and assess the short, medium and long term evolution of physical, biological, socio-economic and institutional factors should be adopted.
- Process monitoring, concerning the surveillance of progress among all groups of actors involved in formulating and implementing action programmes at the various levels, and identification of opportunities for improvement. Indicators should permit an assessment of the quality of ongoing processes and of the commitment of the various actor groups to the formulation and implementation of action programmes.
- Impact monitoring to assess the effectiveness of measures to combat desertification. This concerns the surveillance and quantification of how the physical, biological and socio-economic environment responds to various NAP strategies, policies and measures.

Desertification monitoring allows defining environmentally sensitive areas (ESA). Monitoring may be performed using ground and satellite observations, requires interviews and questionnaires for identification of social, economic and institutional problems, and that the development of good databases, including spatial tools, be developed.

Three general types of ESA definitions can be distinguished, based on the stage of land degradation:

- Critical ESAs: Areas already highly degraded through past misuse, presenting a threat to the environment of the surrounding areas.
- Fragile ESAs: Areas in which any change in the delicate balance of natural and human activity is likely to bring about desertification. Lack of care in adopting

preventive measures and impacts of (predicted) climate change could shift these to critical ESAs.

- Potential ESAs: Vulnerable areas that could be threatened by desertification under significant climate change, if land use is impoverished or where offsite impacts may produce severe problems elsewhere under variable land use or socio-economic conditions. This would also include abandoned land which is not properly managed.

The production of a set of vulnerability and impact indicators should constitute an important step towards creating a Desertification Monitoring System (Enne and Zucca 2000). In affected countries these would have the dual function of:

- Providing a diagnosis, integrated in space and time, of the state of natural resources and of populations and institutional conditions of the affected regions;
- Supporting the decision-making process relative to defining and implementing measures and practices for preventing, mitigating or reversing the desertification processes.

Setting up of a desertification monitoring system should include (i) areas where remedial programmes have already been started, (ii) monitoring the progress of the programmes and assessing their usefulness; and (iii) disseminating related information to all concerned. The assembly and evaluation of information should be a continuous process, providing a feedback mechanism for national planning and action. To carry this out, national action should be considered to establish tools for assessing desertification:

a) Indicating the relative seriousness of the situation for all the regions affected, with a view to establishing priorities and degrees of urgency;
b) Standardizing the monitoring facilities and methods in regions affected or likely to become so;
c) Improving networks of climatological, meteorological and hydrological stations in regions exposed to desertification so as to permit more detailed and sustained monitoring and assessment of climatic and hydrological conditions in relation to the desertification process. Satellite imagery techniques should be employed where appropriate. National meteorological and hydrological services should provide ongoing assessments;
d) Observing atmospheric processes, the state of vegetation and soil cover, dust transport, shifting of sand dunes, the distribution, migration and abundance of wildlife, the condition of livestock, the condition of crops and crop yields, and changes in water availability and use;
e) Mapping to get easier spatial information on processes.

Databases should be accessible to various types of users, including stakeholders and researchers, and not be exclusive to the officers in charge of surveillance programmes. Related data needs to be appropriately publicized with a view of raising the awareness of, and focussing the attention of the population and the society.

The intensity and characteristics of desertification processes varies in different regions of the world. Vulnerability also varies as identified in Fig. 4.13. Necessarily, the measures to combat desertification, including preventing vulnerable areas becoming threatened by these processes are very diverse and are discussed in Chapters 9 and 10 of this book. Information is available in numerous research papers and books such as those referenced above. The challenge is to create information suitable for users and stakeholders relative to their practices and management issues, and for decision-makers and policy-makers aimed at creating appropriate conditions for avoiding, preventing and fighting desertification.

There is a need to implement an education process focussed on the users and stakeholders, aimed at making them actors in the development processes that overcome the desertification problems. Another challenge is to develop educational programs (as outlined in Chapter 12) that build up the awareness of the society about desertification and man-made water scarcity. The UNCCD includes in its website good educational kits that could be broadly used when translated into local languages (http://www.unesco.org/mab/ekocd/intro_home2.html and http://www.unccd.int/publicinfo/unescoKit/unescoKit.php).

4.2.4 Social and Political Constraints and Issues

As has been discussed earlier, desertification not only affects the physical environment but also the society, particularly the quality of life and the productivity that maintains human settlements. There is close association between where the emphasis of the economy is placed and the balance of societal parameters that give reasonable appreciation to the value of natural resources as a social good and as a factor for enhancing the quality of life. This societal dimension of the quality of natural resources is probably the most important from the point of justifying the need for adopting a strategy for combating desertification. Preserving environmental resources is as important for the population as the proper and sustainable use of those resources.

Desertification and the factors contributing to it do not only spring from single-person actions or lack of information. Rather desertification results from the collective response of communities and social groups to development opportunities provided by the low priority and importance societies tend to give to environmental protection concerns. When emphasis is placed on urban development and related activities, the surrounding rural areas which have poor natural resources do not receive appropriate support to withstand the adverse effects and changes of recent decades.

In territories where water resources are scarce, soil resources are poor and vegetation is fragile, natural resources are not able to cope with the demands of the modern society. In other words, the former equilibrium has been broken and the society has not yet been able to establish new equilibrium conditions. There is a lack of technologies and investment for progressively converting former land and water uses to present needs. A market economy does not function properly in these conditions.

Land degradation occurs affecting the population and causing the aforementioned problems of poverty, malnutrition, low levels of health, illiteracy and low educational levels, and migration of the younger generation because new opportunities are lacking. Then weak human resources add to poor natural resources and the problems are reinforced. Land degradation then turns into desertification.

In areas without rich human resources, local institutions are generally weak and public participation is low. However, changes in economic and social conditions in those areas would require that social driving forces stimulate new investments and employment, adaptation technologies, and health and education facilities that could provide for the dynamics of positive changes. Strong local institutions and public participation are also necessary to call for political action, mainly related to development policies. This includes the promotion of an effective democratic governance of natural resources. Otherwise, land degradation becomes self reinforcing because social driving forces are not strong enough to support the desirable changes. We find another circle of weakness leading and aggravating desertification (Fig. 4.14).

Desertification became a prominent environmental issue. However, it is necessary to clearly recognize how societal factors – social, cultural, economic, institutional and political – really influence and are influenced by desertification. Treating desertification solely as an environmental problem is definitely not enough. Measures to combat desertification must consider strengthening the societal driving forces. Nevertheless, this has to be performed with respect to the cultural, religious and political values of each region or country.

Fig. 4.14 The relationship of poor natural and human resources and social driving forces in a desertification process

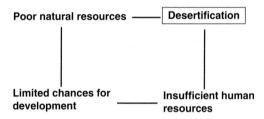

4.2.5 Issues to Combat Desertification

From the above, it is evident that combating desertification is complex and difficult (Fig. 4.15). As the UNCCD states, it "includes activities which are part of the integrated development of land in arid, semi-arid and dry sub-humid areas for sustainable development". Problems of a physical nature relate to agriculture, forest and rangelands; land degradation affects soils in various ways, namely through water or wind erosion; man-induced water scarcity may refer to water quantity and quality, as well as to surface and groundwater; revegetation may focus on protection or production objectives; land uses are diverse and refer to various activities. Social problems should give priority to poverty eradication, health care and education, food security and improved nutritional conditions. To support these initiatives it is necessary to

Fig. 4.15 Principal issues
and activities for breaking the
vicious cycle of
desertification

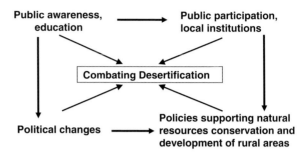

develop an economic base that generates employment and incomes for the population from agricultural productivity, that encourages use of traditional products and
industries, attracts new industries to use the local raw materials, and strengthens the
local service industries and means of communication. Implementing these activities
requires that local institutions become driving forces for development. They need
the support of public participation and for the people to be involved in the decision-
making processes that build and encourage implementation of the required policies,
including those for democratic governance of natural resources (Fig. 4.15). These
institutional activities need to include rural and urban development, development of
production and conservation technologies, implementation of water allocation and
water quality standards, credit, investment and taxation among other things. Given
the complexity of issues that may be required, the success of combating desertification relies on developing public awareness and education on the problem and
on the possibilities for its solution. A necessary adjunct to these activities may be
the support of political and policy changes, as well as development of public participation and strengthening of local institutions. The required policies for resource
conservation and rural development may then be built and implemented with the
participation and support of local actors and stakeholders.

Combating desertification requires not only local but external driving forces.
Political and policy changes are not effective if activities are restricted to the local
level; the engagement of regional or national governments is required. Policies that
give priorities for investment or provide favourable taxes for those target areas
need National awareness and support. Education and awareness raising programs
need support and participation from beyond the local area, as well as programs
for strengthening local public participation and institution building. These factors
explain why NGOs are active in these non popular areas and activities. However the
main need for these external actors is that they play a role of catalysers for only a
limited time. Development will not be sustained when creating dependencies; it is
essential to have the support and participation of interested, local populations.

Various examples of successful developments and cases are presented at the
UNCCD website (http://www.unccd.int/knowledge/menu.php). Special reference
is given to the role of women, to African and Asian experiences and to the use
of traditional technologies. Traditional systems are seen as having several important attributes that inhibit desertification and thus make them sustainable users

and managers of dryland resources. Traditional systems, however, require updated knowledge, mainly when they need to be integrated into, but not replaced by modern ones.

National Action Programmes (NAP) are one of the key instruments in the implementation of the Convention. They are strengthened by Action Programmes on Sub-regional (SRAP) and Regional (RAP) levels. National Action Programmes are developed in the framework of a participative approach involving the local communities and they spell out the practical steps and measures to be taken to combat desertification in specific ecosystems. Appropriate information is available from http://www.unccd.int/actionprogrammes/menu.php.

Chapter 5
Conceptual Thinking in Coping with Water Scarcity

Abstract There are important conceptual and ownership issues that can over-ride the physical and engineering aspects of coping with water scarcity. Water has a very high value for support of life, but it is often not fully valued by the community of water users. How do we place a value on water? What are its social, environmental, economic, and cultural values? Who owns the water? Who has the right to collect, store, allocate, distribute and sell or lease water? Who has the authority, and how do they proceed, to set the price to be paid for water? How are ownership and accessibility managed where water crosses political boundaries?

5.1 Introduction

When thinking of coping with water scarcity there are many issues to be considered other than the direct technical matters dealing with collection of water and prevention of losses. The human dimension is of great importance. Over its history humankind has developed cultural traditions, social structures and institutions which have a huge impact on the availability and use of water. For example in some traditions there is a concept of ownership whereas in others the idea of ownership of a product of a natural process makes no sense.

This chapter attempts to discuss some of these "non technical" issues and their impact on water resources availability. To those from a particular culture, the traditions of another place, and their influence and constraints on water resource use may appear strange and even unacceptable. However in dealing with cultural issues great care is needed, particularly to understand how the current situation developed historically. To advocate change of some practices may be offensive to the local population, no matter what the logic from a water resource point of view. Therefore advocacy of change must be approached with great sensitivity and empathy. It may be that no change will be possible during the life of the current generation. Other traditions however, may have no more basis than "accepted practice" and change of these to increase benefits to all concerned may only be a matter of logical argument and demonstration.

This chapter attempts to consider some of the key ideas that influence water resources development, and therefore also impact strategies for coping with water

L.S. Pereira et al., *Coping with Water Scarcity*, DOI 10.1007/978-1-4020-9579-5_5,   77
© Springer Science+Business Media B.V. 2009

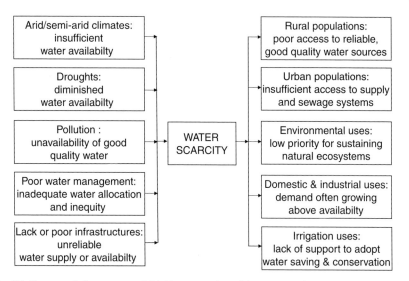

Fig. 5.1 Framework for conceptual thinking on coping with water scarcity

scarcity (Fig. 5.1). As has been mentioned in previous chapters, water scarcity may be due to natural causes – aridity and droughts – or to man-made causes. Man-made scarcity is produced through a wide range of problems that may include: (1) pollution or contamination by upstream users – urban, industrial, mining – causing degradation of that water before it can be accessed by downstream users, (2) poor water management such that there are strong inequities in allocations both in time and space between potential users, and (3) insufficient or inadequate infrastructure for water collection, storage, conveyance and distribution, especially during periods of stress, and when infrastructure maintenance is poor.

Water scarcity problems are different among various groups of the population and users, e.g. rural *vs.* urban populations, environmental *vs.* human and economic uses, or domestic and industrial *vs.* irrigation uses. A key issue is controlling the demand for access to water, since the demand for water is everywhere growing very rapidly, including in water scarce areas.

5.2 Social Value of Water

5.2.1 Water for Life

Water is essential to life and therefore it affects lifestyles. Conversely in areas of water scarcity the availability of water is very much affected by the lifestyles of the local population. Land use and animal husbandry practices have considerable influence on water use. In water scarce regions there is a need to capture some of the precipitation for household use and to retain as much as possible of the remainder in the soil to promote crop or forage growth. Runoff from roofs and surfaces which have been compacted by the passage of many feet (animal and human) can be directed to storage locations. However crop and grazing lands need to be kept tilled, or at least in a non-compacted state to promote rapid infiltration of precipitation whenever it occurs, and to retain water in the soil, and to avoid soil water evaporation.

5.2.2 Differences Between Urban and Rural Needs

In urban areas there is almost always some form of municipal water supply, either in the form of a piped system or using tankers. The major concern of municipal water managers is to capture water, usually outside the urban boundaries, and to store it for later use. Roof water can be used in-house and roofs need to be constructed to facilitate capture of the water. Where the traditional roofing is not suited to easy capture of its runoff, for example where the roof comprises thatched grasses and leaves, there may be a need to consider a change of roof style. This will usually require some difficult decisions and expenses, but if the roof provides the only water supply for life support, a difficult (both culturally and economically) decision may need to be made, or some expensive alternative water source may need to be found. Land surface runoff can be directed to sumps from where it can be extracted for use or encouraged to percolate to an aquifer from which it can later be pumped as needed.

Many of the poorer segments of urban societies need to collect water from a standpipe or from a tanker. In these situations the distance the water must be carried to the point of use (residence) is usually less than 200 m. However, under these circumstances, where both water supply and sewerage systems are precarious or non existent, health problems may be more important and warrant more attention and priority than the labour and cost aspects of the water collection.

In rural regions, often, there is no piped water supply. There may be tanker delivery of water, or more commonly, each household needs to collect its own water, usually from a communal well or from a stream or lake (Fig. 5.2). Here much more labour is required to obtain water – to lift water from a well and to carry it a considerable distance from the water source. Water collection may consume a significant proportion of the time and physical labour of the household. Quite often the collected water is of poor quality and also requires time and fuel, generally wood, for boiling it before use.

Drought periods tend to bring more hardship to rural households than to urban dwellers. Urban supply systems, though usually far from perfect, often include reservoirs that store water from periods when it is available, and incorporate opportunities for managers to restrict supply during drought and to conserve at least enough to maintain drinking water supplies. However in rural areas, when a managing authority is lacking and/or the village authority is weak, it is *everyone for himself* and water tends to continue to be used without restriction. As the drought continues the usual supply source may fail completely. To survive, the population

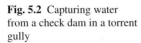
Fig. 5.2 Capturing water from a check dam in a torrent gully

must carry water to their houses from distant sources or, under extreme conditions, move near to a water source (perhaps a city). For these reasons drought usually has more severe impacts on the rural population than on city dwellers. However in extreme drought, if city supply systems fail, the poorer urban dwellers are likely to suffer more than rural populations because water will only be brought into the city at great expense, because sources are quite remote. The prices demanded for water under these circumstances are likely to be beyond the reach of the poor.

Almost all water-scarce regions of the world have at least brief, infrequent episodes of surface runoff, and systems need to be in place to capture this water when it occurs. The systems must be passive (i.e. not needing any human operation at the time of the event) and ready for acceptance of water at all times. The random, brief occurrence of precipitation means that human operation is unlikely to occur successfully in every runoff event and so systems needing human operation will only catch part of the possible available water. The quality of captured water may be an issue but quality can usually be improved relatively simply.

The key issue in regions of water scarcity is to maximise the capture and conservation of any available water (including its protection from disease vectors and preservation of its quality) and to adopt simple but effective water treatment to avoid water born diseases.

5.2.3 Differences Between Arid Zones and Non-arid Areas

Arid zones are readily recognised as regions of water scarcity but many non-arid regions also have scarcity of water supply, usually due to seasonality of precipitation, the density of the population and the nature of the accepted water use practices. In the wetter areas, however, water scarcity can be overcome by changes in water management, for example by encouragement of capture and consumption of roof runoff and re-use of water. In arid zones the scarcity is more likely to be of an absolute kind, with water in all its forms being very limited. Wise management of the scarce resource can significantly improve conditions and lifestyles, but in the arid zone per capita water use will always be lower than it could potentially be in the wetter areas. In the arid zone water has a very high price. In monetary terms the price may not be significant but the collection, storage and supply of water in the arid zone will always require either a high level of technology and/or collection and transport of water from some remote location.

Most communities have adapted their water collection and use practices to their environmental situation over many years. However recent ideas and simple, inexpensive technologies mean that these time honoured practices could realistically be improved upon in many situations. Water availability could be increased and the real cost (money, labour, management of extreme shortage) to communities of their water supply could be reduced. Flexible but realistic thinking and willingness to consider new ideas is required.

5.2.4 Social Effects of Water Supply/Collection Practices

In many areas of water scarcity collection of water consumes a large part of household labour, and therefore of the time of one or more of the household members. While this labour is a major cost to the household, it also provides much opportunity for social interaction for the water collectors. Change, to reduce the labour in water collection, would inevitably change the social structure of a community. This must be recognised as a consequence of change of water management practices. The social change that results may be resisted, resented and may lead to upheaval within a community. It is possible that improvements to water management, which may benefit all, may be opposed by community leadership because they fear social disruption or they do not wish to see the members of the water collection group freed from the constrictions of their water collection activities. Hence any proposals to change water collection or water management needs to be preceded and accompanied by an educational program aimed at preparing all levels of the community for the changes that will occur, highlighting the direct (increased supply at lower cost) and indirect (less labour consumption by water collection, more free time for the water collectors) benefits for the whole community. Arid zone communities are often isolated and conservative, resistant to change, particularly when that change may impact on the social structure.

5.3 Environmental Value of Water

Water is a major factor in shaping the natural environment. The amount of precipitation and its distribution in time has a major influence on the vegetation of the region. However it also influences the species of fauna to be found in the region and has a large, but long term influence on the shape of the landscape. Any changes in the water regime will inevitably lead to changes in these other components of the local environment. This is particularly true of the effects of water extraction on the flow regimes of rivers. There is continual community debate over the desirability of maintaining *natural* conditions in rivers. However the definition of these natural conditions is very variable and dependent on the social, political as well as scientific agenda of the personnel involved.

It has been clearly shown that change of flow regime, by extraction of water or by impounding water, can have large effects on the whole of the downstream river valley. This is true for ephemeral, usually dry streams as well as for perennially flowing rivers. In most cases the initial effects are small because (1) early in the development the flow regime is only slightly changed, and (2) time is required for the impacts to become apparent. However, as the water resource is developed further, the impacts become greater (Table 5.1), usually affecting both surface and groundwater downstream of the point of development, together with any wetlands and the riparian vegetation.

There are many examples of very large environmental impacts of development of water resources. In general these have not been foreseen, either because at the time of the development there was insufficient scientific understanding of water resources processes, or more recently because the planning was piecemeal, considering only the direct benefits of the development and being unaware of, or ignoring possible detrimental effects downstream. Some of the larger impacts (they could be termed disasters) have been (1) the large reduction of inflows to the Aral Sea, and the lowering of its surface by over 20 m and the reduction of its surface area by more than 50%

Table 5.1 Environmental concerns in relation to progressive water abstractions

State	Relative time	Activities
1	0	Natural conditions – no water abstractions
2	0.4	Minor water use – no noticeable environmental effects
3	0.6	Medium water use – some municipal or irrigation abstractions
4	0.6	First thinking about environment requirements
5	0.7	Larger water use – with some polluting effluents
6	0.8	Environmental flow laws introduced as demanded by the people – usually with no enforcement provisions or penalties
7	0.9	Larger water use – usually municipal or irrigation – with serious pollution impacts
8	0.95	Strong environmental laws with enforcement provision and penalties, Environmental flow requirements set at levels to reflect usage at this time or a little earlier (state 7). Environmental flows not sufficient to sustain natural environment. Actual enforcement weak.
9	1.0	Very strong political will with strong enforcement as demanded by the people. Still environmental flows not sufficient to sustain the natural environment. Environmental flow requirements reflect usage at about state 5. To achieve these flows much recycling must occur – all waste water returned to the river must have high level of treatment.

Fig. 5.3 The Aral Sea
shrinking from 1957 to 1999
(source: SIC-ICWC)

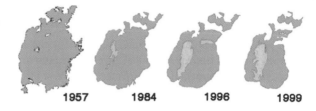

(Fig. 5.3). Here natural inflows have been diverted to consumptive use by irrigation and industrial uses. (2) In the Mediterranean Sea adjacent to the Nile delta the fish catch has fallen dramatically as a result of trapping of nutrients in the form of suspended solids in the Aswan High Dam. There are numerous writings suggesting that the High Dam has brought benefit to Egypt in the form of increased food production, but has cost the nation dearly in reduction of protein supply (fish), natural supply of agricultural nutrients, and has greatly increased the incidence of serious diseases such as schistosomiasis. (3) In Australia development of water resources from rivers that pass through the arid zone on their way to the ocean has led to salinisation of the land surface and river waters and devastating blooms of cyanobacters (blue-green algae) which are toxic to mammals.

Societies have been struggling to decide what priority they put on environmental values. The steps that have been taken over a period of 30 years in some developed countries can be summarised as shown in Table 5.1. There are lobby groups in developed nations which advocate returning the environment to some kind of pre-human condition, though it is usually not possible to define this condition. At the opposite extreme are those who say man is part of the environment and the changes resulting from human activity are *natural* and so they should just be accepted. Between the two extremes, the majority seem to be of the view that as much of the natural environment as possible should be preserved whilst allowing for the life support of humanity. Unfortunately there is no clear definition of the *needs* of humanity and as both population and expected standards of living have risen inexorably the scope for maintenance of remnants of pristine natural environments continues to diminish.

Much attention has been given to *environmental flows* in recent years. It has been accepted by many that preservation of the *natural* ecosystem of river valleys – wetlands, riparian flora and fauna and all the properties that go with a natural river system – is a desirable aim. To achieve this, the concept of *environmental flows* has emerged. There are many definitions and criteria for environmental flows, but in essence they are seen as flows needed to preserve the *natural environment*. Inevitably, where any development has occurred the flows cannot be as they were under natural conditions since the development activity inevitably has disrupted the natural conditions to a greater or lesser extent (Table 5.1). Compromises are needed. For example in a semi arid region where the natural flows are used for irrigation of cash crops it is inevitable that the demand for irrigation water will, if not now, then some time in the future, exceed the natural flow of the river. Hence to preserve any flow for environmental benefits requires that restrictions must be applied to the diversion of flows for irrigation. Where irrigation provides employment and generates large financial benefits for the region very strong measures, often at national and international level, are required to ensure any consideration is given to environmental flows. Water then has a monetary value as well as a non quantifiable *environmental* value. The monetary value of the environmental water can easily be quantified as being at least the value of crop production foregone.

Preservation of flow for environmental benefits is continually under attack by those who consider they could profitably use the water for other social benefits. Unfortunately there is as yet no agreement on what characteristics of the natural flows the environmental flows should reproduce. For example what proportion of the natural mean annual flow, what size and frequency of floods, what length of no-flow intervals should characterise the environmental flows? Most attempts to be environmentally sympathetic by governments have been limited to specifying simple variables such as a proportion of the mean annual flow that is to be reserved for the environment each year. However environmentalists point out that seasonality of flows, temperature of the water and the frequency of overbank flows are probably much more important criteria. There is obviously a long way to go in reaching satisfactory specification of the characteristics of environmental flows and even further to go in gaining widespread acceptance of the compromises that will inevitably be required on all sides. However, the earlier these issues are addressed the less the ultimate cost of restoration and compensation.

5.4 Landscape and Cultural Value of Water

Water has considerable value in the landscape. It provides visual attraction and beauty and is of considerable importance for the life patterns of many creatures, particularly migratory birds.

Numerous ethnic and cultural groups of people also attach special significance to some water bodies, such as rivers, wetlands, lakes and springs.

Development of the water resources of a region inevitably affects all of these values which derive from the undeveloped natural state of the waters. As discussed in the previous section, any water development will change the natural conditions and impact upon some or all of these values. The time for thinking about these impacts is before any development occurs, but the nature of development, usually beginning on a very small scale, is such that these issues are never considered until it is almost too late, with the state of degradation only being repaired or returned to a more acceptable, though usually far from natural state, with great effort and expense. There is a need to recognise this human frailty and attempt to require genuine environmental impact assessments to be conducted before any action is taken to make use of a water source. For almost all water bodies in the world it is already too late for this, but for any future activities the likely consequences need to be examined. Then there may be a possibility of involving the community in seeking environmentally sympathetic development compromises that assist in retaining at least the more significant landscape and cultural values. At present there is still a dominance of the *development is always good* attitude around the world which must inevitably increase the difficulties to be faced by future generations.

The need to consider landscape values as part of values of water is a major issue for water use in water scarce regions. As well as environmental and visual values there is a need to consider the preservation of ancestral water use systems, which are well known heritage items in many parts of the world. Examples are: the *tabias* and *jessour* systems of Southern Tunisia (Fig. 5.4), the *qanats* of Iran, or the Nasca aqueducts of the pre-Inca period in Peru (Fig. 5.5), which nowadays are still of interest for water supply for local populations and have also become tourist attractions.

Fig. 5.4 A water harvesting "jessour" system marking the landscape in arid southern Tunisia

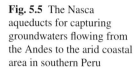

Fig. 5.5 The Nasca
aqueducts for capturing
groundwaters flowing from
the Andes to the arid coastal
area in southern Peru

5.5 Economic Value of Water

5.5.1 Safe Water Supply has a Cost

Water has great value to all life. However it is often viewed as a naturally occurring
commodity which is freely available to all. This sentiment has led to the view that
water should be available free of any charge, which is reasonable thinking if there
is no development of the water resource and individuals collect their own supply.
However in the 21st century Earth's population has increased to a point where it is
no longer possible, nor desirable, for individuals to collect their own supply from the
natural source. Except in areas of very sparse population water supply systems have
been developed, which involve some form of collection and delivery. The collection
and delivery activities incur costs and so all water supply systems require funding,
both of a capital nature to set up the system initially, but also ongoing funding to
meet operational and maintenance costs.

It is probably unfortunate that in many countries the political view has pre-
vailed that the population should not be charged for water supply. As political and
economic conditions have changed, most of these *free water* schemes have been

unsustainable. The result has been that about half the earth's population does not have in-house delivery of safe water. On the other hand, the World Bank points out that electricity is delivered to more than 80% of all households, world-wide. The difference is that everyone accepts that electricity is produced at a cost, and most people are willing to pay for energy provision.

Repeated resolutions to provide running potable water to the majority of households, world-wide, after the Mar del Plata conference in 1978 have failed, primarily because there has been no firm political will to charge realistic prices for the supply of water. Until educational programs can convince the population at large, and the political leaders in particular, that sustainability of water supply can only be achieved when users pay realistic prices for the supply, large sectors of earth's population will continue to pay the price of living without a safe, in-house water supply. This price is very high in terms of the labour involved in collecting water each day and in terms of poor health due to use of unsanitary water and poor hygiene standards which sometimes result from insufficient water availability. Experience in regions with safe, in-house water supply, shows that the cost to individuals of not having a safe delivery system is far higher than the real cost of construction and operation of an adequate system.

The problem of inferior, inadequate household water supply in most areas of the world is not one of physical water scarcity but one of political determination to charge a sufficient price for supply to cover the real, total cost of the supply system. Aid agencies and NGOs have provided less and less funds for development of water supply schemes as they have observed that Governments refuse to charge for water supply at a level that will allow for maintenance of the system plus repayment of the initial capital cost. No matter what the political philosophy of a country, such policies are unsustainable, as evidenced by the supply of electricity to most households, but of water to only a minority of households.

5.5.2 Water Pricing

How should the price of water be determined? This is a very difficult question, especially when the initial supply, in the form of precipitation, is totally free, and many subsistence households have little opportunity to exchange labour or goods for money to pay water charges.

However this argument does not provide a basis for supplying water free of charge. How do households pay for electricity? A large amount of imaginative work is needed to develop charging systems that fit into the local culture and customs. Households which do not have a supply currently pay a large monetary price for deliveries of water by cart, or in terms of labour (mostly by the women and children) to collect and carry the household water. Imaginative programs can be developed to harness these large uncosted and unaccounted energy resources to make water supply systems safe and sustainable and much more efficient without creating social problems. Solutions require imagination and political will and leadership. Unfortunately, in our modern world the politicians and other leaders have no direct experience of the labour required and health effects of water gathering and the water supply (life support) needs of their poor constituents rarely attract their attention.

It is often suggested that real pricing of water cannot be implemented because the poor could not afford to pay. This argument involves prejudiced thinking and a refusal to face reality. The poor find methods of paying for electricity. Realistic pricing mechanisms require measurement of water consumption but the excuse is brought out that water meters are too expensive for the poor. However 80% of households worldwide can afford to pay for an electricity meter. Which is more important for life – water or electricity? Tariff structures can easily be developed to ensure the poor can afford the minimum living requirements but that those who use large quantities pay a premium price. There should be no reduction of tariff for large (or bulk) users since one of the aims of pricing is to reduce consumption. However such a tariff structure causes problems for irrigators who are very large consumers. Hence tariffs for irrigation need to be different from those for urban households, but they still need to be economically realistic. That is, all the water supplier's capital and operational costs should be covered, at the very least. If income of the water collector/distributor is too low, the inevitable result will be neglect of maintenance and eventual collapse of the scheme. There are many examples in water resources development history.

A major problem with the idea of a free market for water is that there will always be uses which can afford to pay a high price for water. For example most urban household users would have no difficulty competing for water supply on the basis of price alone. In Australia typical charges for potable water in urban areas are $1 to $1.50/kl whereas irrigators typically pay $30 to $80/Ml for untreated water. However high prices for irrigation water not only affect the irrigator but eventually affect the urban dweller since the price of food will need to rise or growing of food will cease. The world, and local economies have adapted to low food prices and so radical changes in irrigation water tariffs could disrupt whole economies. Careful, multi-disciplinary thinking is required to develop sustainably structured water tariffs and to integrate these into the broader national and international economic environments.

5.6 Priorities for Water Allocation

5.6.1 Who Owns Water?

The major problem of water allocation is in reaching the initial agreement or acceptance of ownership or the right to allocate. In addition, water allocation has to be done when recognizing the full value of water and not only the costs associated with supply and the benefits generated by the use of water. The scheme in Fig. 5.6 summarizes this conceptual approach.

There is no universal agreement on the status of water as a commodity. It is similar to the status of land. Can it be owned? Should it be a common trust to be used according to agreements? Who has a right to ownership or to be a party to

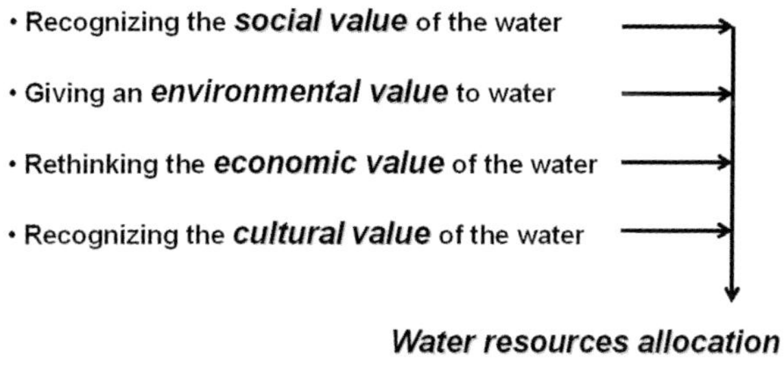

Fig. 5.6 Conceptualizing the relationship between water value and water allocation

• Recognizing the *social value* of the water   ⟶

• Giving an *environmental value* to water   ⟶

• Rethinking the *economic value* of the water   ⟶

• Recognizing the *cultural value* of the water   ⟶

Water resources allocation

a use agreement? These are questions which have strong anthropological and cultural links for which solutions (or disputes) have developed through history. Cultural and anthropological understanding has been over-ridden in many instances by water scarcity, because in recent years water use has escalated, driven by increased population, rising standards of living and the economic advantages of water availability.

There is not space here for a broad discussion of cultural traditions of water rights and ownership. However in recent history many of these cultural links have been swept aside by the political and economic philosophies of majority population groups and in most places water resources have become a political issue both within and across political boundaries. The European experience is that water rights and ownership have been brought under State control by legislation and regulation. Some of the legislated changes have introduced systems of control which are quite different from traditional systems, but which take account of user *rights*, and give some opportunity to users to influence the regulations and controls.

Understanding and development of strategies for coping with water scarcity must take account of this wide range of external forces. Strategies must be developed which are compatible with local political and cultural realities. One particular coping strategy may appear obvious to one cultural group but may be totally unacceptable to another, perhaps for reasons which have their origins in ancient history, but which appear to pragmatists to have no relevance today. In the long term a strategy such as this will not be sustainable. Successful strategies will be those which recognise and either accommodate or over time, by education, change the cultural and political environment. It must be emphasised that for any strategy to be successful it needs to be in agreement with the cultural traditions of the people and be acceptable by the communities they are designed to serve. This may imply that strategies vary within the same country. This is the case of Chile, where the indigenous communities of the arid North have a different legislation and rights structure from their neighbours. These accepted differences reflect the cultural and traditional differences between the groups, thus allowing for peaceful water allocation decisions, and allowing the indigenous community to apply its traditional rules and to celebrate its appropriate ceremonies. Other examples also exist in Andean areas.

Our original question was who has the authority to allocate water? In modern sovereign states water is generally accepted as being linked to territory. It is linked to the location on which it settles from the atmosphere. But what of trans-boundary rivers? Does the upstream state have a right to consumptively use all of the water in the river? As technology develops who will have the right to trap water from the atmosphere and prevent its *natural* travel to precipitate on a different sovereign state or on a neighbour's land?

5.6.2 Water Ownership

Is water to be shared equally by all, to be owned by the state, or to be used freely by those who care to access it? All these, and many more systems underpin the water law/water allocation systems currently operating in various parts of the world. Considering Fig. 5.6, it can be seen that differences are governed by the concepts behind the value of water, which relate to the socio-economic and political frame of each country or region.

First in, first served systems may be fine where the resource is much larger than the demands made upon it. However few parts of the world still have this luxury. Similarly a totally unregulated market can only work where it is acceptable for large sections of the population to be denied any water – but genocide is clearly not acceptable. The intermittent and random nature of natural flow of rivers, quite different from the approximately constant rate of consumption demanded for most human activities means that large scale intervention is required to smooth out the natural fluctuations and make the resource more useable. Large scale projects to provide this constancy of supply usually require state intervention and provide opportunities for states to appropriate ownership to themselves and to enact laws to regulate allocation of water. Ownership of this type usually provides opportunities to raise funds or charge for water. However, as we have seen, some states or regimes have decided for various political, cultural or religious reasons not to charge for allocation or supply of water, but as a result have found it very difficult to provide adequate water supply systems for their citizens.

There can be no ideal ownership system for water. The system adopted must be compatible with and serve the needs of the constitution and population of each locality. The real need is to develop a system that is equitable and is seen by the majority of potential water users to be so. The present trend in Europe and other regions is not presented here as something to be copied by others, but as an example of what is being done in some locations. Each country or cultural region needs to choose a system that fits easily into its cultural and political framework and best serves its clients, the water using population. Adoption of a system that does not serve all the people will not be sustainable, since water is a life support necessity. Dissatisfaction with imposed rules and regulations for water use and allocation will lead to widespread public unrest and ultimately to removal of the regulators by peaceful or perhaps even violent methods.

In areas of water scarcity ownership/allocation systems will always be a focus of attention and potential conflict. Systems that are seen to be fair now, rarely contain provision for managing the increased demands of the future (due to population increase and higher lifestyle expectations). Economic development and long term investment require certainty regarding future water supplies, but any system that provides for revision of allocations as population and demand increases introduces enough uncertainty to stifle investment and curtail development.

In all areas of water scarcity there is a need to build institutional frameworks and allocation systems that allow clear statements and guarantees of future (perhaps reduced, but by forecast amounts) supplies so that potential investors can adequately plan and estimate long term returns. Without long term planning and assurances of future water availability development will be stunted.

Therefore it would appear that in regions of water scarcity ownership of water is probably not the key issue. Rather political stability and water institutions which can provide guarantees of future amounts of water to be supplied, and of its price, are probably far more important.

5.6.3 Water for Human Life

The needs for access to water supplies are obvious but it is worthwhile to summarise them here.

Firstly water is needed for drinking and food preparation. This water needs to be of high quality but the amount needed is not large. Under most conditions a supply of as little as 10 l/person/day would not restrict activities. Secondly water is required for cleaning and basic hygiene. Again the amounts required are not large but tend to increase greatly with general standard of living and inversely with the price of water. The quality requirements of this water, for bathing, cleansing of utensils and living space and for transport of wastes is not high. In modern society very large quantities of water are used as a transport medium for wastes. While this may be an effective means of waste removal it does not make sense for regions of water scarcity. Use of methods of solids removal requiring little or no water should be adopted.

Food production can be the largest user of water. However in regions of water scarcity there is a need to find crops that can survive rain-fed conditions or which need minimal irrigation. For example it makes no sense to grow rice in areas of water scarcity. However rice is currently grown in some water-scarce countries to obtain foreign exchange and for local political reasons, but the corresponding high water demand when compared with what may be required for other irrigation crops makes no sense at all. An example is rice and sugarcane production in the arid coastal area of Northern Peru, which replaced the traditional pre-Colombian crops at the time of colonisation and are kept until today. There is a strong case in such locations to revert to traditional crops or to find substitute crops which need only the same water as the traditional crops.

Research is providing indications of various kinds of food that can be produced with less water. There are also possible dietary changes towards food that grows with low water requirements. However making changes from high water demand to low water demand crops, depends, among other things, on the socio-economic conditions of such regions, on land ownership structures and on political issues. Finding new solutions for these kinds of issues is beyond technological developments and requires socio-political approaches.

5.6.4 Water for Industry

As was stated above, it makes little sense to site high water demand industries in areas of water scarcity. However this situation does occur, as a result of historical developments and availability of raw materials and energy supplies, or due to cultural and personal and business profit reasons. Effort is needed to change this situation, either by moving the industry, which could have high social impacts due to consequent unemployment, or by implementing a comprehensive water recycling program to reduce the water demand and water pollution and contamination.

Implementation of recycling schemes is often resisted; however strategies involving real supply and demand pricing, environmental restrictions and public education need to be used together to achieve change. The adoption of the principle of polluter pays has proven effective in many countries and has led to the progressive adoption of water recycling with consequent decreases in demand.

Most industries can reduce their water use but there is always a cost. However with realistic water pricing it may be cheaper to use less water but at the same time to raise employment. In most parts of the world there is little real incentive to save water; *it is a free renewable resource*. If the real economic value of water was charged, many users, particularly industries, would quickly change their water-use practices.

5.6.5 Irrigation

Irrigation is the human activity which consumes most water. Irrigation water use is primarily for production of raw materials of food and fibre. Irrigation farmers must compete with their fellow irrigators and with dry-land farmers and as a result they can generally only operate where water input costs are low. In general, irrigation water is not treated in any way so the costs are those of collection, conveyance, distribution, and application of the water plus taxes. In some countries government tax raising measures are linked to water usage. Where water is collected, stored and then distributed to irrigators by government agencies the taxes are a form of cost recovery for the capital and operating expenses of the scheme.

The present trend in many parts of the world is to tax not only to cover the operation and maintenance costs but also to recover the investments that were involved in the building of the irrigation systems. However on a price basis alone irrigation has difficulty competing for water because it needs large volumes and the returns from sale of most agricultural products is low, and often distorted by social and political considerations. As a result most irrigation schemes are subsidized from more broadly based taxation systems. These subsidizations can usually be justified in terms of overall social equity. However the unfortunate result of these subsidy schemes, where the user does not pay the real price for the water used is that the water users themselves may cause distortions in the economy. Over time very uneconomic or environmentally damaging practices may become entrenched and when a new policy of water pricing is introduced it may cause the abandonment of irrigation with large social and economic consequences.

Common sense may suggest certain traditions or practices should be stopped or changed but these changes may involve sufficient upheaval and social dislocation that there is not the political will to force a change of activities. The present trend in Europe is for subsidies to be reduced. The European case is paradigmatic: the subsidies to farmers are no longer related to production, and therefore to the irrigation activities. The impacts of this change are smaller than expected because world market prices for agricultural products, particularly cereals, have increased because not only has demand for food increased but also considerable demand for conversion of crops into biofuels. Probably impacts of subsidies for irrigation where they still exist could be lesser if they were cut. For example it makes no sense to encourage irrigated rice growing in an arid area where irrigation farms are going out of production due to rising, saline water tables. Yet in the Murray-Darling Basin in Australia rice growing for export is encouraged by low water costs and an ideal climate, but it is causing huge environmental damage, and particularly damage to the farms of the neighbours of the rice growers. When it is suggested water prices should rise to pay a larger part of operating costs (but no contribution to scheme debt) the irrigators pressure their local politicians for assistance. At the same time they pressure their politicians to be active on the international stage to demand entry to markets for their rice, markets which already have adequate supplies from sources where rice-growing is less costly and much less damaging to the environment. They also pressure politicians to assist their hapless neighbours with their water logging and salinity problems. Simply charging the real price for water, to cover all collection, storage and distribution costs, and the cost of repairing environmental damage would reduce or even stop these questionable practices. But social upheaval of a very small part of the population would also occur.

Similar problems occur in many irrigation regions of the world. Not only does irrigation provide for national food and fibre needs, and export earnings but it also directly and indirectly provides employment in rural areas. These benefits, plus the mistaken notion that fresh water is a free gift of nature, means that irrigation is often seen by politicians solely as a solution to problems, not as what it really is, a potentially beneficial practice which needs very careful social, economic, environmental and political management.

Without careful management irrigation destroys itself and/or its neighbours by water logging, salinity, pests or pesticides. Improving irrigation management is costly and requires that solutions be found that are well suited to the local conditions. This implies the need for a strong technological input to irrigated farming. Large commercial farmers find it relatively easy and usually profitable to adopt innovative ideas applied to their irrigation methods, the scheduling of irrigations, and water conservation practices which may reduce their water demand. These technological changes which may include the use of improved crop varieties, good seeds and plants, rational control of diseases and pests, appropriate use of fertilizers including fertigation (application of fertilizers with the irrigation water), overall leading to higher yields and high quality products. This is possible because farming under such conditions relies on high capital inputs, which are difficult for small farmers, whose production relies on labour, not capital. These farmers have low access to innovation and technological support and therefore often have few opportunities to gain access to ideas for careful water management, water saving and controlling of environmental impacts. Therefore, for small farmers, the State or the society, need to provide both technical and financial support if water management is to be improved.

Irrigation is not sustainable if water supplies are not reliable. Therefore decisions to permit or encourage irrigation imply an effort to provide an adequate, reliable supply. Large investment is usually needed to set up irrigation systems and so a long term view must be adopted, with recognition that success will attract further potential entrants who will also expect a share of the *naturally available, free water*.

The major need for development of irrigation in areas of water scarcity is to minimise water use. Effort is needed to find attractive economic crops needing minimal water, to find and use application methods that minimise loss of water by evaporation from the soil or percolation of water beyond the depth of the root zone and to minimise losses of water from storage and delivery systems. This requires that conveyance and delivery of water occur in concert with crop demand and not following hydraulic considerations alone. For example, when canal systems are poorly managed, it is common for good irrigation land adjacent to delivery canals to be waterlogged, indicating severe leakage from the delivery system and *loss* of both productive land and water. If the real price is paid for water, investors may be willing to lend funds for leak reduction and maintenance and to improve delivery operation. If leaks cannot be reduced it may become uneconomic to continue use of a particular delivery system. Land holders taking delivery from such a system would then need to change their land use practices. In parts of Australia open channel delivery systems are being replaced by pipes, capital costs being covered by water savings in 8–10 years, and widespread environmental benefits being provided free of cost. However, this kind of change is unlikely to be possible in regions where farmers have access to only very small land areas, often of less than 1 ha. Under these circumstances only some centrally organised development, such as can usually only be organised at government level (but with users participation), is required to effect significant change.

Use of irrigation implies water scarcity – at least for part of the year and for the crops that land holders wish to grow. However in some water scarce areas irrigation may not be an optimal use for the available water, but there will always be potential users desirous of irrigating. In these cases, if alternative employment is available and any social disruption can be accommodated, there is a need to discourage irrigation by suitable pricing mechanisms or by regulation. Price changes, made gradually, can often be easily implemented and would have the added benefit that they would discourage waste and inefficient water use. However the rate of price change must be large enough to have real effects. It must be much larger than the inflation rate, and implementing such price increases is difficult. During a period of dramatic change such as this, there is a need to provide support and encouragement to farmers to move from their traditional high-water demand cropping and irrigation practices to modern, reduced demand systems and technologies, particularly where farms are small and the farmers are poor and have limited education.

5.6.6 Water Self Reliance

In a sense this is the oldest system of water management. Individuals collected water to meet their own needs. Modern developments, particularly urbanisation and intensive irrigation farming have led to development of broad scale water collection and distribution enterprises, many of them government run. However in areas of water scarcity there is a need to change thinking back, not to total self reliance, but to taking responsibility for your own water. This may mean greater emphasis on self reliance but it should also lead to development of cooperative institutions and a legal framework with sufficient flexibility to adapt to changing circumstances.

The legal system needs to serve the interests of both the majority and individuals so that no-one is disadvantaged for the gain of others. This is a very idealistic requirement but as demands for access to a finite resource continue to increase such a system will be necessary to avoid ongoing resentment, conflict and violence.

5.6.7 Gender Issues

This is a delicate subject as it is impacted by cultural and traditional sensitivities. However, it is important because in most parts of the world water collection is carried out by women and children, but funding, institutional and infrastructure development are largely in the control of men. For many families in regions of water scarcity the major expenditure of labour is on water collection, either by animal power or by family members. In many of these situations labour use could be reduced by development of innovative water collection schemes. Unfortunately if those who provide the labour are permitted little or no input to the thinking about possible innovation the chances of finding more effective, less laborious water collection methods are limited.

There is a need for widespread education on all aspects of water – on methods of collection and storage, water quality and hygiene, the need to guard against contamination etc. In the 21st century many of the traditional methods of water collection and storage are very inefficient and wasteful of labour – mainly female labour.

One of the approaches that could encourage education on water issues is emphasis on the benefits for the whole family of developing more efficient, less labour intensive methods of water collection and storage. The cost of small plastic pipes is little more than the cost of electric wiring. Why do most homes which do not have a connection to a safe water supply have connection to electricity supply? Is this a gender issue? It may be. As has been discussed earlier there is little logic in most households being connected to an electricity supply, be less than half being connected to a potable water supply. The causes are complex and undoubtedly involve issues of political will and capital financing, but perhaps also issues of division of labour and sexism.

5.6.8 Planning for Optimal Water and Land Use

In many political jurisdictions water and land go together. Water in rivers is often shared between land holders, either equal shares for each individual, or on some proportional basis, such as an amount related to the size of the landholding or the amount of landholding potentially irrigable. However, no matter what the system, human nature is such that any system adopted will be seen to benefit some and disadvantage others. Therefore there will always be pressures to change whatever system is currently in place.

When opportunity is available, such as during development of an irrigation system, or when major political changes are occurring, it is wise to examine the existing system and perhaps to attempt to devise a fairer system. Since there can be no perfect system, an adopted system needs to be seen to be fair by the majority of potential users, not just by actual current users. This means the systems adopted will be different from place to place and from time to time, depending on the cultural and political environment. No system should be regarded as fixed for all time. Change is often needed to allow for changes in population, changes in wealth and education and changes in farming methods and dominant crop types grown. However when change occurs there is a need for equity to be seen to have been achieved, and change to bring advantage to the politically or economically powerful needs to be resisted. After all water is supplied naturally and is the source of life for all.

5.7 International Issues – Treaties Between Sovereign States

River basins and aquifers are no respecters of arbitrarily drawn political boundaries. Some water resource management systems are contained within single political jurisdictions (notably island states) but other water managers may need to work with resources which originate in another state, or that is precipitated within their state but naturally supplies another state. As demands for fresh water increase there is a growing desire to utilise larger proportions of any available resource. This inevitably leads states to be reluctant to allow water to leave the area of their jurisdiction, or alternately to expect that rates of flow that have historically crossed a boundary into their jurisdiction will continue to flow into their area in the future.

It is to the advantage of every state to monitor the flows crossing its boundaries and also to enter into water resources agreements with its neighbours. Agreements or treaties preferably should be written in terms of proportions of historically measured flows rather than in absolute amounts, with provision for monitoring by both parties. Since sovereign states are usually reluctant to permit neighbours to have free access to their territory, monitoring of water resources quantities in a neighbouring state's part of a basin or aquifer is usually not possible. However there can be no substitute for long term factual information (which is accepted by both sides) on the movement of water across a boundary. Such information can provide an invaluable basis for future discussions, negotiations or appeals for fair dealing.

Water has in the past been a cause of serious international disputes and is likely to become a frequent cause of tensions in the future, particularly in regions of increasing water scarcity. For example access to the water resources of the Jordan River Valley has been a (perhaps minor) contributor to tensions between Israel, Jordan, Syria and Palestine. The demands on this limited water resource must increase as population and living standard expectations rise. In recent years agreements have been reached on development of the resource. However continued discussion, flexibility, goodwill and cooperation will be required in the future as the expectations and demands for water grow differently in each of these states.

Another obvious example is the increasing demand for water in the Turkish and Syrian parts of the Euphrates–Tigris basin. A large proportion of the Euphrates–Tigris water resource originates from precipitation within Turkey, but this basin provides most of the water resources of the downstream state of Iraq and some of that of Syria. If the people of Turkey expect to make full use of the water precipitated on their territory means no water passing into the downstream states will be greatly reduced. In the case of Khabour river, located midway the Euphrates and the Tigris in Mesopotamia, the summer flow rate decreased from 20 to less than $2 \, \mathrm{m^3/s}$, which is causing the exhaustion of the Ras-El-Ain aquifer that became the main source of irrigation water. In addition, farmers located downstream, where that aquifer is not accessible, are emigrating from the valley where their ancestors were living for centuries. Discussions (and hopefully agreements) are needed now and development in the future will need to be on a cooperative basis if hardship and conflicts are to be avoided.

In Central Asia the disaster of the Aral Sea is now being addressed by cooperation between the states through which the Syr Darya and Amu Darya flow. This is a good example of international cooperation, among the five Central Asia countries, although real benefits for the Aral Sea and its surrounding population are yet to be observed.

5.8 Conclusion

It has been shown that there are many issues to be considered in coping with water scarcity, in addition to the obvious technological issues of where to find more water and how to make more use of the water we have. Concepts of ownership of the water resource are very important. These affect individuals' rights to access water and issues such as who will manage the resource, and their legal authority. Since water supply infrastructure is large and expensive, investment is usually needed for its development. Investors are unlikely to be interested unless there is some chance of long term stability of management of the resource, backed up by a stable legal system and some positive economic return. The legal system needs to take account of local traditions and customs and have widespread acceptance in the community. Any legal system for apportioning rights of access to water resources needs to incorporate provision for the number of users who have a right of access to water to increase over time. A given of the 21st century is that population increases and expected standards of living rise, both of which put increased pressure on management systems which have been developed to cope with water scarcity.

Chapter 6
Surface Water Use and Harvesting

Abstract Ideas for maximising the availability of surface water for human use are proposed. These ideas are presented to support planning and management initiatives to provide more water for domestic or industrial supply or for crop growth for production of food and fibre. These refer to both large and small scale projects, including water harvesting. Attention is given to reservoir planning and management, as well as to control of water losses and non-beneficial uses of water. Particular attention is given to environmental and health issues in relation to the use of surface water.

6.1 Large and Small Scale Projects

6.1.1 Definitions

"Large" and "small" are relative terms. However it is important to define them because systems management of water resources is usually different depending on the scale of the entity being managed. "Large" usually refers to projects where large volumes of water are involved and supply from the system is consumed by many users. However it can also refer to projects which have large capital requirements and sometimes to projects which operate over large regions. "Small", on the other hand typically refers to projects where the beneficiaries are limited in number, say to one or a few households or farmers. They are usually projects where the capital requirements are quite limited.

6.1.2 Objectives of Water Use and Harvesting

In general terms the usual objectives for water management in regions of water scarcity are (1) to ensure enough water for survival at all times, and after this requirement is met to maximise the possible benefits which the remaining water can provide, (2) for household comfort and well-being, (3) for food and fibre production and (4) for protection of the natural environment. Several issues for achieving these objectives need to be considered (Fig. 6.1).

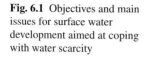

Fig. 6.1 Objectives and main issues for surface water development aimed at coping with water scarcity

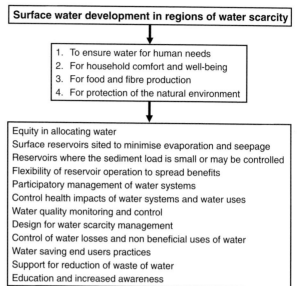

These objectives are almost always in conflict with each other and so compromises are usually required. The conflicts are increased by the numerous different sub-objectives of the individual water users and land holders. For example it is not possible to maximise the availability of water to households while maintaining the completely natural flow of the river from which household water is drawn. Similarly householders may wish to have a secure supply of water for the whole year, but irrigators wish to maximise their water use by crops during the growing season and are willing to accept some water shortage and hardship in the non-growing season.

Conflicts will always occur because there are many users, with diverse interests and aims, all needing to obtain their supply from a single, limited (scarce) source. There is large potential for inequities of supply and for those with power to dominate and take a large share of the limited resource. Similarly inequities and less than optimal water use tends to be encouraged by inflexible political, institutional, or management systems. As with most technologies, efficient water resources use and management schemes can change rapidly and rigid institutional arrangements can often stand in the way of the changes that may be needed to develop or encourage the "best" methods of allocating water to the many potential users.

Waste of water and inefficient water use are usually considered to be states to be avoided. However in times of drought a wasteful or inefficient water use system can often, with considerable effort and cost, be temporarily improved to allow greater use to be made of the limited water available. On the other hand a totally water-use efficient system may contain no flexibility for temporarily increasing supplies or for surviving with a lesser supply, and may fail completely during a drought. Decisions concerning how "efficiently" a system should be expected to operate, and whether or not the "inefficiency" can provide a drought "safety net" must depend on local cultural and social expectations and the physical layout of the resource and of supply points. There is considerable scope here for innovative management, both in preserving the less than efficient long term practices and in changing practices during droughts. It is probably better if this situation is deliberate and widely discussed and accepted in the community rather than being the accidental result of ineffective administration.

Water scarcity should encourage lateral thinking and flexibility in approaching water supply needs. If water is scarce, should we continue to grow our staple food, which has a high water demand, or could we grow an alternate, higher value crop which consumes less water, and purchase our staple food from elsewhere? Can we develop a taste and tolerance for a basic food which can be grown with less water? Is the water demanding industry really needed in this water-scarce region? Can we develop an industry which has more modest water requirements? Can industry recycle its water so that its actual consumption of our precious resource is reduced to near zero?

6.2 Reservoir Management

6.2.1 Need for Reservoirs

Most water supply schemes need to incorporate reservoirs. These may be surface storages or aquifers. The function of the storages is to smooth out the natural variability of the hydrological system to allow human activity to be supported by a constant, or a regular, seasonally varying supply. Where water is scarce it is most unlikely it will be possible to take water on demand from the natural system. In times of high flow (either surface or subsurface) it is often possible to extract whatever water is required. However in drier times the natural flow is likely to be significantly lower than the expected extraction rate. Hence surface or subsurface reservoirs serve as temporary storages, capturing high flows whose water can then be available for use during periods of low natural flow.

Reservoirs need to be sited to minimise non-beneficial leakage or removal of captured water. For example a sub-surface reservoir (aquifer) should be located where there can be large inflow but minimal outflow of water. Making effort to encourage recharge of an aquifer which quickly transmits water to the sea makes little sense. One needs to be found where water is trapped in the local region, and which can then be extracted for later use. Surface reservoirs need to be sited to minimise evaporation and seepage. This means the geology needs to be investigated to attempt to ensure seepage will be small. Evaporation can realistically only be minimised by keeping the surface area of the reservoir as small as possible. Reservoirs should be located to have a maximum volume/surface area ratio; otherwise a large fraction of the captured water will be lost by evaporation. In a region of Australia where the annual pan evaporation exceeds 3 m, one of the reservoirs providing water for a major mine and a town of 30,000 people makes beneficial use of only 20% of all

Fig. 6.2 Sediment trap
upstream of a small reservoir

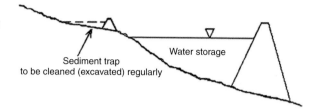

Sediment trap
to be cleaned (excavated) regularly

Water storage

its stored water, the remainder (80%) evaporates and seeps out. If evaporation could be reduced by just 25% the available water supply could be doubled. However at this time, no practical strategies are available for reducing evaporation from large reservoirs. The only possibilities are at the planning stage to choose a site for the reservoir where the storage will be deep and have minimum surface area. This is not always a realistic possibility, as in the example mentioned above.

Silting is another opponent of maximizing water use. Over time, surface reservoirs capture almost all the sediment carried by inflowing streams. Care is needed to site reservoirs where the sediment load of streams is small or provide a sediment trap upstream of the reservoir (Fig. 6.2). The sediment trap will need regular cleaning if it is to function effectively. This is quite possible for a small reservoir which is supplied from only one stream but for large reservoirs it is usually unrealistic because inflow rates may be very large and from many streams. In this case the only means of preventing reservoir silting is to improve soil conservation practices by good land management and channel protection upstream.

6.2.2 Water Scarcity Management

There are many opportunities for management of water in an environment of general shortage. Some of these possibilities for good management of a scarce resource are dependent on availability of capital for investment in fixed infrastructure. These opportunities usually need to be grasped at the planning stage. Their capital intensive nature means they are fairly inflexible and therefore are unlikely to be changed significantly after construction. This means that good long term planning is needed, but this planning needs to avoid focusing on a narrow range of water use opportunities. The chances of major socio-economic or technological changes in the short-to-medium term are high and therefore narrow range or single use capital infrastructure may be a serious impediment to beneficial water management in an as yet unforeseen socio-economic and technological environment. For example there is a move away from having large storage reservoirs as the major investment of new water development projects. Moves are towards larger numbers of smaller scale water storages since these may offer more management options and particularly localised management, which is rarely possible with projects dependent on a single large storage. In overall terms a single large reservoir may provide for maximum control of water, but ultimately the low flexibility of operation of such a system when management is centralized and non-participative often means that only a very limited sector of the population can benefit, whereas with multiple small systems the potential for flexible management and spread of benefit to a wider sector of the community may be larger. With smaller systems the local beneficiaries have a sense of ownership and often willingly contribute to maintenance. Very large reservoirs are usually remote and the users often feel alienated from any control and hence are reluctant to contribute any labour for maintenance or ideas for changes needed in operational modes.

It is unfortunate that in the past large reservoir schemes have been imposed on communities by governments that were ill informed, or were seeking notoriety, or due to importation of development approaches unsuited to the local conditions. For example water development projects, such as for irrigation, where concepts and technologies are imported from different social, cultural, economic and technological environments can produce a shock to the local communities and users. Their reaction may then be not so much to reject the development, as to ignore it. These schemes have often languished with a loss of the aid which funded the development and with large scale devastation of the local ecology and sometimes left the local people far worse off than they were prior to the misguided investment. In all water resource developments the local people need to see benefits from the proposal and to have some control. If they have no sense of "ownership" of the project ideas the chances of worthwhile returns in any form from the investment are small.

Another benefit of smaller systems is that stratification and eutrophication problems can be more easily managed in small reservoirs than large ones. Overriding all these considerations must be the topography of the location. Some landscapes lend themselves to multiple reservoirs, others do not. Large reservoir development depends on having a suitable dam site in the region.

At the operational management level there can be many opportunities for improving water availability and for improving security of supply. These two benefits are usually considered to be opposed – to increase security means more water must be held in reserve and so supply must be decreased. In a general sense this is true but with careful consideration of all the factors which influence supply, sensible flexibility of management can lead to a significant increase in one of these benefits, with no reduction of the other. Understanding of natural processes such as the seasonal distribution of precipitation has enabled the development of statistical techniques which better reflect the natural processes and which can be used in flexible modes to mirror local operations and increase the benefits to be obtained from a resource. For example conservative thinking combined with only simple analytical techniques often suggests that the best management policy is to keep a reservoir as near to full as possible at all times. However it may be that more adventurous management, which may increase water use by 25% (and at the same time reduce losses of water from the reservoir by seepage and evaporation by 10%) may lead to crop failure one year in 10. The overall increase in productivity from such management may be anywhere from 15 to 50%. However, introduction of such a scheme would need to be accompanied by an ongoing education program to encourage water users to build up a reserve (in the form of storage of crop or as a cash balance in a bank) from their increased returns, to allow them to survive in the 10% of years of no income.

There are many reservoirs around the world which are full or near-full most of the time, yet almost all reservoirs were originally designed to become near-empty more than once during their life. Unfortunately there is too little communication between reservoir designers and reservoir operators, and much of that communication is via politicians who tend to see empty reservoirs as some indicator of evil, rather than the near empty reservoir as an item of infrastructure providing optimal performance. Of course, if the design assumptions were wrong, continuing emptiness of a reservoir is not an indicator of good performance. A whole science of "risk management" has developed in recent years. Its aim is to analyse risky operations so that all risks and potential benefits are considered and the best overall outcome is achieved. The key to good risk management is consultation with all interested parties and complete openness in all decision making. Success of reservoir planning, management and operation depends very much on education of, and ongoing communication with affected parties, be they politicians, water users, system operators, local landholders and business operators or neighbours.

Water resources systems are often seen by politicians and the population as preventers of drought. They are not, and water resources planners and managers must educate their "masters" and "customers" to this fact. Water resources projects can supply drinking water over drought periods but they are rarely, if ever designed to supply irrigation water during prolonged droughts. To supply the huge volumes needed for irrigation during a drought would require reservoirs much larger than are physical possibilities in most parts of the world. The main function of reservoirs is to smooth out the normal climatic variations, such as to provide water captured during the wet season for whole year production, or to allow a continuous flow to meet domestic and farming needs on rivers where flow only occurs in two or three flood events each year. The aim of water resources projects is to improve human comfort and productivity in regions of water scarcity. However there must be the recognition that every few years all regions will experience much drier conditions than usual. During these droughts water availability will be very restricted and the normal productivity of the land will be suspended. However, the very much increased productivity that the water resources project has provided in other years (perhaps 8 or 9 out of 10) will more than cover the losses of the drought, provided all water users manage their own resources (bank balance, food bank, stock numbers) sensibly, with this reality as a central feature of their management strategy. Adopting special reservoir and system planning for droughts well before any drought occurs should help in facing the problem and minimizing economic losses when such events inevitably occur.

Continual education programs are needed to remind the affected water users (usually the whole population) and especially irrigators and other large users of the water, of these facts. Unfortunately human nature is to assume the current "ideal" conditions are permanent. They are not. A worse drought will occur. If no provision has been made for this inevitability (e.g. via bank savings) widespread hardship must occur in the very next drought.

Improved analysis of water resources statistics and development of optimal management strategies has been mentioned here. The detail of these approaches is not covered in this book but readers should refer to standard texts on this topic (e.g. Salas 1993, McMahon 1993). Analysis and management strategies are continually improving and so the above list may be appropriate in 2007, but better materials are likely to be available in the years to come.

6.2.3 Operation of Single and Multiple Reservoir Systems

Here we should use the word reservoir in its broadest sense. That is we will include all water storages including surface reservoirs, natural lakes and ground-water aquifers. These reservoirs may be replenished naturally, or they may be pumped systems or even storages built up from desalting of brackish or sea water.

Without doubt the operation of several linked water sources, rather than a single reservoir system, offers the water manager many options and flexibilities with numerous opportunities to maximise the potential availability of water. This applies to all water resource situations, not just to regions where water is scarce. However the cost of this increased flexibility of management is a very large increase in the complexity of the operation of the system and a very much increased possibility that management of the system may be sub-optimal. However there are very large advantages to be gained from multi-reservoir systems, particularly where the reservoirs within the system have different water sources. For example a system with a river-fed surface reservoir, whose storage level depends on recent precipitation and a groundwater system which varies only with long term variations in precipitation has many advantages. High rates of extraction can be obtained from the surface reservoir but only for limited periods, whereas the groundwater can provide a medium level of extraction for very long periods. Similar advantages can be obtained from inclusion of desalinated water, treated waste water, or urban stormwater. Surface water use usually requires a dam with high capital cost but supply may be at low cost, especially if all water can be gravity fed. Groundwater supply usually requires much lower capital cost but may involve larger costs for pumping. It may then be efficient to design a system for surface water supply only, with high-cost groundwater only to be used in emergencies, such as prolonged droughts, when the surface supply is exhausted.

Multiple reservoir systems also offer the advantages of redundancy, meaning that if there is a failure in one part of the system, supply can continue from elsewhere. Provided there are multiple water delivery pathways the system will not be shut down by any single failure of a reservoir, a pipeline, a control valve, or the electricity supply, and therefore the security of supply can be very high.

6.2.4 Groundwater Recharge

Where groundwater is an important source of water there can be considerable advantages in encouraging recharge of the aquifer (see Section 7.5). Care is needed to protect aquifer recharge areas from land use changes that can decrease the recharge. It is well established that increase in quality of vegetation, e.g. by reafforestation, improvement of grazing vegetation or in some cases by introduction of selected cropping, generally increases infiltration, but also increases transpiration and therefore reduces recharge and runoff. Therefore care is needed to allow for the effects of such surface changes on the groundwater availability or to ensure that the vegetation characteristics of the recharge area are not changed. Change from native vegetation to cropping may lead to increased recharge because cropping may involve leaving the soil fallow for part of the year. However the actual effects in any location will depend on the vegetation and crop types and rotations, the soils and the climate of the region.

Surface reservoirs can be used advantageously to increase recharge by holding water over a recharge area for some time, or by diverting water to spread it widely over a recharge area. However care is needed to ensure this does not encourage increased evaporation and a large net loss of water from the system. Precipitation is the source of all fresh water and in most situations the surface and groundwater resources are interdependent. Encouragement of recharge may mean a reduction in available surface water further downstream. Conversely impounding surface water may diminish recharge direct from the stream-bed downstream of the reservoir and hence may reduce the groundwater resource.

6.2.5 Design and Management of Water Resource Systems

Design and management of water resource systems in areas of water scarcity needs particular characteristics. Since water demand in such regions will always exceed supply, and since social, economic and technological conditions are continually changing the systems need to be developed for maximum flexibility of operation. Installation of a single large reservoir to serve a large number of diverse uses is unlikely to fully provide for the real needs. In a region where new ideas and adventurous attempts to increase productivity per unit of water need to be encouraged the flexibility of operation that is offered by multiple smaller reservoirs may provide considerable advantages. There will always be some inflexibility of thinking among those given responsibility for water releases and if this responsibility lays with a single manager or management body this inflexibility tends to become entrenched. However with numerous managers it is likely that one or more will be flexible and encourage experimentation. When the results of flexible thinking are seen, users of water controlled by other managers will demand more progressive practices. The participation of users, hopefully through their Water Users Associations (WUA), may constitute a valuable contribution for improved and more flexible management (e.g. Abernethy 2001). However, WUA are not a solution for every problem and may be of low efficacy when the water system is conceived with approaches that are far from the cultural and social behaviour of the served communities and populations.

Another need for water projects is the careful integration of design and management. Management flexibility will always be limited by the fixed assets of the system, and so there is a need for managers (and water users) to be involved with design as much and as early as possible.

Management plans need to be part of the original design. Similarly ongoing education programs for both managers and users also need to be set up as part of the initial design. Participatory management of a resource by all interested water users should begin at the planning stage. Such arrangements can be very cumbersome and so some kind of consensus or representation scheme needs to be devised that is sympathetic to local cultural expectations. It is common for education programs to be run at the inception of a new capital project. However it is difficult to maintain enthusiasm for such programs once a project begins operation. Without an ongoing commitment to education projects tend to have fixed operational programs and slowly run down, for example as has occurred with the "warabandi" system on the Indian sub-continent, during its 100 years of operation (Bandaragoda 1998). Continuing education is needed for newcomers to the project and to continually introduce new ideas and new technologies to the community. Flexible thinking regarding water needs to be part of school education since the children will be the users and managers of the next generation. It is much easier to educate the young than the middle aged who have become fixed in their ways.

As time passes and management (release rules) of reservoirs changes little by little, there is going to be an ongoing need to re-analyse the whole management system. Unfortunately this is a rare practice. However a system is usually designed to maximise water use with a given set of release rules. If these rules change over time, and they usually will, the system will then probably make less than optimal use of the available resource. A large example of this is the "warabandi" system of water sharing in Pakistan, which, over its 80 years of operation, has developed severe, entrenched inequities. The system's current operating rules should periodically be re-analysed so that adjustments can be made to these rules, to make better use of the water. Wastage as a result of ad hoc changes to operating rules, without a complete re-analysis of the system and involving the users is common. The work needed to re-analyse the system is quite small, but persuading users to adopt slightly different practices to maximise benefits from the system is a much larger task which must not be (but often is) omitted from the implementation plan.

6.3 Control of Water Losses and Non Beneficial Uses of Water

In regions of water scarcity there should be strong motivation to prevent waste of the precious resource. There are many ways in which water is lost or wasted. Some of these are relatively easy to prevent. In particular most of the operational losses that occur once water comes within the sphere of man's control should be preventable. Other losses, such as evaporation and removal of flowing water by infiltration into stream beds are, in general, very difficult to prevent. The major types of wastages and possible strategies for their control will now be discussed.

6.3.1 Location of Losses

When it is thought that significant volumes of water are being lost unaccountably from a system it is worthwhile making some effort to determine exactly where those losses are occurring. This requires some means of measurement of flows at a number of locations within the system.

Reservoirs are a common site for major losses. To check whether significant losses are occurring there is a need to measure, or estimate fairly accurately, all the observed movements of water and to conduct a simple water budget for a couple of years. Evaporation can be estimated by operating an evaporation pan adjacent to the storage, and making allowance for the higher loss from the pan than from the lake due to differences in energy transfer within the lake and through the sides and base of the pan. If the sum of all the inputs and outputs of water do not match the observed change in reservoir storage volume there must be some unaccounted movement of water. It is usually assumed that there will be a few percent of inflows lost by seepage, but much larger, unexpected seepage losses are possible. It may be that the seepage water moves to an aquifer from which water can be extracted downslope. In this case this seepage water may not be a loss, but just a relocation of the resource.

Conveyance and distribution systems are often the source of large wastage of water, but some of this wastage can be prevented. In many situations water is transported to the point of use via the river bed downstream of a dam. In regions of water scarcity these river beds will usually take up huge quantities of water. Similarly, unlined delivery canals may also leak large volumes. Canals can be lined with clay materials, buried sheeting, synthetic membranes or concrete. However, unless great care is taken during construction and with maintenance, the leakage is usually reduced only by a small amount, perhaps by 20–40% (Rushton 1986), and any form of lining is very expensive. However it may be that most leakage occurs from a short length of a canal and in this situation remedial action such as some form of lining of this short section may make economic sense. Where water has very high value it may be economic to deliver it by pipeline. Provided the pipeline is maintained properly, leakage can be practically eliminated. In several countries open delivery channels are now being replaced with pipelines, even for very large flows. For example in the Wimmera region in South East Australia more than 80% of water that entered the several hundred kilometre long delivery channels never reached the users. The whole flow to this region is now being piped, with the benefits of the saved water repaying all capital costs in just a few years. Added uncosted side benefits are reductions in waterlogging and salinisation along the delivery route and improved water quality at the point of use.

Water use practices are a major source of wastage. Factory production processes and cooling systems which do not recycle can be very wasteful of water. In many parts of the world recycling of factory process water is now mandatory. The high cost of recycling systems themselves encourage reduction of losses by forcing factory managers to adopt processes which induce minimal water use and pollution. Evaporative cooling systems avoid problems of raising the temperature of stream water, but they consume large volumes of water. Where water is scarce evaporative cooling is unlikely to be an optimal practice since heat is dissipated by the change of state of water from liquid to vapour, and the vapour is lost from the system.

There are many opportunities to save water by changing irrigation practices. In traditional irrigation enterprises flood and furrow irrigation are very common. In both cases, when not upgraded, it is difficult to properly control the amount of water application, and so, much water flows beyond the root zone, runs off the field and large amounts of water are evaporated. Farmers need to be encouraged to change their practices since it is possible, using upgraded and appropriate irrigation techniques, to obtain equally high yielding crops with less water application (e.g. a study on cotton irrigation improvement by Dalton et al. 2001).

Similarly sprinkler irrigation can be very wasteful if used in hot, arid and windy conditions. Wherever possible irrigation water should be applied in windless conditions. Drip irrigation tends to have the least waste of water but it needs careful management and maintenance, is expensive to install and is not suited to all crops. Appropriate selection, design and management of farm irrigation systems is required for water saving and maximum water productivity as analysed in Section 10.6.

One form of encouragement of the use of water saving practices is to stop subsidising water supplies and to charge the full cost of collection and distribution of the water plus the cost of maintenance and operation of the system. These aspects are dealt with in Chapter 10.

6.3.2 Reduction of Evaporation

There is no effective economic way to reduce evaporation from large water bodies. At the planning stage reservoir locations need to be chosen to minimise evaporation – by ensuring the volume to surface area ratio for the reservoir is a maximum.

Evaporation can be reduced from small, vertical sided reservoirs by providing a roof to shield the reservoir from solar radiation or by covering its surface by light coloured floating blocks. Solar radiation is the main source of energy for evaporation so any shielding can be helpful. Floating covers are only effective if they are not porous and if their upper surface can be kept dry. If water splashes onto the surface of floating blocks there will be little or no reduction of the evaporation. Monomolecular cetyl alcohol, which forms a layer one molecule thick across the surface can be very effective but the integrity of the protective layer is spoiled by any wind. Similarly oils can be used but they can introduce toxicity problems, so that care is needed to choose materials appropriate for particular applications.

The most effective means of evaporation reduction is to minimise the water surface exposed to the atmosphere and to the sun. Deep storages are to be preferred. Subsurface storage can be very effective for evaporation reduction. However with subsurface storage there will always be seepage flows and costs in recovering the water and so these must be weighed against the gains from evaporation reduction.

Evaporation from the land surface can be reduced by mulching the surface or by tillage to keep the surface capillaries broken. Similarly, infrequent watering to completely fill the root zone of the soil can use less water than more frequent small applications which only partially wet the root zone. Each water application wets the surface and encourages evaporation at least from the top few centimetres of the soil.

6.3.3 Support for Reduction of Waste of Water

In regions of water scarcity, reduction of wastage of water should be part of the thinking and awareness of every individual. It needs to be taught in families to small children and in all formal education systems. In many societies the connection between simple traditional activities and waste of a precious resource is not realised. There is a need for continuous education programs to ensure everyone believes the loss of a single drop of water is a cause for sadness and concern. Most villagers are aware of how precious every drop is, but there is a need to educate city dwellers, and even some village folk. Commitment is needed from society leaders to ensure ongoing educational programs. These education programs also need to encourage less use of water in all activities – so-called demand management. Excessive or unnecessary use of water is also a problem, and could be labelled antisocial. A requirement to pay the full cost of water supply can reinforce education programs. Unfortunately governments in both poor and wealthier nations are often reluctant to charge the real cost of supplying water and miss the opportunity to combine educational programs with real, "on the ground" policy actions. Households for which all water must be carried several kilometres by its members are usually very parsimonious in their water use. However city dwellers who receive piped water either free or at little cost are effectively encouraged by the system to be profligate water users (wasters).

6.4 Water Harvesting

Water harvesting refers to methods used to collect water (1) from sources where the water is widely dispersed and quickly changes location or form and becomes unavailable, or (2) that is occurring in quantities and at locations where it is unusable unless some intervention is practiced to gather the water to locations where it can provide benefits. There is a number of water harvesting techniques that are practiced in many of the water scarce areas of the world. However there are many other techniques that have developed from local necessity, in sympathy with the local physical conditions and customs, that could be adapted to benefit populations in other water scarce locations. A large number of such practices have been described in a number of publications (e.g. Ennabli 1993, FAO 1994, Sharma 2001, Oweis et al. 2004, Martinéz et al. 2009). A number of the widely practiced techniques will now be briefly discussed.

6.4.1 Rainwater Collection

Rainwater can provide a considerable water resource, not only in humid regions but also in semi arid and arid regions. Large volumes of water flow from roofs. In many regions roof water has not been collected because traditional roofing materials did not permit easy collection, and storage of collected water was difficult

Fig. 6.3 Rainwater
harvesting for village
domestic water supply in
Cape Verde (courtesy by
Angela Moreno)

and expensive. However in recent years the ready availability of some form of roof
sheeting and innovative ideas for water storage have made roof water a serious water
resource consideration. A large number of websites produce relevant information on
the subject, including for design (e.g. Thomas and Martinson 2007). An example of
a cooperative village rainwater harvesting for households is shown in Fig. 6.3. However there is a great need to encourage its use and to teach simple, low-cost means of
collecting water from roofs, and constructing suitable storage facilities. For example in the dry north east of Thailand storage containers (jars) were constructed by
householders from concrete reinforced with wire netting to hold volumes ranging
from 100 up to 3 000 l. Roofs were modified slightly to facilitate collection of all
but the first few litres from each rainfall event. These innovations radically changed
the lifestyles of the majority of inhabitants of the region, by providing enough water
for in-house uses over the dry season.

Measures are needed to keep insects (mosquitoes and flies) away from the stored
water to avoid increases in diseases carried by these creatures (malaria, dengue
fever). An example of an improved rainwater collection area including a filter is
shown in Fig. 6.4. This can be achieved by covering the openings of storage jars

Fig. 6.4 Rainwater collection
area with a filter as used in
Northwest Brazil

with mosquito netting or by carefully disinfecting the stored water (with bleach). Similarly some attention is needed to avoid bacterial infections that can result from bird faeces and dust depositions. Disinfection can deal with this problem also. Care is also needed to ensure the collected water does not concentrate dangerous salts which can be contained in dust near various mining activities (lead, arsenic, etc.).

6.4.2 Terracing

Terracing can be used to collect water for two purposes. Firstly a horizontal surface reduces runoff and maximises the infiltration of water into the soil. If the soil surface is kept tilled and free of vegetation except for the desirable crop, almost all rain falling on the terrace will be used for crop growth. In regions of low rainfall, soil water can be stored over long periods provided the soil surface is kept tilled or mulched and vegetation free. Thus it may be possible to store 2 years rainfall to obtain one cereal crop. If there are several terraces down a hillslope it may be possible to grow a crop on alternate terraces each year. Secondly terraces can be used as temporary water storages to reduce flow velocities of surface water and prevent erosion. The slow flowing water on the terrace may then be directed into a storage facility (excavated hole or small dam) for storage and subsequent use. Without the terraces and their associated flat slopes small storage facilities usually become filled with silt in a very short time. Terracing helps keep the soil where it is needed and useful and provides silt free water for storage. Figure 6.5 shows old Inca "andene" terraces, which are still used at present in the Andes (Llerena et al. 2004).

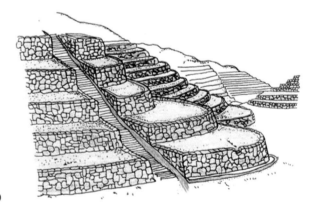

Fig. 6.5 Inca "andene" terraces used in the Andes (courtesy by Antonio Enciso)

6.4.3 Small Dams

In regions of high evaporation, small (and shallow) dams are usually a cause of loss of water resources. However in groundwater recharge areas, small dams, or even low embankments across floodways, can be used to increase aquifer recharge. On grasslands used for grazing the encouragement of infiltration by low embankments which reduce flow velocities and hold water for a few hours after rain events can increase soil water and greatly increase the time after rain that water is available to vegetation roots. By this means it may be possible to double annual biomass production in semi arid regions.

A major difficulty in management of small water supply dams is siltation. Small dams will usually be constructed on small headwater streams. These are often steep and can carry large amounts of silt. If the land surface upstream of the dam is in its natural state the silt and suspended solids load may be small, but usually the upstream area will already have been heavily affected by grazing animals and human activity. Most human activity increases erosion and sediment loads in watercourses. Exceptions occur where large efforts are made to keep erosion to a minimum.

If no attention is given to sediment control most small headwater dams will fill with sediment within a few years, with almost total loss of the water source and the effort and/or capital invested in the construction. The most effective erosion prevention is complete protection of the upstream land from human activity (including grazing of their livestock animals). This is usually unrealistic. Next best is to take intensive measures to prevent surface and stream channel erosion. Third best is to provide a sediment trap upstream of the reservoir (Fig. 6.2). This will usually take the form of a basin where flow velocity will be reduced almost to zero so that sediment will be deposited before the water flows into the main storage. Sediment traps need to be quite large, to reduce velocities sufficiently for suspended materials to be deposited. It may be possible to achieve this with a wide horizontal section in the approach channel. The sediment trap must be constructed to allow easy maintenance. The sediment will need to be removed periodically (perhaps as often as annually) and so the trap must be arranged so that locally available removal equipment can be used to clear the basin with minimum effort and cost. It is obvious that a sizeable proportion of the total capital cost of the small dam must be invested in the sediment trap (perhaps 20% of the total cost) and that a large proportion of the ongoing maintenance cost (perhaps 80%) will be used for clearing the sediment trap.

Animal and human access to the edges of reservoirs behind small dams is another serious cause of damage to the storage capacity and to the water quality. While animals wading and wallowing at the edge of the reservoir may be thought to be kind to the animals, this activity will quickly destroy the water resource. Continual stirring up of edge sediment by the animals reduces the water storage capacity, causes the stored water to be continually turbid, and introduces dangerous bacteria and nutrients. If possible small reservoirs should be securely fenced and water should be supplied to both humans and animals at valved collection points and drinking troughs outside the fence. Power to lift water to these points can be supplied by animals or a simple wind mill.

Without careful consideration of sediment management and provision of resources for this activity most small headwater dams will be condemned to early failure. In most cases combinations of the above measures will be needed to prevent loss of the reservoir by siltation.

6.4.4 Runoff Enhancement

In places of water scarcity where the availability of water needs to be increased it is possible to enhance runoff from rainfall events by partially sealing the soil surface. This can be done by applying surface sealing materials or by compacting the soil surface. Application of surface sealants (such as bitumen) is expensive and provides increased runoff for only a limited time. The sealing materials usually weather rapidly and are penetrated (from below) by vegetation so that their integrity as a surface seal is usually lost quite quickly.

Compaction of the surface requires labour and perhaps some machinery but it tends to be longer lasting. It may be necessary to import aggregates to mix with the surface layer (perhaps to a depth of 20 mm) to provide a high density material at the surface and some cementing agent to prevent the surface layer being removed by wind and water. The surface can be compacted by one or two passes of machinery, by using hand tamping or by the passage of many feet, particularly hoofs of sheep and goats. Compacted natural soils, or modified soils can have very low infiltration characteristics and provide up to 80% of all rainfall from a storm as runoff. Even small areas can deliver significant quantities of water. For example a 12.5 mm rainfall event could produce 1000 l of water from a 100 m^2 compacted area.

Prepared runoff enhancement surfaces need protection from damage and general maintenance to preserve their impermeable characteristics. There is also a need for collection of the runoff in a suitable container from which it will not be lost by seepage or evaporation. The modified surface and any delivery channels need to have flat, but definite gradients to move water to the collection point without causing erosion of any of the surfaces. Care is needed to choose enhancement materials which are non toxic and do not produce toxins as they break down over time.

Runoff enhancement is often practiced to grow trees and vegetables. Water can be gathered from a semi sealed surface to flow towards a tree or small vegetable plot to infiltrate the soil in the vicinity of the plant roots (Fig. 6.6). In this way water availability for crops can be increased many fold over the natural rainfall to

Fig. 6.6 Runoff basins water harvesting at Tal Haddya, Syria, showing the runoff collection area upstream of the planted crop

provide sufficient for growing essential life-support or high value crops in regions of water scarcity. The "jessour" system shown in Fig. 5.4 is another example of runoff enhancement for trees cultivation in arid areas. Various runoff enhancement arrangements practiced in North Africa and West Asia are described by Ennabli (1993), FAO (1994), and Oweis et al. (2004) among other.

6.4.5 Runoff Collection

In undulating to flat terrain runoff may be collected by small earth bunds, namely prepared with a simple plough, thus enhancing infiltration near to crop roots. Various systems are described in literature (e.g. FAO 1994, Oweis et al. 2004) such as the typical crescents used in semi-arid areas of North Africa (Fig. 6.7).

Where runoff collects into large numbers of very small stream channels, the available water resource may be quite large, but its diversity makes it difficult to use. Under these circumstances it is often possible to capture part or all of the flow in a small channel and divert it to another channel. Diversion channels which run almost parallel to the contours can be used to carry this diverted flow to a different location. If flows from a few of these channels can be gathered there may be enough water to fill an excavated hole or small dam which is deep enough to prevent loss of all the water by evaporation. The small diversion channels, following the contours, with a slight slope, can often be constructed using hand tools, or animal power, or with simple earth moving equipment such as a road grader blade. In this way it may be possible to capture flow from several stream channels, each draining a few tens of hectares, to gather water from two or three hundred hectares to a single location. Five millimetres runoff from 200 ha represents 100,000 l of water. There are many landscapes where there is small relief where many small streams flow. However when these streams reach the almost horizontal surrounding land the water is dispersed and quickly infiltrates. If excavated holes were dug in the undulating terrain above the flat area, and water was diverted via near horizontal catch drains to these holes, a relatively large water resource could be available.

Fig. 6.7 Runoff collection for an individual tree, Central Tunisia

It is also possible to use the water that is temporarily held in horizontal catch drains and then infiltrates to grow a row or two of a crop, in the bed (e.g. Khouri et al. 1995, Prinz 1996).

6.4.6 Flood Spreading

Flood flows are a feature of all landscapes, including regions of water scarcity. A very large part of the annual flows may occur in one or two floods, but the flow is often so large that the water passes through the region and can not be used where the rain fell. Some advantage can be gained from these large flows by encouraging them to spread across flat areas. If water can be retained on flat surfaces for a day or two the upper soil layers may be saturated or water may percolate downwards to replenish the local aquifer. In both these circumstances the water thus "harvested" is available for later use, in the first case for growth of crops or to support grazing and in the second case for whatever purpose groundwater is used (UNDP 1990, Prinz 1996; Oweis et al. 1999).

Where the terrain is suitable it may be possible to restrict flow in the river channel and force water to flow over the floodplain or into an old floodway. Flood spreading by these means can most easily be achieved where the upstream-downstream gradient of the river valley is quite small. With small longitudinal gradients water forced onto the floodplain will tend to flow in the downstream direction very slowly, allowing maximum time for infiltration to occur.

Channel flow can be restricted by placement of large rocks, dams and erodible dykes in the bed or by securing a log in the stream bed. Overbank flow velocities can be reduced by construction of low banks (perhaps 30 cm high) across the direction of flow (Fig. 6.8). Low, wide banks will become stabilised by natural vegetation and are likely to need minimal maintenance. However large maintenance is required after a large flood destroys the constructed infrastructure, an inevitable occurrence once every few years. The benefits of flood spreading (Cordery 2004), where the terrain is suitable, can be quite large in terms of increased biomass production and increased groundwater resources.

6.4.7 Water Holes and Ponds

Natural water holes and ponds can be exploited for water supply purposes. There is a need to ascertain the source of the water – is the hole fed by groundwater or by small surface water flows after each rainfall event? If the supply is from groundwater it may be possible to treat the pond as a well and increase the extractable water by making the pond deeper and increasing the hydraulic gradient towards the pond. If the water is supplied by surface flows in a stream bed there are several possible ways to increase the water availability. One means could be to raise an embankment across the stream bed downstream of the water hole to increase the volume of water

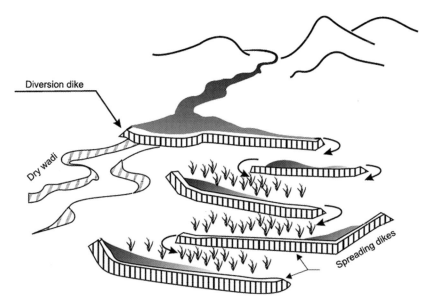

Fig. 6.8 Spate irrigation by spreading flood water over the cultivated land (from Oweis et al. 1999)

captured during each flow event. This is a kind of small dam. The embankment will need to be carefully designed and constructed to prevent its destruction in the next significant flow in the stream.

It is possible that the water hole exists because there are water tight strata below. Raising the water level may not increase the volume stored, but may supply water to the permeable strata above the natural water level of the hole. In such a case increasing the available resource may only be possible by excavating a larger hole in the water tight stratum (very expensive) or by pumping the excess water to another storage location during the brief period the water is above its natural level.

The water supply obtainable from a natural pond can often be increased by raising an embankment around the pond, or by increasing the depth of the pond. Increasing the surface area of the pond usually only increases the removal of water by evaporation, but increasing the depth of a pond can often provide significant benefits. If an embankment is to be raised, excavating materials from the pond bed and thus deepening the pond can provide the added benefit of increasing storage capacity without increasing evaporation loss. It must always be remembered that in areas of high potential evaporation, depth of water is the key to effective water storage.

If drinking water is to be supplied to stock it is better to gravitate or pump the water to a drinking trough with a water level controlled shut-off valve than to allow the animals to wade into the edge of a pond. Wading animals cause erosion, turbidity and generally contaminate the water, and so water holes and ponds should be fenced off.

6.4.8 Tanks

The word "tank" has several different meanings. In some countries, particularly in south Asia, a tank is a water hole. It is often an enhanced natural pond or it may be an excavated hole. These have been dealt with in the previous section.

In other countries a tank means a constructed water container. It is usually fabricated from sheet steel, cement concrete or plastic. Such containers are expensive, relative to the volume of water they store, but they have an important set of advantages over larger, landscape-type storages. They are of particular importance where the resource is very limited and needs to be protected from evaporation and/or contamination.

As discussed elsewhere it is not possible to prevent large evaporation losses and some seepage from ponds and lakes. However small tanks, with capacity anywhere from 1000 l to 10,000 l can be totally enclosed, almost completely preventing losses. Such storage containers are worthwhile where limited amounts of roofwater are collected or where water is desalted at considerable expense. Constructed containers also offer easy prevention of contamination since access to the stored water can be controlled. This includes prevention of access by mosquitoes and other water sourced disease vectors.

For similar reasons it is also important to allow water to be removed from the storage tank only via a valve outlet. Dipping of containers into the tank or any other form of water removal inevitably leads to contamination of the water by chemicals, bacteria, insects and the catalyst effect of light. While ultra violet light has a purifying effect on water, general sunlight encourages growth of many undesirable water residents. There are significant advantages in keeping water storage containers totally enclosed.

6.5 Environmental and Health Issues

6.5.1 Overview

As population increases and ever larger demands are made for water resources there is a need to consider the environment of which the water resources are a part. Any human intervention, such as withdrawal of water for life support, diversion for irrigation, storage for use in dry periods, or enhancing runoff in arid regions impacts the natural flow regimes. When populations were much smaller than today and the bulk of these populations were subsistence farmers or hunter-gatherers these interventions usually had little effect on regional water resources, although there were some exceptions. However recent population increases, and changes of lifestyle and technology have meant water demands have increased to a point where alterations to natural water regimes have been large enough to seriously impact previously existing water uses and to restrict continuation of newly introduced water resource utilisation.

Developments of river basin water resources have occurred piecemeal, without basin-wide planning, and the result has often been widespread environmental damage. Devastation of the Aral Sea and the surrounding region as a by-product of upstream irrigation development is an obvious gigantic example, but there are hundreds, perhaps thousands of other examples. These range from consumption of all flow leading to decimation of all aquatic species downstream, to trapping of sediment in reservoirs and subsequent erosion of stream beds below the reservoir to consume some of the water flow energy, to salinisation and desertification of huge areas and salinisation of once fresh water due to unwise irrigation and forest clearing, etc. The list is very long.

As stated above some of these unexpected environmental effects impact on the activities which caused them but others only have impacts downstream, perhaps on the other side of a political boundary. While the benefits of the new developments may be considerable for some, are the overall benefits positive? Recent times are characterised by change but is devastation of a water resource a desirable outcome? Huge financial resources are currently being consumed in attempts to overcome, or at least reduce, man's damaging impacts on his own life support system – his fresh water resources. Unless great care and wise, broadscale thinking is applied to water resources, the near future will be characterised by rapidly diminishing water availability per person. The Water Framework Directive adopted in the European Union is a good example for many other regions and countries to consider, to assist them to face these problems and avoid rapidly approaching crises. The problems are most serious in the semi-arid and sub-humid regions of the earth where large land resources are available but water is very limited. Perhaps in the near future, continuation of current unwise developments in these regions will so damage the water regimes that continuation of present lifestyles will be impossible.

Therefore any thinking about increasing water availability and water use as discussed above must be considered in the context of environmental impacts. While the political reality is that if people are short of water then the (usually very limited) local resource will be developed to meet their needs. This approach is usually justified as being a humanitarian necessity to meet a particular need. The intention is to "only meet this existing need and to decide that needs such as this will not be permitted to develop elsewhere". Unfortunately meeting these local crises becomes the norm and the desperate needs for water by the increasing population becomes the focus of attention. The effects of the development of supplies on the overall hydrology of the region becomes a non-issue, until disaster strikes and the whole system fails because the demands for water exceed the total resource of the region. History gives many examples of civilisations which developed around a water resource which was used to irrigate crops for the consumption of the local population. Many of these developments eventually failed due to waterlogging or salinisation or both. The major irrigation dependent societies to have survived over the long term are some in China and in India, and those of the Nile valley. In the Nile the water and the valley soils have very low salt content, and before the construction of Aswan High Dam the soils were leached by annual floods (Biswas 1980).

As suggested above, there are always going to be political and social pressures to overdevelop water resources, often as a "temporary" strategy to alleviate social problems which have other causes. However these "temporary" activities rarely have a closure date and the less than acceptable environmental damages become the cause of widespread disaster. Overstressing of water resources, even for short-term emergency social needs should be resisted. Repair of even short-term damage is almost always a long-term problem and continuation of the "short-term" measures usually leads to irreversible damage. For example, once the soils of a valley flood plain become salinised it is almost impossible to remove the salt. Leaching is not possible, even if enough water is available, because the salty water table is then close to the surface. Drainage of the high groundwater is not an easy possibility – to where to drain it? – into the river? Then all downstream water uses will be affected, lost in some cases. Pipe the salt water to the sea? Huge cost. Grow trees to lower the water table? Too late for that, since except for some very specialized species trees will not grow in the salinised soils. In the Pakistan Punjab a 100 year old, salinised irrigation system was rehabilitated in the 1970s. From a local economic viewpoint, this rehabilitation investment made little sense, and with the difficulties of managing water applications the system is again in danger of failing. Short term, and indeed any, politically expedient pressures on water resources need to be avoided.

Apart from these general issues there are many specific environmental issues which need to be considered in regions of water scarcity. All can be managed, but without suitable management all will bring any water resource developments undone. The more important ones are discussed below. A general framework of the items which must be overtly considered in any water resource development project is presented in Fig. 6.9.

6.5.2 Protection of Stored Water for Drinking

Stored water needs protection from evaporation and seepage loss and from contamination. Losses are discussed in 6.3 above. Contamination occurs mainly from human or other animal contact. Water related diseases are spread more easily and more frequently than people realise and therefore care to control health impacts is a must in all water projects (Table 6.1). Drinking water requirements and respective practices are dealt with in related reference books and papers, mainly the guidelines produced by WHO (2006a). If water is to be used for drinking and food preparation no direct or indirect contact with any animals should be allowed. This includes use of buckets to extract water, which have been touched by human hands. Any human or animal contact with the water, or with any surfaces which contact the water, such as buckets or water jars, or lifting ropes or wires, have potential to transmit bacteria to the water which can then cause disease in the water users. Therefore it is much better to use a pump to extract water from below ground storages and a tap in the base of an above ground tank. This applies to surface storages and tanks as well as to wells. Animals and people need to be kept away from surface storages by surrounding them with secure fencing. Animal water can be supplied in a drinking trough to which water can be pumped using a windmill or hand pump. Allowing

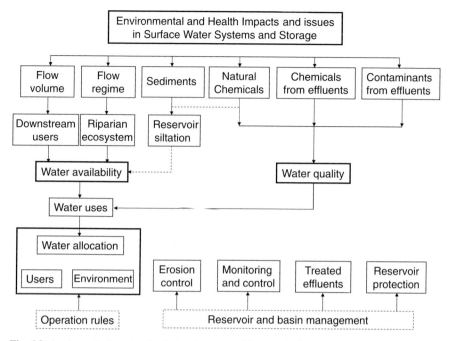

Fig. 6.9 A schematic framework relating impacts and issues relative to environment and health

grazing or work animals to drink directly from a surface pond brings contamination but also causes erosion of the edges, causes the water to be continually turbid and eventually leads to silting and loss of storage capacity.

Birds should also be discouraged from accessing water stored for drinking purposes. It is impossible to exclude them, unless the surface is small enough to be covered. However domestic birds and large flocks of wild birds need to be kept away by fencing and frightening.

Similarly stored water needs protection from disease vectors such as mosquitoes, flies and molluscs, which are summarized in Table 6.1 and discussed in Section 6.5.6. The simplest insect protection is to keep the stored water enclosed. This is particularly important for water stored near accommodation, such as in rain-water storage tanks. Having adequate water supply close to one's house is important, but if frequent incidences of malaria are the result it is probably worthwhile locating the storage at some distance. Similar problems occur in urban areas where treated water is stored near or within the city before distribution. These storages also need protection since they can provide ideal breeding places for mosquitoes and flies, which can then attack the dense local population.

6.5.3 Sediments

Most human activity leads to soil erosion, at much greater rates than occurs naturally. Most eroded materials eventually find their way to stream channels and into

Table 6.1 Main water related diseases (adapted from WHO 2004, 2006b)

Disease	Impacts/causes	Control
Diarrhoea	• 1.8 million deaths per year (including cholera); 90% are children under 5 • 88% of diarrhoeal disease is attributed to unsafe water supply, inadequate sanitation and hygiene.	• Improved water supply reduces diarrhoea morbidity by 6–25% • Improved sanitation reduces diarrhoea morbidity by 32%. • Hygiene interventions including hygiene education can lead to a reduction by up to 45%. • Improvements in drinking-water quality through household water treatment can lead to a reduction of diarrhoea episodes up to 39%.
Malaria	• 1.3 million people die of malaria each year, 90% are children under 5. • 396 million episodes of malaria every year, most in Africa south of the Sahara. • Intensified irrigation, dams and other water related projects highly contribute to this disease burden.	• Better management of water resources reduces transmission of malaria and other vector-borne diseases.
Schistosomiasis	• 160 million people are infected with schistosomiasis. • Tens of thousands of deaths every year, mainly in sub-Saharan Africa. • It strongly relates to unsanitary excreta disposal and absence of sources of safe water. • Many people are infected by wading in water – at the edge of tanks, in canals and paddy fields	• Basic sanitation reduces the disease by up to 77%. • Man-made reservoirs and poorly designed irrigation schemes are main drivers of schistosomiasis expansion and intensification. Much control can be achieved by drying all irrigation canals for a few weeks every year
Trachoma	• 500 million people are at risk from trachoma. • 146 million are threatened by blindness and 6 million people are visually impaired. • It is strongly related to lack of face washing, often due to absence of nearby safe water.	• Improving access to safe water sources and better hygiene practices can reduce trachoma morbidity by 27%.
Intestinal helminths (Ascariasis, Trichuriasis, Hookworm disease)	• 133 million people suffer from high intensity infections with consequences such as cognitive impairment, massive dysentery, anaemia. • 9400 deaths per year.	• Access to safe water and sanitation facilities and better hygiene practice can reduce morbidity from ascariasis by 29% and hookworm by 4%.

water storage reservoirs and ponds. The best strategy for dealing with sediments is to prevent erosion. Human activity, including farming and herding, involve removal of vegetation and disturbance of the soil surface. Subsequent wind and rainfall set surface soil grains in motion and those moved by water all travel to the natural drainage system, the river. Increase in sediment delivery changes the character of the river. The sediments build up on the river bed, reducing the channel cross-section and changing the hydraulic gradient. Natural energy processes will then work to develop a new state of equilibrium between bed-slope, channel cross section, discharge and sediment movement. The river channel may widen to accommodate the risen bed level, thus introducing more sediments from eroding banks, and/or the gradient may increase, to provide energy to move the increased sediment load. Whatever occurs the character of the river channel will change, certainly downstream, but possibly upstream as well, as a result of changed gradients. Property of landholders downstream may be greatly affected. Banks may be eroded and the river may even change its course. During floods large amounts of sediment may be deposited on the floodplain, which may be beneficial, but is more likely, at least in the short term, to be unwelcome. This discussion does not apply just to perennial rivers. It applies equally to rivers which flow for only short periods, such as the wadies, which for most of the time are characterised by dry channels.

The more obvious problems produced by sediments are loss of fertile soils and the loss of storage capacity of reservoirs. It is not uncommon for storages to be completely filled with sediment within a few years of construction. Loss of a storage in this manner is a very serious matter. All investment in construction of the storage is lost, but more seriously, one of the few possible locations in a region for construction of a storage is also lost, probably for ever. Where large amounts of water are available it may be possible to manage silt inflows with spillway gates and sediment sluices, but in areas of water scarcity this is not possible.

As discussed in Section 6.4.2 above the best means of protection of a water resource from sediment problems is protection of the land surface from erosion. This is the only measure that can be taken for reservoirs that have large catchments. For small reservoirs it may be possible to construct a sediment trap upstream of the reservoir (Fig. 6.2). The sediment trap will consume some of the available water resource but this may be a small price to pay for increasing the reservoir life from just a few years to decades. After an inflow event it may be possible to save the water held in a sediment trap. Within a few days most suspended sediments will be deposited on the bottom of the trap and the clear surface water can be pumped to the main reservoir. If this is not done a water volume equal to the capacity of the sediment trap will be lost from each runoff event. The shallow character of the sediment trap means the water held will all be removed by evaporation within a few weeks, probably before the occurrence of the next runoff event. Since most runoff events are very small, perhaps producing little more than the capacity of the sediment trap, transfer of sediment trap water to the main storage after each runoff event could significantly increase water availability.

This practice is not viable in areas where erosion and sediment yields are very high such as the Loess Plateau in central-west China. Here one of the only means of

Fig. 6.10 Sediment trapped in a check dam in an ephemeral water course provided excellent agricultural soil in Cape Verde. Note the surface has recently been worked by agricultural machinery showing its excellent agricultural characteristics. The dam was arranged to allow high flows to bypass this trapped soil to avoid eroding it away, while at the same time trapping sediments from smaller flows

capturing surface water is to construct rock dykes across valleys. In some areas these fill with sediment and provide a flat surface for crop growing (Fig. 6.10), particularly trees with deep roots to reach the water stored in the trapped sediments.

6.5.4 Water Quality – Chemical and Bacterial

Water quality issues are of utmost importance where the water is for human consumption. Bacterial contamination is of great importance for human consumption whereas chemical quality is of importance for almost all uses. In regions of water scarcity where there is little or no choice of water source it is important to protect the quality of the available water.

Most bacterial contamination results from contacts between the water and animals, particularly humans. Human and animal wastes need to be totally separated from water resources wherever possible. It is quite common for much of the bacterial contamination of water to occur at or after its collection from its natural source. For example water collected in jars becomes contaminated by hands which lift the jars or which haul on ropes used to raise the water to the jars. Similarly at the collection point any spilled water which drains back to the water body is almost certain to contain bacterial contaminants. Great care (and education) is needed to prevent these and many other simple causes of drinking water contamination. Table 6.2 summarises the main aspects that characterise improved (safe) and unimproved drinking water and sanitation systems.

In many situations household wastewater is returned to the water course without any treatment. When this wastewater becomes a significant part of the total flow (say more than about 5%) the bacterial quality of the resource quickly declines, to a point where it is of little value to any user and it can become a serious health hazard. It is imperative that human wastes not be allowed to enter streams, but that dry pit toilets be used or that simple water treatment systems such as septic tanks

Table 6.2 Improved and unimproved drinking water and sanitation facilities (adapted from WHO 2006b)

	Improved	Unimproved
Drinking Water Sources	• Piped water into dwelling, plot or yard • Public tap/standpipe • Tubewell/borehole • Protected dug well • Protected spring • Rainwater collection	• Unprotected dug well • Unprotected spring • Cart with small tank/drum • Bottled water • Tanker-truck • Surface water (river, dam, lake, pond, stream, canal, irrigation channels)

Sanitation Facilities*	• Flush or pour–flush to:
	
	  – piped sewer system
	  – septic tank
	  – pit latrine
	
	• Ventilated improved pit latrine
	• Pit latrine with slab
	• Composting toilet
		• Flush or pour–flush to elsewhere
		• Pit latrine without slab or open pit
		• Bucket
		• Hanging toilet or hanging latrine
		• No facilities or bush or field

* Only facilities which are not shared or are not public are considered improved.

be interposed between households and the stream system. Without universal use of such methods the scarce water resource of a region can be completely destroyed by a few careless or selfish households.

The problem of safe access to drinking water and to safe sanitation facilities is enormous, affecting billions of people – at least 1.1 billion people have no access to safe drinking water, and 84% of these live in rural areas (WHO 2006b) as summarized in Fig. 6.11. Therefore these problems require careful attention when dealing with surface water (and groundwater) quality: there is the need to minimize/avoid health problems when accessing water, and it is also essential that sanitation problems are not passed on to affect downstream populations and users, especially when water is scarce.

For drinking water purposes bacterially contaminated water can be disinfected by introduction of a small amount of bleach or by boiling. However if it is the household's responsibility to carry out its own disinfection, then health problems are likely to be frequent because disinfection is expensive and needs careful attention to detail, meaning that the protection system is likely to be much less than perfect.

Chemical contamination has many causes. The most common are contact between the water and deposits of chemicals on or within the soil and rocks over and through which the water flows. Also common is the contamination with water discharged from mines and from industry without appropriate treatment as often occurs in developing countries. Various salts are highly soluble in water. Unfortunately, around the world, there are communities whose water supply is contaminated with naturally occurring toxins such as arsenic, nitrates and fluorides, to name just three. Chemical contamination has been reported to be frequent in Latin-America (Antón

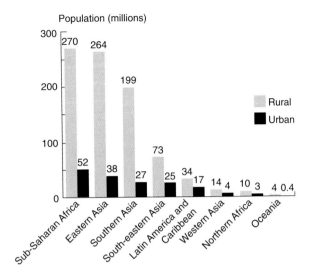

Fig. 6.11 Rural and urban population (millions) without access to an improved drinking water source in 2004 in developing regions (source: WHO 2006b)

and Diaz Delgado 2000). Chemical contamination often occurs in water supplies at levels too low to have an objectionable taste, but high enough to cause severe health effects in the long term.

Taking arsenic as an example of chemical contamination problems, WHO (2004) has reported that in Bangladesh, between 28 and 35 million people consume drinking-water with elevated levels of arsenic and the number of related cases of skin lesions is estimated at 1.5 million. Arsenic contamination of groundwater has been found in many countries, including Argentina, Bangladesh, Chile, China, India, Mexico, Thailand and the United States. It should be noted that the effects of low-level drinking water arsenic contamination has long term effects which may not appear in the consumers for many years. However once the effects are observed (skin lesions, bone deformities, etc) they cannot be reversed. Those affected can receive treatment to alleviate some of the symptoms, but the damage done by the deposition of arsenic in bone, nerve and other tissues tends to be permanent. Members of the general population have no means of knowing whether or not their drinking water is totally safe. For example in Bangladesh the arsenic contaminated groundwater resource was developed as an alternative to bacterially contaminated water. Unfortunately the new resource was not subjected to detailed chemical analysis and it has only been realised 10–20 years after implementation of the new scheme that the supplied water is slowly poisoning all the users. It is imperative that governments in all countries (including the developed nations) implement chemical analysis of all major water supply sources to attempt to prevent similar disasters occurring again (in other places from arsenic or any other chemical poisons). Testing for some of the poisons is quite difficult and therefore needs to be a government responsibility.

In situations where chemical contamination is a possibility the population needs to be encouraged to make alternate arrangements for drinking water, such as collecting roof runoff, perhaps in pots and jars. In the arid zone and regions with long seasonal dry spells quite large storage is required – enough to supply 10 l/day/household member for the duration of the dry season, i.e., perhaps 2500 l/person.

Fine sediments, particularly those of colloidal size which produce turbidity, are often a source of contamination. The sediments themselves may be harmless but many forms of contaminants, both bacterial and dissolved salts often adhere to the microscopic particles. Allowing water to stand for a few days and then removing for consumption only the clear water from the top can avoid many of the dangers. More effective is filtering the water through sand or a membrane. Filtering can be very effective in removing all particulate contamination but it is expensive and needs use of some specialized equipment. However filtering is totally ineffective in removing dissolved contaminants such as various salts (NaCl, $CaCO_3$, arsenic) which have not adhered to the surfaces of particulates.

As has been mentioned earlier, stored water, particularly if located near a residence, needs protection from insects and other vermin. This can be achieved by either totally enclosing the stored water or by installing and carefully maintaining insect screens.

Contaminants have many undesirable effects in addition to those affecting human health. For example some chemical and bacterial constituents can have adverse effects on the appearance of lakes (due to eutrophication), on industrial processes, and as mentioned elsewhere in this book, all kinds of salts have influence on irrigated crops. Some trace salts can be highly advantageous to crops but most salts, when they are present in irrigation supplies above some threshold level, can be very detrimental to most crops. As a result effort is needed to prevent these salts entering irrigation supplies. Removal of salts from any water (including irrigation supplies) is very costly, since it can only be achieved by chemical precipitation or by desalination techniques, both of which are expensive and usually unrealistic for the very large volumes required for irrigation.

6.5.5 The Riparian Ecosystem and Biodiversity

Wherever there are significant population densities, or irrigation is practiced, the natural flow regime and the range of species inhabiting the rivers will be changed. Over historical time the aquatic flora and fauna have developed in concert with the flow regime. When the flow regime changes some of the species will become stressed and die off. Other species which prefer the new flow regime will then colonise the river bed and flood-plain. These changes may be seen as desirable or undesirable. In most locations, not only are the aquatic species adapted to the flow regime, but other species, including humans, have adapted their lifestyles to be at least partially dependent on the existing environmental conditions. For example where villagers are dependent on catching fish or crustaceans to provide their protein, but flows are stopped by a dam, the villagers will need to change their lifestyle dramatically to survive. All water resources development, water diversion, or use of streams to transport and dilute wastes causes changes to the environment, usually in ways that adversely affect downstream populations.

In recent years it has been realised that many of the changes produced by significant modification of flow regimes are undesirable. Policies have been developing to reduce the changes in the flow regimes such that the general environment downstream will not be drastically modified. At present it is quite unclear how much alteration is acceptable. Large debates are occurring over how close to the pre-development regime the post-development flows should be. In most situations the aquatic environments are not sufficiently understood for clear guidelines to be developed. Post development flows are being set at about 25% of the pre-development situation, with similar variability, but this is an arbitrary figure which has no scientific basis. Presumably it will be enough to support some species, but others will disappear.

Environmental flows, the flows needed to sustain the naturally occurring species (flora and fauna), are very difficult to define and are currently the subject of vigorous scientific and political debates. For many streams maintenance of the flows necessary to support the natural systems means virtually no development can occur. In other regimes, particularly in semi arid and arid regions very little is known about possible effects of developing the very limited water resources. It is quite possible that every region would be very different. In some areas there may be little or no effect but in most arid regions any removal of the very scarce resource is likely to have large effects. These effects may not be immediately apparent but in the long term they will be noticed, but by then it will probably be impossible to reverse the effects even if there was a desire to do so. For example nesting places of migratory birds in wetlands may be endangered so that the population may be progressively reduced each year. This reduction may not be noticed for some years and then by the time it is decided some remedial action is needed and changes are implemented the birds may be extinct. Similar problems can occur with many other arid zone species, particularly because they are rarely seen and therefore their life cycle is not yet understood.

Care is needed to avoid the attitude that human survival is paramount. What effects does extinction of other species have on humanity? In most cases we do not know. Therefore great care, and exercise of the precautionary principle, needs to be encouraged.

In all regions of water scarcity, not just arid regions, the problem is quite critical. Human survival in the region may require that all water is appropriated for "beneficial use". In these situations the local population must decide whether or not the downstream environment should be preserved in a near-natural state, or whether it can be sacrificed for the development. Does it matter? There is a need to educate those who make such decisions and assist them to evaluate the total benefits and losses of such changes in local biodiversity. Most of the problems produced by such developments in the past were the result of either ignorance or piecemeal local development. However the consequences can be very large. Examples of huge environmental damage are the disappearance of fresh water inflows to the Aral Sea as a result of upstream irrigation and the hugely reduced flows of water to the sea. Reduction of freshwater flows to oceans causes changes in marine life, allows brackish water further inland along estuaries, changes river mouth species, disrupts intake of fresh water supplies for a range of supply schemes, damages or destroys mangrove forests and allows pollutants to concentrate in estuaries where previously they were dispersed into the ocean. A typical problem area is Chatt-el-Arab in southern Iraq, where the decrease of flow in the Tigris and Euphrates Rivers is a result of dams constructed in Turkey and increased water use along the length of the rivers before they reach the estuary.

Relaxing requirements for environmental flows in regions of water scarcity to allow the local population to survive will usually provide only short-term benefits. The degradation produced by cutting off all flows will eventually be so large as to discourage continued settlement in the region. An example is the draining of the marshes at the head of the Persian Gulf as mentioned above.

6.5.6 Water Borne Diseases

Water borne diseases are endemic to many parts of the world. Many common ailments are transmitted via water by unhygienic or unsanitary practices. Thirsty, uneducated people are likely to drink highly contaminated water. The need to only drink from safe sources needs to be continually emphasised to children in their earliest years. It is well known that simple cleanliness, particularly having clean hands whenever they come in contact with stored or transported water is very important. Myriad bacteria are transferred from dirty hands to water to mouth and gut to develop into one or another form of enteric infection. These infections can be manifest in the form of a slight reduction in an individuals normal comfort level, to the other extreme of serious illnesses such as cholera, typhoid, hepatitis, dysentery or one of many other potentially deadly diseases. These problems become more serious as water becomes increasingly scarce because there is then thought to be less water available for washing. The opposite is true. As the resource becomes more precious the need for cleanliness, and in particular for hand washing before any contact with water becomes imperative. Where water is very scarce this precautionary behaviour is rarely practiced and as a result stomach upsets, and worse, are part of the daily expectations, which significantly lower enjoyment of life and eventually shorten it. Prevention of these problems is quite possible but it needs continued education, publicity and discipline.

Education concerning water consumption and hygiene needs to begin with the very young at home, and should be reinforced by continual reminders in all school curricula, beginning in the first days of school attendance. This topic, with suggestions for implementation, is discussed in Chapter 12.

Any form of water resource development causes changes in natural conditions. Many of these changes offer opportunities for multiplication of numerous disease vectors, which may have produced little or no problems under natural conditions, but with the changes could have devastating effects. An example would be development of a tube-well to raise water to the surface. When the water is temporarily held in a surface tank or reservoir it provides opportunity for breeding of mosquitoes and flies. These opportunities were not available when the water was stored only in the aquifer. In regions where malaria, dengue fever or similar insect transmitted diseases are endemic, storage of water on the surface needs to be accompanied by precautionary measures to prevent the surface water becoming a breeding site for these disease transmitters.

In the tropics, almost all water resources development tends to increase the opportunities for some type of undesirable, and in many cases deadly, disease vector. This means that an integral part of any tropical water resource development plan, particularly in areas of water scarcity, needs to be an overt, up-front plan to prevent any increase in these preventable diseases. This means that planners must investigate what diseases could possibly increase with the development and then implement measures for their prevention. In most cases this will mean changing the method of implementation of the scheme and probably raise the cost of the project. But with innovative ideas the changes and extra costs need not be large. Failure to do this results in people's lives being changed for the better by increased availability of water, but for the worse by increased morbidity. In some earlier developments the increase in disease and shortening of life expectancy has been (and continues to be) so large that even now it would probably be better if the developments were abandoned. Examples are to be found in power production reservoirs, and in many irrigation developments in regions where river blindness and schistosomiasis are endemic. Unfortunately, attempts to control disease problems after they become obvious on completion of the project always costs much more than measures built into the original plans for the project – if indeed post construction control is possible.

The more common preventable insect transmitted diseases resulting from water resources development, mainly in the tropics, are malaria, dengue, encephalitis, river blindness, and schistosomiasis. All of these can be kept in check by a combination of biological/entomological understanding and engineering works and management. It is beyond the scope of this book to discuss preventive measures, but these are available in many other references. However it is important that planners, decision makers and financiers include disease prevention as an integral part of any water resource development project, especially in tropical regions.

6.6 Conclusion

In regions where water is scarce the small local population has usually been effective in deriving the water it needs for an acceptable lifestyle. However as populations increase there is often a need to

- Make better (optimal) use of the usual water sources
- Find more water
- Carefully protect all the water.

Protection is needed to prevent, or at least minimize evaporation and seepage of stored water and to prevent physical, chemical and biological contamination. The principles of how to collect more water, how to make better use of what is collected and how to protect it from spoliation have been discussed in this chapter and some examples have been given of practical ways of conserving water supplies. Some of the principles involve cultural practices and beliefs, and hence change to these, requiring sensitivity and understanding, will only occur slowly. The majority of situations where water development could be worthwhile usually require some adjustments to the project so that the aims and desires of the local people can be met. Planning and implementation of these adjustments, mainly to prevent loss of the investment due to water losses and wastes, or to avoid proliferation of life threatening diseases usually requires only application of simple common sense and logic, and it should be possible, with some determination and will to quickly and economically implement these anywhere.

Chapter 7
Groundwater Use and Recharge

Abstract Principal aspects of groundwater use and recharge are reviewed in the perspective of achieving sustainable groundwater development, i.e. avoiding groundwater degradation. The analysis concerns both major aquifers and well fields and minor aquifers of local importance, such as in islands. It includes issues for exploitation and management, as well as requirements for aquifer monitoring and control and maintenance of wells, pumps and other facilities. Attention is paid to the effects and environmental impacts of aquifer over-exploitation as well as to artificial recharge and the need for conjunctive use of surface- and groundwater.

7.1 Introduction

Groundwater plays an important role in alleviating water scarcity problems due to its inherent physical and storage characteristics. Aquifers have some specific characteristics that distinguish them from other water sources (Box 7.1) and make them quite unique in their usefulness. As such, the aquifers can be extremely beneficial in supplementing other sources from which there may be diminished yield due to drought and dry spells and can constitute useful strategic reserves for coping with water scarcity. However, for groundwater to play such a key role under conditions of water scarcity an increased understanding of the aquifer system is required. Its operation and management would demand greater attention than under normal circumstances because it must be remembered that it will never be a sufficiently abundant resource to provide for all ongoing demands.

The main requirements for effective groundwater use under conditions of water scarcity include:

- Full assessment of the resource.
- Development of proper operational and water allocation plans.
- Awareness and consideration concerning groundwater quantity and quality protection, and enforcement of the necessary legal and institutional measures.
- Application of integrated water resources management towards securing the sustainability of the source.

L.S. Pereira et al., *Coping with Water Scarcity*, DOI 10.1007/978-1-4020-9579-5_7,

Box 7.1 Value of groundwater resources in coping with water scarcity

- Groundwater may be found in areas where surface water is absent or difficult to mobilize.
- In the case of temporal or seasonal scarcity of water, an aquifer can be operated as a seasonal water storage reservoir tapped during the dry season and refilled during the rainy season.
- Groundwater can be used as a supplementary source for blending with water from other non-conventional sources, and in conjunctive operation with other sources.
- Aquifers can be used for the storage and additional treatment of recycled water.
- Sediments in surface water are removed when water enters an aquifer and the quality of the water is improved.
- Aquifers often have a much larger storage capacity than surface reservoirs.
- In aquifers, the transport of water from the areas of recharge to the areas of extraction is natural and may involve very large distances.
- Groundwater reacts slowly to seasonal and medium term climate variability, acting as a buffer for such variations.

In arid and semi-arid regions, where water scarcity is endemic, groundwater plays an immense role in meeting domestic and irrigation demands. In these regions massive use of groundwater has been practiced for some time now for large cities and long-established irrigated agricultural developments. Examples can be found in the South and Southwest United States, Northwest Mexico, the Mexico City, Israel and Palestine, the Arabian Peninsula, the large areas of India, Pakistan and China. Settlements in oases of arid and desert areas of North Africa, the Near East and northern Asia rely on groundwater use.

In these areas, although the value of groundwater in the development of agriculture and in meeting domestic and industrial demand has been considerable, concerns about the sustainability of groundwater production and the extent of overexploitation and "mining" are relevant. This is due to the fact that a good assessment of groundwater recharge is quite difficult and in most cases it is at best based on assumptions with large uncertainties. These concerns are more pronounced in arid regions and where large groundwater developments have occurred in recent times, as for the Arabic Peninsula and Libya, where developments are based on tapping fossil groundwater.

Groundwater is the largest source of stored fresh water and is often the only source of water in arid and semi-arid regions. Large aquifers underlie North Africa and the Arabian Peninsula. Though costly to develop, several countries could share the benefits of these aquifers, but exploitation strategies still need to be developed together with agreements on potential abstraction. Many countries are already

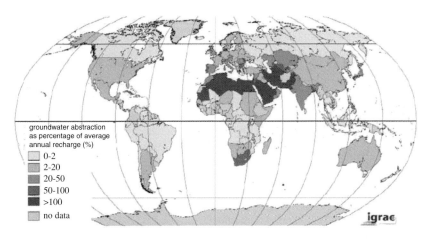

Fig. 7.1 Groundwater abstraction rate as a percentage of average recharge (source: IGRAC 2004)

heavily dependent on groundwater and face severe problems of depletion, since water withdrawals are already exceeding renewable yields (Fig. 7.1).

Technological developments and improved understanding of groundwater hydraulics and hydrogeology of aquifers have allowed large-scale exploitation of groundwater. This however, has led to severe problems with the availability of water and the quality of groundwater in arid and semi-arid regions where natural recharge is less than the growing demand, or where the use of groundwater is the main way for coping with drought. In using groundwater for meeting temporary drought conditions or to cope with more permanent water scarcity problems, exploitation should be such that the sustainability of the system is not impaired. For this to be achieved, the groundwater reservoir characteristics need to be fully understood together with the discharge, recharge, and storage potential of the aquifer, for both renewable and non-renewable systems. This knowledge should lead to the development of a rational exploitation policy for the groundwater in storage, and the establishment of operational procedures for wells, boreholes and well fields. Management considerations such as economic, legal, environmental, and institutional issues should also be integrated into the policies to ensure the sustainability of the groundwater resource.

7.2 Major Aquifers and Well Fields

7.2.1 Groundwater Reservoir Characteristics

The groundwater reservoirs could be igneous, metamorphic or sedimentary, depending on the origin of the geological formation. The lithology determines the type of aquifer such as sandstone, limestone or basalt, whilst its porosity determines how water is transmitted such as through intergranular spaces, fractures, channels or karstic pathways. These are further defined by their physiographic setting such as plateau, alluvial fill or fan.

The extent and conditions for natural recharge of aquifers when these are exploited, is of great importance. This is most pertinent for areas of low rainfall, such as in arid and semi-arid regions. The exploitation of aquifers has to be planned on the basis of their "safe yield" or "perennial yield" (see Box 7.2 for pertinent aquifer definitions). Evaluations have to be performed so as to know whether the aquifers contain renewable or non-renewable water. If the water is renewable, at what rate this is renewed? If the groundwater is non renewable it is said that the aquifer contains "fossil" water, as appears to be the case for the Saharan aquifers tapped by the Great Man-Made River project in Libya (Margat and Saad 1984, Otchet 2000).

Box 7.2 Some pertinent aquifer definitions

- **Porosity** is the ratio of the volume of voids to the volume of aquifer; it depends on many factors such as the cementation of the pores, the fracturing, the packing of the grains, their shape, size and their arrangement.
- The **effective porosity**, or **specific yield**, or **storage coefficient** of an aquifer refer to the amount of water that can be drained by gravity and is expressed as a percentage.
- The **hydraulic conductivity** or **permeability** is the physical property of an aquifer to transmit water and it is expressed in terms of velocity. It is related to the effective porosity.
- **Transmissivity** of an aquifer is the product of the average permeability and the saturated thickness of an aquifer.
- Aquifer **pumping test** is the controlled pumping of a well for a specified time during which the rate of pumping and the drop of the water level, during and the recovery after ceasing of pumping are monitored; numerous techniques for evaluation of pumping test results for estimation of permeability and specific yield are available in the literature.
- **Safe yield** of an aquifer can be defined as the water that can be abstracted continuously from an aquifer without undesirable results.
- **Perennial yield** is the flow of water that can be abstracted from a given aquifer without producing results which lead to an adverse situation.
- **Fossil water** is the groundwater that entered the aquifer as recharge in past geologic periods, under different climatic conditions, and is not renewable.

The porosity of an aquifer determines the storage and to some extent the release of groundwater by an aquifer. Of greater importance is the effective porosity or specific yield of an aquifer since it controls the quantity of water that can be released by an aquifer. Another important physical parameter of an aquifer, which is related to the effective porosity, is the hydraulic conductivity or permeability, indicating how easily the water can be transmitted through the aquifer. Both physical properties can be evaluated in the laboratory using undisturbed aquifer samples or by the analysis of pumping tests on wells or boreholes.

The groundwater storage and exploitation potential of an aquifer is not controlled solely by the lithological conditions. The setting of the aquifer, its structure and geometry should be such as to retain the water and make it available for exploitation. Thus, the dimensions and thickness, the stratigraphy or variation of lithologic units, both vertically and laterally, and the geologic and hydrologic boundaries, together with the effective porosity determine the storage and water bearing capacity of an aquifer.

The hydraulic factors associated with the occurrence of groundwater in aquifers and which affect uses of the water are:

- The quantities of water available for natural or artificial recharge (climate, abundance of recharge, infiltration areas).
- The hydrogeological properties, related to the movement of groundwater (hydraulic conductivity, transmissivity, specific yield, storage coefficient, as defined in Box 7.2).
- The hydrological boundaries of the aquifer (coast, contact with waters of inferior quality, contact with a stream or lake).

Depending on whether an aquifer forms a water-table under atmospheric pressure, or it is surrounded by strata of markedly less permeability, the aquifer is called an "unconfined, water-table, or a phreatic aquifer" in the first case, or "confined aquifer" in the latter case. The latter may be artesian if the water naturally rises to the surface due to pressure within the aquifer.

The use that is made of the aquifer, and the quantities of water that may be available in the future, depend on hydro-meteorological factors such as the rainfall, evapotranspiration, and runoff, and on the ease with which the aquifer is recharged. Recharge relates to the infiltration and percolation characteristics of the unsaturated zone and on the distance to the saturated zone. Floodwaters and runoff waters, especially in arid regions, are only partly available for recharge since, usually, their occurrence is brief and there is insufficient opportunity for them to infiltrate to the aquifer.

The chemical quality of groundwater is related to the water bearing materials of the aquifer and the amount of soluble substances they contain. Furthermore, the duration of contact of the groundwater with the lithology of the aquifer influences the concentration of dissolved solids. In coastal aquifers, quality may be affected by sea intrusion, while point and non-point source pollutants can affect phreatic groundwater.

In arid and semi-arid areas, high evapotranspiration is the main cause of increase of salt concentrations in the unsaturated zone. This salt may be transported by water percolating to the aquifer. Deep aquifers often have a high concentration of dissolved solids whilst shallow systems may have low salt content because the latter receive more continuous contributions of recharge from rainfall and surface run-off, frequently replacing volumes extracted or discharging from the aquifer.

The exploitation of a phreatic aquifer and the increased local use of groundwater for irrigation may affect the quality of the groundwater, especially in areas of high evapotranspiration, due to the accumulation of salts at the surface and their subsequent leaching into the aquifer. Fertilizers and pesticides used in agricultural activities also affect the quality of the local groundwater when proper crop management and good agricultural practices are not adopted. Similarly, pollutants from industrial and urban areas, including heavy metals and organic substances, often produce point source contamination of aquifers when raw wastewater percolates to the groundwater.

The sustainable aquifers are those that receive recharge mainly from surface water, rainfall, streams and lakes in excess of their exploitation. This recharge could be continuous or temporary and intermittent depending on the source of supply.

7.2.2 Discharge, Recharge and Storage of Aquifers

The outflow from an aquifer is discharged through springs, through subsurface flow into the sea, lakes, surface streams or other aquifer systems, rises into surface depressions when the water table is high, and passes into the atmosphere by evaporation and evapotranspiration. Groundwater contributes to the surface flow of streams and is the main contributor to base-flow and to dry-weather flow. Wells and boreholes, drains, *qanats* and similar, are artificial means for extracting groundwater. In exploited aquifers, they usually account for most of the discharge.

The recharge, which is the downward movement of water, is controlled by the supply of water infiltrating, the vertical permeability between the ground surface and the water table, and the depth to the water table. Subsurface flow originating in upland areas may add to the aquifer recharge. Because the residence time of groundwater within the aquifer is normally very large, periods of drought only affect aquifers after some time whereas for surface resources the impact of drought is much quicker.

The term "sustainability" used in reference to groundwater resources means the use of these resources in such quantities and under such conditions that would allow their renewal at a rate greater than or equal to their use. Theoretically, all water resources are renewable. The rate of renewal though, differs for the various types of aquifers. This renewal may occur within periods as short as one year in the case of a shallow minor phreatic aquifer in areas where there is rainfall, to many years as in the case of deep artesian aquifers. In the case of fossil water as found underneath the Sahara and the Arabian Peninsula, this could only be measured in geologic time. Age-dating techniques for the water in these large groundwater reservoirs show that these groundwaters infiltrated during the ice age (30,000 years ago). Thus one should distinguish between renewable and non-renewable groundwater resources.

Sustainability thus is a quantitative term and has the meaning of non-depletion of the reserves within the life span of water-works, say 40–50 years, rather than geologic time. In this respect the concept of "safe yield" of an aquifer needs to be applied. This defines the quantities of groundwater that can be used which will not create any undesirable results (Box 7.2). Undesirable results could be the lowering of the water table to depths that would be too large for economic pumping, or the deterioration of the groundwater quality (as discussed in Section 7.4). However, well planned "groundwater mining" is possible under very specific conditions, and may be economically and socially acceptable to cope with extremely severe droughts. The term groundwater mining is used when the conscious and planned abstraction rate greatly exceeds aquifer recharge (UNCED 1992).

The development of groundwater reserves depends on the special climatic conditions in an area, the water balance, the spatial distribution of precipitation, and the long-term recharge. An outline of the types of groundwater reserves is given in Box 7.3. The groundwater reserves are also subjected to temporal and spatial variations that may vary on an annual or inter-annual basis. This particularity of groundwater, as opposed to other minerals, plays a decisive role in its exploitation since withdrawal should not exceed the aquifer annual or inter-annual recharge. The opposite practice, such as an extended over-exploitation, would lead to the depletion of reserves and the drying up of major parts of the aquifer and its springs. This may result in irreparable damage of the aquifer system, either directly or indirectly, by drawing in water of inferior quality such as seawater as is the case of coastal aquifers, or the compaction of the aquifer granular structure, leading to land subsidence, as discussed further in Section 7.4.

Box 7.3 Types of groundwater reserves and their exploitation

- Reserves in **live storage** which are naturally discharged and could be exploited by pumping.
- Reserves in **dead storage** which are below the level of natural discharge and which could be exploited only after the live storage is exhausted.
- **Local reserves** that are pumped by a well or a number of wells and their pumping creates a local cone of depression.
- Reserves in **irreversible storage** whose pumping leads to a permanent consolidation of the aquifer and loss of water storage space, leading to land subsidence.

The rational development and management of water resources needs to be based on long-term planning and advanced understanding of their quantitative and qualitative temporal and spatial variations, both at a national scale and to the scale of the individual hydrologic catchment and aquifer. On the other hand, the exploitation of non-renewable aquifer systems does not differ from any other mining operation and any such policy should be carefully evaluated on its own merits. Mining of groundwater might be acceptable if environmental values are not compromised and the rights of future generations are duly considered. Such a policy has to be carefully examined beforehand so as not to jeopardize the long term interests of the community and should focus on:

- Economically feasible capital investment, the local and community present and future costs and benefits due to the groundwater development.

- The foreseeable consequences arising from the progressive reduction in the availability of water, including ecological impacts.
- Equity considerations with respect to future generations.

When tapping an aquifer, the pumping schedule and quantities available largely depend on the groundwater in storage, both in terms of the type of reserves (Box 7.3) and the yield capacity of the aquifer (Box 7.4). For an idealized aquifer, in the long run, inflow and outflow must balance.

Box 7.4 The yield capacity potential of groundwater

- **Mining yield** is the quantity extracted in excess of the recharge and which leads to the depletion of the aquifer.
- **Perennial yield** is the extraction that is carried out under specified conditions and which does not create any undesirable repercussions on the groundwater reserves.
- **Deferred perennial yield** consists of two different pumping yields. The starting yield is greater than the rate of the perennial yield and results in the depression of the water level. This may provide water at lower cost without any detrimental effects on the aquifer. In fact this may reduce initial losses that occur at high water-table conditions such as by evapotranspiration or subterranean flow to the sea. When the water level is lowered to a predetermined level, then the rate of abstraction is adjusted to the rate of the perennial yield.
- **Maximum perennial yield.** For this to be accomplished there is need for maximizing the recharge potential and the aquifer must be treated as one unit. It assumes that surface water is conjunctively used with the operation of the aquifer, and that there is a rational distribution of water demand.
- **Sustainable yield.** This is the maximum extraction rate that could be sustained by the natural recharge, irrespective of any short-term variations of recharge.
- **Optimal yield.** This is related not only to the safe yield of the aquifer but also to the optimization of the operation and management of the aquifer and includes conjunctive use aspects where these can be applied, and artificial recharge.

A groundwater system holds a certain amount of water in storage. The storage – annual discharge ratio of an aquifer, in the dimension of years is usually quite high compared to the same ratio for a river, where it is usually very small. This feature makes groundwater very attractive in the case of droughts and for temporary water shortage conditions.

7.2.3 Exploitation of Groundwater Storage

The exploitation of the groundwater storage should be a gradual process, starting from an investigation of the aquifer and its characteristics, followed by an evaluation of its potential, and ending with the design and operation of the wells and well-fields. Normally, an aquifer is the first source to be developed by farmers and other private individuals. This usually occurs well before any major water distribution scheme is put into operation because of the ease of access and the relatively low cost of developing a groundwater source. Therefore, in practice, the investigations that are carried out for increased exploitation of an aquifer aim to assess the safe yield and additional potential of the source.

The necessary groundwater investigations required are outlined in many publications and are beyond the scope of this text. The important issue is that before a major water scheme is implemented all the necessary investigations, surveys and data collection needed for the evaluation of the available potential should be undertaken (USBR 1985).

An exploratory drilling program, which is based on a number of other investigations, is the most direct method for gathering information about a groundwater reservoir, and provides a basis for designing operational boreholes and well fields after pumping tests. The location and spacing of wells and boreholes, their depths, diameters, casing and other design specifications (gravel-pack, screen size, and location of pump) have to be determined through the exploratory drilling and pumping test program. Groundwater quality considerations usually determine the type of casing and grain size of filters to be installed. The optimum design of a well field is quite a complicated process that normally requires the use of mathematical simulation models to evaluate the field and to minimize interference effects between the wells and to provide results to enable yield maximization.

In water scarcity regions, groundwater often constitutes the main source of water. During droughts, exploitation of large aquifers should be undertaken with care, to avoid irreparable damage to the aquifer and to the quality of the groundwater. On many occasions and in many countries the efforts made to cope with water scarcity are an integral part of efforts made to conserve groundwater and prevent over-exploitation.

Obviously, exploitation in excess of the natural recharge will result in the lowering of the reserves and the piezometric head surface. This in turn will have an impact on the yield of the wells and greater energy costs will be required for pumping. Depending on the hydraulic depression created, water of inferior quality may intrude the aquifer, such as seawater or inferior quality water from deeper horizons. Similarly, though depending on the texture of the aquifer material, land subsidence conditions may develop that could result in irreparable damage to the aquifer and even cause damage to properties at the ground surface (see Section 7.4).

The exploitable reserves of groundwater are those that correspond to the regulated reserves, or live storage. These are included between the highest and lowest water-table level of each hydrologic year and are only a fraction of the geologic reserves that are between the lowest water level mentioned above and the bottom of the aquifer. Groundwater source development under conditions of water scarcity requires careful quantification of the aquifer potential and the optimization of its use. The exploitation policy should be so formulated as to avoid depletion of the reserves beyond a predefined level that should be defined as the minimum acceptable level which would be expected to cause minimal problems. The main issues that should be considered for a groundwater exploitation policy under water scarcity are listed in Box 7.5.

Box 7.5 Main issues for aquifer exploitation under water scarcity

- The groundwater yield potential needs to be evaluated.
- The exploitation policy should be designed so as to safeguard the sustainability of the aquifer.
- The distribution of wells should be optimally designed to avoid interference between the wells and the development of water level depressions due to concentrated pumping.
- Water conservation measures should be implemented, such as less water-demanding cropping patterns, improved irrigation systems, and demand management strategies.
- Schemes for rationing of water need to be established.

7.2.4 Management Considerations

Managing groundwater reservoirs under the stress conditions of meeting the demand, particularly on the occasion of lack of sufficient water resources, is a very difficult task. Some of the main issues for groundwater management in water scarcity regions are listed in Box 7.6.

Commonly, in areas where there is any kind of water scarcity, including a drought, an aquifer is already being exploited. This exploitation normally goes back a long time, having developed gradually and often haphazardly and without serious regulation and control of the aquifer system. The quantities extracted and the distribution of boreholes is usually far from optimal. This leads to excessive local depressions of the water table, interference of one pumping well with another, and salt water intrusion in parts of the aquifer, whilst elsewhere in the same aquifer there may be undesirable losses of fresh water to the sea or to other systems.

Under drought conditions, the available groundwater reserves are often relied upon to meet a temporary lack of water resources. Furthermore, during such periods the competition and demand for water are very high and managing an aquifer so as to meet only part of the water demand, as is desirable in the policy of sustainable use of the aquifer, is quite difficult.

The optimal utilization of a groundwater resource needs to be based on long-term planning. Advanced knowledge is needed of the likely temporal and spatial variations of the quantity and quality characteristics of the groundwater. To gain this knowledge data should be assembled on historical sequences of droughts and extended dry spells, so that related use and conservation measures can be built into the exploitation policy. The inclusion of the contingency of reduced recharge in the policy is imperative in regions facing various levels of water scarcity. This is especially important to demonstrate the dangers and hopefully attenuate the pressure for increased groundwater exploitation to meet the growing demand.

Box 7.6 Main issues for groundwater management in water scarcity regions

- The lack of sufficiency of water resources makes the management of groundwater reservoirs particularly difficult. Therefore, a good assessment of the demand is required.
- Normally, aquifers under water scarcity conditions are heavily exploited, and their management is often far from optimal. A good knowledge of the overall water resource in the region is necessary.
- The high competition and demand for water makes management of an aquifer for sustainable use quite difficult. Priorities for allocation among different users should therefore be clearly understood.
- Advanced knowledge of the quantitative and qualitative characteristics of the aquifer is a prerequisite for its optimal utilization.
- Contingency plans for periods of drought, when demand is higher and recharge is reduced, are imperative to cope with increased groundwater exploitation.
- Implementation of plans and measures on aquifers in water scarcity regions is more difficult than in regions with relative abundance of water resources. Therefore, appropriate institutional, legal and administrative frameworks are required.
- The optimal development of groundwater should be based on the rational planning of the whole aquifer and should be based on the capacity of its "perennial yield".
- Induced and artificial groundwater recharge should be carried out as a supplement to natural recharge for balancing the supply with the demand.
- Aquifers in water scarcity regions may reach critically low levels quite fast during droughts. Thus, monitoring is essential.
- Economical, social and environmental issues, in addition to technical ones, are of significant importance in managing an aquifer in water scarcity regions.
- The establishment of legislative measures to protect groundwater from pollution and over-exploitation are of paramount importance in regions with deficient water supplies.
- An organizational framework with the authority and personnel able to develop, monitor and control the use of aquifers needs to be established.

For sensible exploitation and optimal development of the groundwater resource, a number of measures can be used such as:

- Legal measures.
- Administrative actions.
- Set up of institutional bodies responsible for the management and operation of waterworks.
- Economic incentives and penalties.

The optimal development of groundwater should be based upon the rational planning over the whole aquifer since actions on any part of the aquifer affect the whole system. The sound management of an aquifer should be based on its capacity, defined in terms of "perennial yield", and on a number of selected targets such as meeting a prescribed demand at a specified water quality and at minimal possible cost. Operationally, the whole aquifer should be considered.

A major concern in the utilization of groundwater is to have sufficient recharge during wet years to satisfy water needs during the dry periods. This can be partially accomplished by induced recharge, by artificial groundwater recharge (see Section 7.5), and also by increasing the availability of storage space in the aquifer by pumping during the dry season.

The utilization of groundwater storage in arid to semi-arid regions should seek to smooth out the large variation of recharge that occurs from year to year. The aim is to preserve as much as possible and to maximize the utilization of the limited quantities of fresh water that are normally available in a stressed environment. It should be noted that in these environments water is often saline or brackish due to the high rates of evaporation. Priority should be given to making the correct evaluation of the groundwater storage and of the quantities available for safe exploitation. Over-exploitation of groundwater in such regions leads to increased salinity. This fact makes it necessary that storage, extraction and salinity should always be jointly considered.

The groundwater quality in coastal aquifers is particularly vulnerable due to the inland propagation of the seawater-freshwater interface. Depletion of the groundwater storage due to over-pumping and a drop of the water table to levels below the mean sea– level, create favourable hydraulic conditions for seawater intrusion. The salt water moves inland in the form of a wedge below the fresh water. This is a slow process that is often only detected after the problem has already become acute. Depleting the storage to levels below the mean sea level for limited periods of time may be possible depending on the characteristics of the aquifer. This should be done with due care and with a continuous monitoring of both the water levels and the quality of the groundwater.

In managing an aquifer, issues other than technical ones, should also be considered. For example economical, legal, environmental and institutional considerations need to be taken into account. The development and maintenance cost of a groundwater reservoir is usually small as compared to surface water resources development. Groundwater becomes costly if the water table is drawn to large depths and the yield is reduced. Water conservation measures can become very effective if the cost of groundwater is made to match the cost of surface water from implemented schemes. The cost of artificial recharge should be included in that for the continued use of the aquifer.

In many countries, the rational management of aquifers is impeded by the legal ownership of the groundwater and existing water rights, which have developed historically, before the development of current understanding of how the system could best be operated and managed. When the concept that groundwater is public property is adopted, then the administrative authority acting in the community's best interests should control and regulate the use of groundwater resources. Under such conditions it is possible to arrive at near optimal use of an aquifer.

In countries or regions suffering from deficient water supplies, the establishment of legislative measures to protect the groundwater from pollution and over-draught is of paramount importance. However, such measures should consider the basic premise that the aquifers may not necessarily coincide with administrative or political boundaries.

Groundwater is or should be a public property, and licensing should regulate its exploitation. Developing, monitoring and controlling the use of aquifers requires an organizational framework able to carry out these duties and responsibilities. The effective implementation of any groundwater policy requires a competent, well-equipped and efficient organization that can operate on the basis of the existing legislative framework. The form that the legal framework takes, whether under a unified water code or in the provision of various separate laws, decrees or regulations is of secondary importance. The same applies to the organizational setup, which will vary from country to country, from a single Authority to a State Ministry. Of importance is good coordination among all interested parties and local organizations. To be effective the institutional setup must be strong and supplied with competent and well-trained staff in such numbers as could cope with the monitoring, studying, regulating and controlling the use of the aquifers.

7.2.5 Aquifer Monitoring and Control

A complete list of required data for monitoring and controlling an aquifer is difficult to compile because of the variety of aquifers and groundwater management problems that may be encountered. The type of data and monitoring practices required would depend on the stage of the aquifer use, its development, the demand that it has to meet, the type of aquifer and its sources of recharge, the climatic conditions, and the importance of the aquifer within the other available water resources. Monitoring and data collection are required for:

- Evaluating the aquifer itself such as its lithology, geometry and hydraulic characteristics.
- Evaluating the "perennial yield" of the aquifer, thus involving climatic data and other water-balance parameters.
- Monitoring the continuous development of pumping and hydraulic conditions such as extraction, water levels and the quality of groundwater.

Monitoring the overall water-balance of the aquifer for adjusting the amount extracted is of immense importance. Groundwater level monitoring provides information on the response of the aquifer to recharge, natural discharge and extraction conditions, and allows improved decisions on new pumping schedules to be developed. Monitoring the chemical and bacterial quality enables the evaluation of the suitability of water for human consumption and for the irrigation of certain sensitive crops. It provides information on the intrusion of undesirable water of low quality, including seawater, into the aquifer, and allows for the dynamic follow up of nitrates, heavy metals and organic components. These may result from the extensive use of fertilizers and pesticides, as well as from point source pollution from industry or town sewage. A basic checklist of monitoring and control data is provided in Box 7.7.

7.2.6 Maintenance of Wells, Pumps and Other Facilities

Wells, pumps, water storage, conveyance, and water control structures should be subjected to periodic inspection and testing. It is common for wells and pumps to be neglected since the nature of deterioration that occurs in a well is not readily identified until the well fails.

The deterioration of a well usually develops slowly to a critical point and then accelerates rapidly to failure. If the problem is identified early on, then rehabilitation may be possible. Failure of wells may occur by encrustation and blocking of the screen area of the casing due to the chemical quality of the groundwater. Encrustation may also occur if the screen is exposed to the atmosphere for prolonged periods. Such exposure may occur during periods of excessive pumping and lowering of the aquifer water level. Under the same conditions, the exposed inflow face of the aquifer against the well may be blocked for the same reasons. Other failures may arise by sand entering the well due to bad design. Sand entry can affect the pumping plant and the depth of the well itself. Items that should be monitored at regular intervals are the yield of the well, the static water level, the depth of the well and the sand or silt content of the pumped water.

During routine lubrication and servicing of pumping facilities, the following should be observed (USBR 1985):

- An increase in sand content of the discharge,
- Decrease in discharge,
- Excessive heating of the motor,
- Excessive oil consumption,
- Excessive vibration,
- Cracking or uneven settlement of the pump pad or foundation, and
- Settlement or cracking of the ground and ground surface gradient around the well.

Remediation measures for these problems need to be taken quickly to avoid development of other problems.

Box 7.7 Check-list on field investigations, monitoring and control for aquifers
Field investigations:

- Developing planimetric, topographic and geologic maps.
- Field mapping of the structure, stratigraphy and lithology of the aquifer.
- Survey of wells, boreholes, springs and other aquifer extraction facilities.
- Establishment of groundwater level and quality monitoring networks.
- Location of sites for pumping tests, observation wells and piezometers.
- Logging of new drill-holes and wells.
- Performing geophysical surveys.
- Performing chemical and bacterial analysis of water samples.
- Collection of aquifer data such as type, thickness, boundaries, permeability, storativity, discharge and recharge.
- Determining groundwater and surface water relationships.

Hydrologic monitoring:

- Recording and analyzing data on static and pumping water level in wells, yield, specific capacity and quality of groundwater.
- Follow-up groundwater table and piezometric levels.
- Monitoring recharge, discharge and inflow contributing areas.
- Follow-up groundwater development and use.
- Recording weather data such as precipitation, temperature, air humidity, sunshine or radiation, wind velocity.
- Investigating special aspects in aquifer behaviour and performance such as seawater intrusion, artificial recharge, and point source pollution.

Control through monitoring:

- New wells, drilling permits, pumping permits.
- Pumping by existing facilities and use of water as compared to permitted quantities.
- Use of fertilizers, pesticides and their levels of presence in groundwater samples.
- Point source pollutants from industry and urban areas.
- Application of legislation and regulations regarding the use of the aquifer.

7.3 Minor Aquifers of Local Importance

7.3.1 Particular Aspects of Minor Aquifers

Items listed above for consideration for major aquifers apply equally well for minor aquifers, but the importance of the problems may be ordered differently. By definition, minor aquifers are of limited extent and have much smaller reserves. Thus, the impacts on a minor aquifer of a prolonged drought, or of an increase of pumping when water demand rises, would be greater in absolute terms than those on a large aquifer, i.e. minor aquifers are more vulnerable to increases in demand.

A minor aquifer would need far more careful management based on studies of the aquifer and data on water level, estimates of recharge and yield potential, and aquifer performance under pumping. However the chances are that a minor aquifer would not have received such detailed attention, so information on its performance would be limited. Thus, under conditions of drought, if a minor aquifer is over-exploited, the repercussions could be quite severe. The community depending on it could early on experience scarcity of water with the shallow wells becoming dry and the flow of springs reducing. Exploitation and management issues for minor aquifers are listed in Box 7.8.

Minor aquifers are often associated with small isolated developments of dunes and dune complexes commonly located in coastal areas. These are often associated with alluvial plains and alluvial deposits along streambeds, with buried river channel deposits, and mostly in arid zones, with piedmont or foot-hill formations such as alluvial fans. Minor aquifers often exist in islands, mainly those of volcanic origin. Small developments of any other geologic formation could similarly be termed minor aquifers.

A minor aquifer is normally an aquifer of small surface area and thus of small recharge capacity. The qualifying term "minor" does not only refer to the size of the formation, but also to the low water transmitting capacity or to the limited water reserves. The latter could be due to a combination of lithologic characteristics, recharge opportunity, geometry and morphology of the aquifer.

If the aquifer has small reserves, then the impact of increased demand would be immediate and equilibrium conditions would be disturbed with any marginal increase of extraction. On the other hand, if the aquifer has a low transmitting capacity and thus low yield potential, it is very likely that it would be a system of low permeability and small specific yield coefficient. This will result in great water level fluctuations and, with the occurrence of a drought, the drop in water level will be quite large. This will have a direct impact on the yield of the wells and the flow of springs, and quite often it would quickly result in shallow wells becoming dry, or in yields from boreholes being greatly reduced. This would necessitate deepening of the wells or their relocation to a thicker part of the aquifer.

Minor aquifers discharging to the sea or to other neighbouring systems will lose most of their water reserves if low recharge conditions prevail over an extended period. On the other hand minor aquifers in contact with water bodies of inferior quality stand to be intruded to a serious degree if increased demand is placed upon them and the hydraulic equilibrium is disturbed. In effect, minor aquifers are very sensitive to climatic conditions and to any extra burden of water demand that might be placed upon them. Similarly, such small systems are more vulnerable to pollution and intrusion of water of inferior quality. Localized aquifers with low storage coefficient and small sustainable yield are more susceptible to the side effects of over-exploitation, particularly when drought occurs.

Box 7.8 Exploitation and management issues for minor aquifers
Exploitation issues:

- Exploitation and management problems for minor aquifers could be more critical than for large aquifers. Therefore, the corresponding operation rules must be well known and accepted by the communities using those aquifers.
- Minor aquifers are more vulnerable to increased demand; similarly, they are more vulnerable to pollution and intrusion of water of inferior quality and susceptible to the side effects of over-exploitation. Rules for protection of the aquifers should be developed and well known by the users.
- Minor aquifers are more sensitive to climatic conditions. Contingency planning to cope with droughts is required.
- More careful management and control are required for minor aquifers since impacts of over-exploitation could be severe. Upper limits for rates of water removal need to be enforced as part of the operation rules.

Management issues:

- Often, the local minor aquifer is the sole reason for existence of a community and for its ongoing prosperity. The involvement of the users in management is therefore essential.
- The sufficiency of minor aquifers in years of drought is of the utmost importance especially in isolated and remote areas such as small islands. The repercussions of water scarcity are more pronounced and abrupt, especially on the occurrence of drought. Contingency plans need to be prepared with anticipation and then to be implemented through appropriate participation of users.
- Good understanding of the geometry and lithological variations as well as of the aquifer characteristics is essential for development of meaningful management practices.
- Monitoring of groundwater levels, water quality trends and estimates of recharge, as well as the control and monitoring of pumping, are highly important.
- A favourable water-balance needs to be maintained particularly during drought. At these times it is important to estimate, ahead of time, the quantities that could be pumped. Monitoring needs to be more carefully and frequently practiced under those conditions.
- Water balance evaluations help the development of water allocation and water conservation measures to be implemented in periods of water crisis. However, it is important to recruit the agreement and commitment of users when water conservation and saving measures and practices need to be implemented.
- Prudent management of the development and use of local minor aquifers is paramount for the well being of the communities that depend on them.

7.3.2 Local Importance of Minor Aquifers and Management Issues

Many communities and settlements have been established, grew and thrived on the basis of their proximity to an available supply of water. The importance of a local aquifer to some communities is that it is the sole reason for existence of that community and its continued prosperity. This is the case for many oases in desert areas.

Many settlements were established on the basis of an existing local minor aquifer. The growth of the settlement on many occasions has been controlled by the potential of such an aquifer. When demand for water increases beyond its capacity, water becomes particularly scarce. Augmentation of the water supply from alternate, perhaps distant sources is then a matter of costs and technical feasibility.

In isolated and remote areas such as small islands, the dependence of a community on a local minor aquifer for its domestic supply and irrigation requirements is total. The sufficiency of the aquifer in years of drought is of the utmost importance, which creates great concern to the authorities and the community members (Margeta 1987, Margeta et al. 1999).

The vulnerability of minor aquifers to deleterious effects from increased extraction or from lower recharge during droughts is high, and communities that rely on them stand to feel the repercussions much more than others that depend on larger systems. In the latter case, the effect of an extended dry spell may not be noticed until much later or may not be felt at all if the dry period is followed by a wet season. Minor aquifers are also more vulnerable when technological developments, such as electrical pumps are introduced to replace the traditional extraction by hand or with help of animals. Then much more water can be easily pumped and it can be extracted from much larger depths. Therefore, wells that have been exploited for centuries may become dry or nearly dry in a short period causing enormous problems to populations that have traditionally been relying on them. This may also occur in shallow wells in large aquifers, but generally the consequences may not become apparent so quickly and may be remedied when the management of the aquifer is improved. For minor aquifers, over-exploitation due to use of more powerful pumping facilities may create a non-reversible situation.

Usually there is a contradiction between the importance of a local minor aquifer and the control and monitoring of its performance for allowing the necessary management actions to be taken. For communities depending on minor aquifers, the control of the use of the aquifer usually develops naturally and any new or excessive use of water is usually a matter of conflict for discussion among the members of the community. For larger communities, local or regional government usually assigns significant interest to the management of local aquifers, reflecting their importance for the well being of the community.

Good understanding of the geometry and lithologic variations as well as of the aquifer characteristics is essential for meaningful management practices. Monitoring of groundwater levels, water quality trends and control of pumping, together with estimation of recharge is quite important. A favourable water balance needs to be maintained, together with evaluations of the quantities that could be pumped in periods of drought. These evaluations will help the development of water allocation and water conservation measures to be implemented during periods of water crisis. Prudent management of local minor aquifers for their protection and use is paramount for the well being of the communities that depend on them (Falkland and Brunel 1993).

7.4 Environmental, Economic and Social Impacts of Aquifer Overexploitation

7.4.1 General

The use of groundwater and the depletion of groundwater storage both induce changes in the hydrological system. Disturbance of the equilibrium conditions and drawing down of the water-table causes changes in the rate and direction of groundwater flow, and may induce invasion by inferior quality water to replace the depleted reserves. Lowering of the groundwater levels may also result in spring and river flow diminution, wetland surface reduction, changes in natural flora and fauna and reduction of biodiversity.

The vulnerability of groundwater areas to adverse impacts of drought or to extended periods of man-made water shortage varies. Some aquifers are more vulnerable than others and, as a consequence, some communities are more vulnerable to the impacts of droughts and man-made water scarcity. Large demand placed on an aquifer in excess of its safe yield for a lengthy period could result in a number of serious environmental impacts (Box 7.9). These impacts need to be anticipated so that proper management of the groundwater reservoir and remedial actions are taken in time and proper pumping regimes are established.

Box 7.9   Main environmental impacts due to aquifer over-exploitation

- Lowering of the groundwater level.
- Increased energy consumption.
- Reduction of well yield.
- Depletion of spring discharges.
- Reduction of stream base-flows.
- Drying of wetlands.
- Disturbance of natural flora and fauna affecting biodiversity.
- Degradation of groundwater quality.
- Seawater intrusion in coastal aquifers.
- Land surface changes due to subsidence.
- Changes in landscapes where water used to support the natural ecosystems.

The impact of drought or of man-made water shortage on groundwater may vary from place to place within a country. Impacts could be highly localized because rainfall is characteristically spatially and temporally variable in low rainfall areas. The geology and type of aquifer contribute to differing regional experiences. Groundwater drought is more a problem of having access to water rather than of absolute water scarcity.

Aquifers in arid areas depend on the occasional storm and flood events that may occur, so their development and use should be quite prudent since their sustainable operation is often questionable. The sustainability of groundwater resources in the arid zone is a serious long-term concern and continuous monitoring and evaluation should guide the management of these aquifers.

In certain arid and hyper-arid regions, groundwater is tapped from aquifers that contain fossil water. A good example is that of northern Africa and Arabian aquifers, referred to earlier. A vast amount of the non-renewable or fossil groundwater, stored over geologic time, is trapped in the Palaeozoic to Mesozoic-Neogene (Nubian) sandstones that underlie wide areas of the Arabian Peninsula and the eastern Sahara desert in Saudi Arabia, Jordan, Egypt, and Libya. The dominance and importance of this resource is paramount in the water-resource planning and strategy in many countries, especially Egypt, Libya, Saudi Arabia, Kuwait, Qatar, and Bahrain. Tapping these fossil aquifers is associated with many problems beyond the huge investments involved and their sustainability. For example salt accumulations in surface soil layers and/or underlying aquifers, which is a typical and difficult problem for irrigation with groundwater in the arid region, cannot be neglected in any long-term development project. In Saudi Arabia this has already caused a substantial depletion of non-renewable groundwater resources. Non-renewable or fossil groundwater resources should generally be carefully managed and saved as a strategic reserve for emergency or short-term use.

As discussed in Chapters 2 and 4, both desertification and water shortage are the result of human intervention and the associated excessive rate of exploitation of water resources. The control of desertification relies on the sustainability of agricultural land. This in turn depends directly on the availability of water resources. The pressure exerted on the available water resources, both surface and groundwater, as related to the total renewable water resource, provides an indication of the sustainability of agricultural land. If the exploitation rate is unfavourable, which is common under conditions of water scarcity, the actual use of the resource will lead to the exhaustion of available supplies and necessitate use of water of marginal quality. It also leads to contamination of water sources by sea intrusion or contamination by water bodies of inferior quality. Similarly, the lack of sufficient water resources also leads to increased pollution of the small available resources. In the same way, the non-sustainable use of water, such as over-exploitation of groundwater and draught of fossil water, may lead to unsustainable land use. Use of water of inferior quality leads to soil salinization, which in turn renders the soils unsuitable for agriculture. The land is then abandoned and prone to erosion and increased desertification.

Water shortage could easily result from groundwater mismanagement, even in regions where shortages could have been avoided and a balance between supply and demand could be achieved. Groundwater abuse and misuse in water scarcity regions has far reaching environmental and economic consequences and comprehension of this is quite crucial in coping under conditions of limited water availability. Effective water resources management is thus essential in maintaining a balance between supply and demand within the renewable water resources potential of a region, to avoid human-induced water shortages.

7.4.2 Groundwater Levels

A combination of factors, including land use, technology (well, borehole, pumping plant), geology, hydrology, hydrogeology, and demography conspire to create problem areas. Simple case studies may refer to the lowering of groundwater levels below shallow wells, leaving them dry, or to the reduction of recharge due to the building of a new dam upstream of an aquifer recharge area. Other examples refer to concentration of pumping near the coast creating local up-coning and seawater intrusion. Yet others refer to groundwater level drop due to increased pumping, caused by the sudden expansion of population at certain locations in response to tourism development or work opportunity. The examples indicate that vulnerability varies over space and time and it relates to the ability of users and communities to access alternative water sources and develop new water supply schemes.

Groundwater level fluctuations monitored over a large span of time reveal information on the groundwater flow regime and water balance of aquifers. More particularly they allow the following types of conclusions to be drawn:

- A groundwater level fluctuation about a steady mean value suggests a situation of water balance normally associated with steady hydrologic conditions.
- The fall of the mean groundwater level when accompanied by a steady reduction of the range of the annual fluctuation, suggests a diminishing natural recharge, something which could be attributed to a prolonged drought, urbanization (reduction of recharge areas) or the diversion of surface flows into surface reservoirs or for irrigation or urban consumption.
- The fall of the mean groundwater level accompanied by an increase in the range of the annual fluctuations, suggests excessive groundwater extraction.
- A steady recovery of the groundwater level could be due to return flow from irrigation, or leakage from wastewater.

In aquifers experiencing over-exploitation, the groundwater levels fall causing depletion of groundwater reserves, a reduction of transmissivity in water-table aquifers, and a reduction of the yield and specific yield of wells. This effect is more pronounced in heterogeneous and fractured rock aquifers. Under such situations, where water levels fall below the depth of springs and shallow wells, these sources become dry or may not be viable during a prolonged water scarcity episode. In such an event, wells may have to be deepened, provided the aquifer extends to greater depths. Furthermore, the existing pumping equipment may need to be replaced if the new depths are beyond their working performance range. Improper maintenance of wells and of the pumping machinery may compound the problem of water inadequacy; especially under conditions of falling water levels (see Section 7.2.6).

7.4.3 Water Quality Deterioration

Over-exploitation of aquifers that are in contact with the sea or other water bodies of inferior quality may result in the deterioration of the aquifer's water quality. Several other causes may also lead to the deterioration of water quality of aquifers including contamination by agriculture, industry and urban water uses. A drop of the water table below the level of the sea disturbs the existing equilibrium and seawater propagates inland rendering the groundwater useless for domestic supplies or irrigation. This process takes some time. Therefore, it is not unusual to lower the water table to levels below the sea level for limited periods of time, varying from months to years, depending on the local hydrogeological conditions. This is common in arid and semi-arid regions where major recharge events do not occur annually, but at irregular, often long time intervals. However continued over-exploitation, may create irreversible conditions for the quality of the groundwater.

In the same way, if groundwater of inferior quality exists in the same area with an exploited aquifer that inferior water may invade if the aquifer is over-used. Wells and groundwater galleries may have to be abandoned if invaded by water of inferior quality unless some form of treatment is introduced before water use. Repeated use of low quality water may seriously damage agricultural soil, vegetation, and animal biodiversity and may cause human health problems.

Over-exploitation of aquifers during periods of water scarcity causes much steeper hydraulic gradients towards wells. These conditions favour increased groundwater contamination from the ground surface especially if the area is agriculturally developed. This quality deterioration could become quite serious and affect the use of groundwater for drinking purposes. The pollutants that cause the most concern are nitrates and to a lesser extent pesticides, heavy metals and organic compounds. Outbreaks of disease may occur as people use unprotected, traditional sources or sources invaded by inferior quality groundwater. In general, scarcity of water causes the search for and development of new sources of water and as the scarcity worsens so does the quality of water considered acceptable for use.

7.4.4 Sea Water Intrusion

Although sea intrusion falls under the heading of groundwater quality deterioration, the extent of this problem is so important for over-exploited coastal aquifers that it merits a separate discussion.

Some 60% of the world population lives within 60 km of the coast. This concentration of human settlements, associated with increasing agricultural and industrial activities along the coast, exerts an excessive pressure on the coastal aquifers. Some seawater intrusion and deterioration of the water quality is inevitable whenever groundwater is abstracted from coastal aquifers. This may become a serious problem especially under water scarcity conditions.

Seawater intrusion is a slow process and considerable time may pass from the onset of the phenomenon until increased salt is noticed in water wells. For this

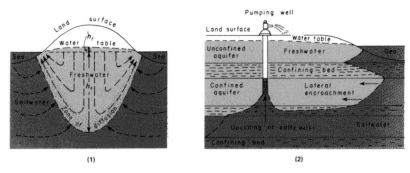

Fig. 7.2 Potential for sea water intrusion and need for cautious exploration: (1) case of fresh water lens above salty water in an island environment, e.g. the main aquifer of Malta; (2) overpumping in a coastal confined aquifer (source: Heath 1983)

reason, seawater propagation inland has to be adequately understood and monitoring programs need to be established to detect it early, so that the situation may be controlled, and if possible reversed.

In aquifers that are in contact with the sea, there is an interface between the fresh water, whose relative density is $1.0 \, kg \, l^{-1}$ and the seawater, which has a relative density of about $1.025 \, kg \, l^{-1}$. Because of this difference in density, the seawater is below the fresh water on the landward side of the coast (Fig. 7.2). The fresh and saline water are miscible, therefore this interface is not sharp but rather a mixing zone exists between the two water bodies. The extent of the mixing zone depends on the hydrodynamics of the system and on the degree of interference to the natural equilibrium conditions, usually caused by exploitation.

Groundwater levels and the depth to the saline wedge can be estimated from the "Badon-Herzberg" equation that is based on a simple application of hydrostatics. This relationship, arising from the relative densities of fresh and saline water, suggests that the depth from the mean sea level to the saline part of the interface, under steady conditions, is about 40 times the height of the water table above mean sea level. Thus, a drop of one meter of the water table would theoretically result in a rise of the saline water by about 40 m. This relationship applies well when the system is under steady conditions and when the dispersion zone is very small compared to the total saturated thickness. The problem becomes quite complex under unsteady conditions and for aquifers with a small thickness.

Sea intrusion problems may occur both on a large scale, such as along the whole front of the aquifer with the coast line, or on a local or small scale, such as for a well or a well field relatively close to the coast as shown in Fig. 7.2. Large-scale propagation of the interface occurs in response to large groundwater abstraction and de-watering of the fresh water from the aquifer. This over-exploitation that induces the sea intrusion endangers the continued use of the aquifer and the water supply. Local or small scale sea intrusion is caused by individual wells or well-fields near the coast, the over-pumping of which causes an up-coning and local deformation of the interface resulting in the gradual increase in salt content of the pumped groundwater.

In coastal aquifers, the hydraulic gradient under natural conditions is towards the sea so that there is a natural outflow of fresh groundwater. Very often, and especially where the topographic elevation is low, which is usual on the coastal plain, this hydraulic gradient is small, and therefore, a little groundwater extraction may be enough to disturb the natural system, reverse the hydraulic gradient, and cause sea-water invasion. The problem becomes quite serious when the coastal aquifer is over-used and, more so, when there is a concentration of wells near the coast. Concentration of wells near the coast is common since normally the aquifer is thickest at this location, the pumping head is small, the yields are large and this is where settlements are located.

Appropriate management approaches to prevent and control seawater intrusion need to be considered as a priority for sustainable management of coastal groundwater resources. This becomes imperative in regions experiencing water scarcity and where heavy exploitation of groundwater is carried out. The difficult dilemma though, is to have to control and reduce extraction at a time when the alternative water resources are limited and the demand is high since this is usually the circumstance in which the problem becomes apparent and acute.

Among the measures that need to be taken are to reduce the pumping and in some cases, to change the location of the wells away from the coast, adopting a more optimal distribution of wells in the aquifer. These are difficult measures to implement due to social, legal and economic reasons. As a result, the situation often becomes worse due to delays in obtaining agreements before measures can be implemented. Other approaches include artificial groundwater recharge, which can be used to establish a high water table near the coastline (see Section 7.5).

7.4.5 Land Subsidence and Land Collapse

Intensive groundwater use, resulting in depletion of aquifer storage and lowering of the water table, under certain aquifer conditions, may cause land subsidence and the development of new sinkholes in the case of carbonate aquifers (Fig. 7.3).

Land subsidence occurs in areas of serious lowering of the water table where the lithology of the underlying aquifer consists of poorly compacted sands and clays. Subsidence is associated with de-watering and consolidation of clays. The dynamics of land subsidence caused by a declining water table depend on the intensity of groundwater abstraction, the extent of decline of the water table, the geological structure of the aquifer, the soil compressibility properties and the thickness of the aquifer (Galloway et al. 1999, Ortega-Guerrero et al. 1999).

Groundwater withdrawal has little effect on aquifers consisting of solid and coherent rocks or unconsolidated gravel beds and sandy rocks that are non-compressible. In contrast, clayey soils, silts and peat have the largest susceptibility to compression and consolidation. Similarly, aquifers exhibiting inter-bedding of quite compressible clayey rocks with sand and gravel or other permeable

Fig. 7.3 Sinkhole formed due to drawdown of the aquifer at Ras-El-Ain, Syria (courtesy by Nicola Lamaddalena)

formations are liable to land subsidence when subjected to substantial groundwater withdrawals. The permeable formations intensify and accelerate the process of soil draining and water depletion, thus leading to the compaction of the clayey component.

Intensive groundwater abstraction from calcareous aquifers may also cause another similar problem, that of land collapse. This phenomenon is due to karstic processes. Sinkhole formation, which is the collapse of land resembling a funnel, is a natural problem in areas underlain by calcareous rocks, such as limestone and gypsum. The rock material is dissolved by the movement of groundwater, and this process is seriously intensified when excessive lowering of the water table occurs.

The intensive use of groundwater and the lowering of the water table re-activate the karstification phenomenon of calcareous rocks. The activation of karstic processes proceeds with the restoration of older karstic features and the creation of new ones. Intensive groundwater use in such an area changes the natural and static hydrodynamic conditions and the directions of flow of groundwater and in some cases of surface water also. The re-activation is caused by the introduction of "fresh" water unsaturated in calcium carbonate, which dissolves the calcium carbonate of the rock material, whether limestone or gypsum. At the same time, the mechanical action by the incoming water adds to the problem. As a consequence, new recharge areas, in terms of sinkholes and caving, are developed, and previous discharge points may become feed locations. The activation of karstic processes poses serious dangers to surface structures and new wells and pumping patterns need to be planned only after all the repercussions are studied in such areas.

7.4.6 Stream Base-Flow Reduction, Drying of Wetlands and Landscape Changes

Excessive groundwater abstraction and lowering of the reserves and the water table may have a major impact on river flow, and particularly on the base-flow, which generally constitutes the natural discharge of the aquifer. This may come about by the reduction of the flow of springs and by modification of the hydraulic conditions controlling the stream aquifer interactions due to the reduction of the groundwater levels.

Water scarcity affects both surface and groundwater because of the intrinsic relationship between them, especially in the case of an unconfined aquifer, and any disturbance to the existing equilibrium of one brings about changes and modifications to the regime of the other. Increased groundwater use from an aquifer should be studied in connection with the surface runoff of the same catchment area and the effect of the development of the one source on the other should be carefully assessed. Normally, a conjunctive use approach should enable maximization of water resources utilization, but the side effects on the general environment due to excessive groundwater use should be considered.

The decrease of the stream base-flow due to a reduced contribution from groundwater has repercussions on downstream users and water impoundment works, as well as on the continued existence of wetlands, and riparian vegetation in the low lands of the river basin. The quality of the surface water resource is also bound to deteriorate due to the absence of the blending effect provided by groundwater outflows into the stream, especially during low flow periods.

The presence of wetlands is normally associated with the presence of surface stream base-flow, or due to a high water table associated with a local lowland area. De-watering of the local aquifer causes a lowering of the water table and a reduction of stream base-flow resulting in the drying of wetlands. Shallow water table aquifers supply water by capillarity to the natural vegetation, or to agricultural crops. A drop of the water table will result in vegetation withering and drying up. This impact is most pronounced in the case of wetlands and riparian vegetation belts. Some local endemic species may be affected or may even disappear. The fauna associated with the affected vegetation and wetlands suffer equally. These include migratory and non-migratory birds which use the affected areas for breeding. Soil fauna and flora are also impacted. Overall, biodiversity may be affected to a large extent.

The same applies to the general landscape that may equally be affected by excessive groundwater use and depletion in areas of unconfined aquifers. Natural ecosystems that produce specific or unique green landscapes in an arid territory, which depends on shallow water tables, wither and die when access to capillary soil water is denied by the lowering of the local water table.

7.4.7 Economic and Social Impacts

The quest for water in areas of water scarcity or under a prolonged drought episode results in intensive groundwater extraction. This increased groundwater pumping produces many undesirable environmental impacts that in turn lead to serious economic consequences.

Capital investment in well construction and pumping plant may be totally lost if the well runs dry, unless it can be deepened. Larger capital investment will be required for wells of increasing depth. Additional costs will be incurred where yields decrease and escalating recurrent costs will be associated with pumping from ever-greater depths. The same apply in the case of intrusion of water of inferior quality with the consequent abandoning of wells.

Great economic and social costs are associated with the abandoning of wells, particularly in minor aquifers, when they become dry or their use becomes unacceptable due to deterioration of water quality. Costs refer to the need to develop other water sources and to changes in the economic activities which were reliant on the use of that groundwater.

Large economic costs are also associated with measures to re-establish equilibrium where the unbalanced exploitation of aquifers led to such deterioration. Social costs are also associated with the restrictions that have to be imposed on small farmers, who are often poor, when access to irrigation water has to be limited. The abandonment of the area by populations migrating to locations where they expect to have better life conditions constitutes a major impact of over-use of groundwater. When springs dry out in rural areas, rural populations are obliged to increase their time investment in collecting water from distant locations, often in conflict with other villages that were previously using that source.

Economic and social costs are also very high when the over-use of aquifers, coupled with uncontrolled land uses, facilitates the contamination of aquifers, so making them inappropriate for domestic and other high quality-requirement uses. Then quite costly water treatment to free water from nitrates, heavy metals and other substances have to be practiced, leading to higher costs for the water delivered to users. Social costs are particularly evident when poor urban and peri-urban populations need to search for other water sources. In the case of degraded surface wells and springs in rural areas, villagers need to search for new sources at some distance, demanding additional time and increasing competition and sometimes conflicts with others.

Large economic and social costs also accrue from land subsidence produced by lowering of water tables. Roads, houses and other infrastructures may be highly affected when uneven ground conditions occur due to differential land subsidence. Costs are associated with the repair of houses and buildings, roads and hydraulic infrastructures. Surface water conduits may have to be totally rebuilt. Additionally, economic activities may be very much disturbed causing serious problems for poor families.

The environmental costs of wetland diminution and landscape change are high but very difficult to assess in economic terms. Impacts are more evident when rare vegetation and animal species are endangered, when nesting conditions for migratory birds are affected, or when the "green" landscape constitutes an almost unique ecosystem in an arid or semi-arid zone. Economic consequences are easier to evaluate when human populations make use of such oases or natural green areas. The environmental impact of any groundwater management plan on wetlands and riparian ecosystems depending on groundwater should be taken into account. Moreover, the added benefits from the recovery of such ecosystems, together with the economic and social impacts as referred above, should be considered when performing any costs-benefit analysis on the improvement of the groundwater quantity and quality in any deteriorated area.

7.5 Artificial Recharge

7.5.1 General

Under scarcity conditions, withdrawals during the dry season often by far exceed the safe yield of aquifers, resulting in the depletion of the reserves and occasionally, allowing water of inferior quality to intrude into the aquifers. As already analysed, flash floods during the rainy season may not have sufficient opportunity to infiltrate the aquifers with the result that much needed water may not be utilized.

Among a number of management interventions that could help improve the situation is that of artificial groundwater recharge. This aims to increase the groundwater potential by artificially inducing increased quantities of surface water to infiltrate the ground and be later available at times of need. It could also help control or even reverse the sea-intrusion propagation in an aquifer caused by long-term or seasonal excessive pumping. This can be accomplished by creating positive groundwater levels through artificial groundwater recharge at selected strategic points in the aquifer. Furthermore, it could be used to improve the quality of pre-treated sewage water, and store it for subsequent development and reuse. In areas where there is seasonal variation of stream flow availability, water can be conserved through artificial groundwater recharge in the wet season for use during the dry season. Furthermore, control and recharge of water in wet years may help reduce the impact of droughts and to some degree alleviate man induced water shortage problems.

Maintaining a head of fresh water above sea level can prevent seawater intrusion. In many such places, artificial groundwater recharge is relied upon for blocking encroachment of seawater, retarding it or even reversing its movement. In such cases the recharge schemes serve as hydraulic mechanisms to control seawater intrusion and to allow improved management of the available groundwater reserves. Artificial groundwater recharge can be used to create a groundwater mound or ridge. A series of spreading grounds or injection wells or a combination of both could be used, depending on the geologic conditions. Such an approach has been practiced extensively on the coast of California. The location of the recharge works and of the fresh water barrier should be far enough inland so as to force the entire seawater wedge back seaward. If too close to the coast the fresh water will separate the salt wedge and force the landward part of it still further inland.

Artificial groundwater recharge schemes may also enable the disposal and storage in the aquifer of treated effluent. This is particularly advantageous during seasons when reuse requirements are minimal. The water can then be recovered for use when it is needed. An added advantage is that wastewater made to infiltrate through the soil is renovated and its quality is considerably improved through the so-called soil - aquifer- treatment (SAT), as referred in Section 8.2.7. This is nothing more than the "trickle filter bed" that is used in standard sewage treatment plants except that the soil in SAT is in its natural state. This filter is more extensive but is equally efficient provided that the soil bed is pervious. In the same manner, bacteria forming around the soil grains feed on the organic matter of the sewage, relieving it of its organic load. Only the upper part of the soil horizon in the unsaturated zone takes part in this process, extending from half to a few metres depth in exceptional cases.

Thus, artificial groundwater recharge may be defined as the planned activity whereby surface water from streams or reservoirs is made to infiltrate the ground, commonly at rates and in quantities many times in excess of natural recharge, providing for an increase in the yield of the resource. Reclaimed sewage water can also be considered as an additional source of water as dealt in Section 8.2.6. The benefits of artificial recharge are summarized in Box 7.10.

Box 7.10 Benefits provided by artificial groundwater recharge

- Counteraction of aquifer over-exploitation (depletion of reserves, localized excessive pumping).
- Combating seawater intrusion in coastal aquifers.
- Increased use of underground storage through the building up of reserves during wet years.
- Renovation and quality improvement of reclaimed sewage effluent.
- Improved economics compared to surface impoundment of water.
- Improved overall groundwater quality, depending on the quality of the recharge-water.

Water resources development anywhere, and especially in areas of water scarcity, requires the consideration of all options available. Among these options is the artificial groundwater recharge that has to be competitive with other alternatives. The benefits and disadvantages need to be assessed on the basis of capital investment required in construction, operation and maintenance costs, aesthetic considerations, amenities, and environmental and social impacts. The augmentation of the groundwater and conservation of water in general, through artificial recharge should be compared with other conventional and non-conventional schemes.

Depending on the quantities of water involved and the aquifer characteristics, artificial groundwater recharge can achieve two important hydraulic effects. These are a rise of the piezometric level and an increase in the available volume of water in the aquifer. The extent of the rise of the piezometric level is controlled by the quantity of water infiltrated, the transmissivity of the aquifer, which controls the dispersion of the water within it, and the storage coefficient or the specific yield of the aquifer, which controls the level of rise. The volumetric effect is related to the specific yield, the transmissivity and the geometry of the aquifer. The bulk of the recharged water spreads out depending on the natural groundwater flow pattern and the rate of recharge that is being accomplished.

The basic and most important factors that need to be taken into consideration for any scheme of artificial recharge are given in Box 7.11. Aquifers that can absorb large quantities of water but do not release them quickly are best suited for artificial recharge. This implies high vertical and moderate horizontal permeability, conditions that are not frequent in nature. Carbonate karstic aquifers may accept large quantities of artificially recharged water but tend to release them very quickly, making them less suitable for this activity. Alluvial aquifers appear to be most suitable, normally being close to sources of surface water from streams, and because they usually present high infiltration capacities coupled with moderate horizontal transmitting capabilities. Most of the artificial recharge works are located on coastal alluvial aquifers mainly because these areas are heavily populated and the local aquifers are extensively exploited. Furthermore, artificial recharge works are the last schemes where excess water can be gainfully utilized before it runs off to the sea. Artificial recharge thus serves the double purpose of augmenting the reserves of water and protecting the aquifer from seawater intrusion.

The artificial recharge scheme to be selected depends on such factors as the relationship between the physical and mechanical properties of the aquifer and the desirable infiltration rates. It also depends on the effect of the aquifer material on the chemical and biological improvement of the water, the ability of the aquifer to store the recharged quantities temporarily and to permit lateral movement of the water at reasonable rates and over adequate retention times.

7.5.2 Methods of Artificial Recharge

The artificial groundwater recharge schemes may be classified under two broad groups:

- The indirect methods through which increased recharge is achieved by locating means of groundwater abstraction as close as possible to areas where surface water is in contact with the aquifer or areas of natural water discharge. In such cases, the natural hydraulic gradient is affected so as to cause increased recharge.
- The direct methods through which surface water is conveyed from lakes, reservoirs, waste water treatment plants or is being diverted from flowing streams to suitable areas of aquifers where it is made to infiltrate to the groundwater from basins, trenches, dry riverbeds, injection wells, pits, etc.

Box 7.11 Hydrogeological and groundwater factors influencing artificial groundwater recharge, and choice of scheme

- The hydrogeological conditions must be amenable to the infiltration and percolation processes associated with artificial recharge.
- The physical and mechanical properties of the pervious media (soil and aquifer) must facilitate desirable infiltration rates.

- The water-bearing deposits must be able to store the recharged water temporarily and subsequently to permit its lateral movement at acceptable rates.
- There must be an efficient means of recovering the recharged water (which is more important in the case of combating sea intrusion since a large proportion of the quantity recharged would otherwise flow to the sea).
- The hydraulic conductivity and coefficient of storage of the aquifer need to be high. These define the ability of the aquifer to allow water to infiltrate and then to disperse, and the capability of the aquifer to temporarily store the infiltrated water.
- The filtration, deposition, precipitation and absorption characteristics of the upper layers of the aquifer. These processes take place during and immediately after infiltration in the recharge basin. They tend to improve the water quality but to reduce the pore dimensions (i.e. become clogged) and lead to smaller infiltration rates depending on the sediments carried by the water being recharged.
- The availability of an adequate source of recharge water of suitable chemical and physical quality.
- The proposed recharge installations must be competitive to other water resource development options and be economically and environmentally sound.

Indirect or induced artificial groundwater recharge consists of abstracting groundwater within a short distance from a flowing stream, lake or impoundment. The pumping lowers the piezometric surface to cause a steeper hydraulic gradient, between the water source and the extraction point, thus inducing increased recharge. The same artificially developed conditions tend to reduce the outflow of local groundwater into streams and surface water storage. The groundwater abstraction facilities could consist of well fields, a gallery or a line of wells. Depending on the hydrogeological parameters of the local aquifer, considerable amounts of surface water could be induced to infiltrate into the aquifer. Higher permeability favours larger quantities to be recovered through pumping wells.

The problems associated with this otherwise straightforward approach are the silting of the river banks and channels where the increased infiltration takes place, the quality of the recovered water, which depends on the state of the surface water, and the possible need for storage facilities to hold the water recovered by pumping. A minor problem that may develop, and which has to be investigated in advance of the implementation of induced recharge, is the additional proportion of groundwater that may be abstracted over and above the induced infiltration of surface water. Excessive natural groundwater pumping may have undesirable effects on local groundwater hydraulic conditions and adversely affect other users if the piezometric head drops.

With the direct methods of recharge, better control can be exercised over both the quantity and the quality of the water. In addition, the recharge activity is performed during the periods of available surface water irrespective of whether there is a need for pumping or not. Surface water diverted from a stream or lake is transported to the selected site and introduced into the aquifer by spreading basins, injection wells, pits and shafts. Direct methods have a number of advantages over indirect methods including the lesser importance of the distance to the raw water source, the opportunity to treat the raw water before recharge, less effort for clean up of the clogging of the spreading grounds, operation of the recharge scheme at will, and the opportunity for strategic location of the recharge works on the basis of environmental, hydrogeological and management considerations.

7.5.3 Artificial Recharge by Spreading

Artificial recharge by spreading is usually carried out when the aquifer extends close to the ground surface. Recharge is accomplished by spreading water over the ground surface or by conveying the raw water to basins and ditches (Fig. 7.4). Use of spreading grounds is the most common method for artificial recharge. The operational efficiency of the spreading grounds depends on the following factors:

- Presence of sufficiently pervious layers between the ground surface and the aquifer.
- Enough thickness and storage capacity of the unsaturated layers above the watertable.
- Appropriate transmissivity of the aquifer horizons.
- Surface water, clean enough to avoid excessive clogging.

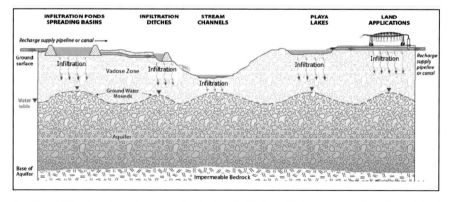

Fig. 7.4 Artificial recharge by spreading basins, infiltration ditches, stream channels, and land depressions (source: Topper et al. 2004)

The quantity of water that can enter the aquifer from spreading grounds depends on three basic factors:

- The infiltration rate at which the water penetrates the ground surface.
- The percolation rate, i.e. the rate at which water can move downward through the unsaturated zone until reaching the aquifer.
- The capacity for horizontal movement of water in the aquifer, which depends on the horizontal permeability and thickness of the aquifer.

The infiltration rate tends to reduce over time due to the clogging of soil pores by sediments carried in the raw water, growth of algae, colloidal swelling, soil dispersion and microbial activity. The infiltration rate may recover when adopting alternate wet and dry periods of spreading or, with much greater cost, by scraping away the clogged surface layer, among other techniques.

A spreading basin is normally constructed with a flat bottom that is covered evenly by small quantities of water. This requires the availability of large surfaces of land for meaningfully sized recharge works. Normally the basins are arranged so that excess water runs into the downstream basins. Retaining basins may be used for settlement of suspended sediments before water enters the spreading basins. The settling of sediments may also be assisted with the addition of coagulation agents. Intermittent operation of the spreading basins may allow the reconstitution of the major part of the initial infiltration capacity.

Variations of the spreading ground technique (Fig. 7.5) consist of the use of ditches that are easier to maintain and for which clogging is a lesser problem since a major part of the sediment is carried out of the ditches. Similarly, flooding dry riverbeds or relatively flat land could be an effective artificial recharge scheme with self-cleaning abilities. Another variation of this approach would be the use of the flat infiltration area for irrigated agriculture where excess water is applied to percolate downwards from the irrigation basins into recharge.

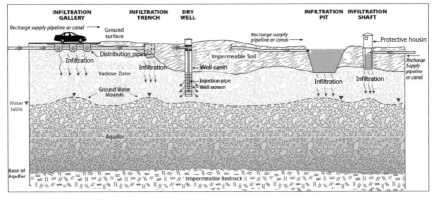

Fig. 7.5 Artificial recharge by infiltration galleries, trenches, dry wells, pits and shafts (source: Topper et al. 2004)

7.5.4 Artificial Recharge by Well Injection

When the aquifer is located at some moderate depth below the ground surface, recharge may be accomplished by introducing water through pits and shafts (Fig. 7.5) and, in the case of the presence of overburden of a large thickness, water can be injected through wells or boreholes which reach the aquifer (Fig. 7.6). This technique is also preferred where land is scarce, when environmental reasons oppose the use of large spreading grounds, when the local hydrogeological conditions do not favour spreading, or wells and pits are already available for use for recharge. In areas where the pervious formations are at shallow depth, recharge can be accomplished by digging pits or shafts. Abandoned gravel pits could also serve as recharge sites.

Injection wells are quite versatile in that they can be used on any type of aquifer. However, they require pre-treatment of the raw water, since these wells easily become clogged, necessitating difficult cleaning operations or even their abandonment and drilling of new injection wells. The recharged water is directly injected into the aquifer without much opportunity for further polishing, and so the technique must be restricted to using high quality injection water. In most cases it should be of drinking quality. Even if such high quality water is used, trouble-free operations cannot be guaranteed for long periods of time.

The injected water is preferably introduced at the bottom of the well since "free fall" of water into the well entraps air bubbles that adhere to the pores of the formation restricting the intake rate. A common practice is to inject water through extraction wells. This allows cleaning of the well and restoration of the capacity for water to flow to the pump. In such a case the water is injected through the pump column and the pump.

The capacity of the wells to absorb the injected water depends on the available conveyance capacity of the surface installations and the well equipment, on whether the aquifer can laterally transmit the water, and on the extent of clogging of the

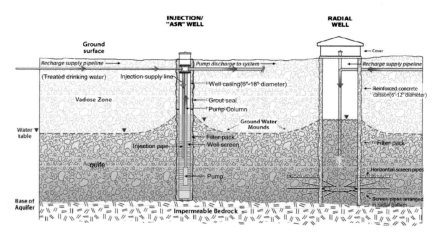

Fig. 7.6 Recharge by injection wells (source: Topper et al. 2004)

borehole walls that is caused by the injection itself. Chemical and bio-chemical processes, admission of air into the aquifer's pores, accumulation of colloidal particles on the walls of the borehole and in the immediate aquifer interstices may be the cause of clogging. When the capacity of the recharge well deteriorates to below an acceptable operational level, some kind of redevelopment is necessary. This can be done by occasional back washing by pumping the same well.

7.5.5 Recharge with Surface and Subsurface Dams

Other types of artificial recharge schemes refer to floodwater retention dams, especially on broad *wadi* floodplains. The purpose of these is to delay the flow of the water and provide the opportunity for recharge into the local aquifer. The structures consist usually of low dams, including earth walls and gabions built to be toppled by floods (Fig. 7.7).

More permanent structures refer to recharge dams, which are proper water impoundment works across stream channels, located at sites overlying aquifers. These are equipped with proper spillways and could be used for artificial groundwater recharge at the site and to release water through outlets towards spreading grounds or other types of recharge schemes. In these cases, the recharge dams act to provide both direct recharge and sediment retention, providing clear water for recharge purposes.

In areas where the pervious formations are at shallow depth, recharge can be accomplished by digging pits or wells, including abandoned gravel pits.

Provided the hydrogeological conditions permit, subsurface dams could be created by a number of approaches such as slurry trenches, cement grout and other chemical injection, reducing the permeability of the subsurface lithology. The excavation and construction of the subsurface impervious wall is performed during the dry season. The subsurface dams control the subsurface flow (Fig. 7.8), impound it underground and provide a source of water during the season of demand. This approached is being used in the semi-arid Northeast Brazil to support agricultural

Fig. 7.7 Gabions recharge dam in a wadi, Central Tunisia

Fig. 7.8 Subsurface dam for shallow aquifer use in crop production (adapted from Silva et al. 2006)

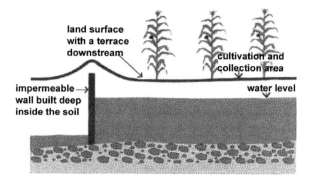

production based on capillary rise from the groundwater when the water table is kept high through subsurface dams (Silva et al. 2006; FUNASA 2007).

7.5.6 Problems and Solutions

A summary of the basic problems in performing artificial groundwater recharge, and some solutions, are listed in Box 7.12.

A major problem in artificial recharge activities is the non-availability of sufficient quantities of water of acceptable quality for recharge with some degree of regularity. The recharge schemes, especially in areas of water shortage, tend to operate on a very occasional, intermittent manner, usually associated with the availability of surplus water that can be used for this purpose. This causes the recharge schemes to remain idle for long stretches of time, requiring proper maintenance before re-activation.

Another major problem that can be expected in the operation of an artificial recharge scheme is its effectiveness over a long period of operation, since, in time, a reduction of the infiltration capacity of the scheme can be expected. This depends almost entirely on the quality of the raw water and the amount of suspended sediments it contains. Thus, measures that reduce the sediment content of the water would reduce the rate of clogging and prolong the efficient use of the scheme. This condition is essential in the case of recharge through injection wells where suspended sediments tend to clog the aquifer formation close to the well.

Box 7.12 Problems in artificial groundwater recharge and solutions
Problems:

- Availability of water
- Quality of available water for recharge
- Regularity of water supply for recharge
- Need for maintenance after lengthy idleness of the scheme

- Reduction of infiltration capacity due to quality and suspended load of water
- Compaction of soil in recharge area during maintenance

Solutions:

- Reduction of sediments for prolonging the efficiency of the recharge scheme
- Control of biological and chemical impurities
- Absence of dissolved air and gases, which reduce the permeability of the aquifer Occasional re-development of the well through pumping for removal of bacterial growth on the walls of injection wells
- Chlorination of injected wells to reduce the bacterial growth
- Intermittent operation of recharge works for reduction of algal growth and other biological activity
- Careful cleaning and maintenance operations so as to avoid soil compaction

The quality of the injected water in wells is very important. Suspended solids, biological and chemical impurities, dissolved air and gases, turbulence and temperature of both the groundwater and injected water all have an effect on the life and efficiency of a well by causing clogging or corrosion of the well screen. A large amount of dissolved air in the recharge water tends to reduce the permeability of the aquifer by "air binding". The injected water should have a temperature only slightly higher than the temperature of the aquifer.

Bacterial growths also tend to develop on the walls of the injection wells and the exposed face of the aquifer. These are normally removed by occasional redevelopment of the well by pumping. Very often, chlorinating the injected water could reduce the bacterial growth but even in this case some re-development of the well by pumping is normally required. Clogging by biological activity depends on the quality of the raw water, the environmental conditions such as temperature and sunshine duration, which promote algal growth, and the type of soil at the ground surface. This problem is usually resolved by the intermittent operation of the recharge works since alternating wet and dry periods of the spreading ground helps renovate it and recover most of its recharge effectiveness.

Artificial recharge of confined aquifers can only be achieved if the confining layer is penetrated through excavation or drilling. Additionally, sufficient pressure exceeding that of the groundwater in the confined aquifer must be exerted so that water can be injected into the aquifer. Normally, artificial recharge in this type of aquifer can only be achieved through injection wells.

The smaller the aquifer, the greater the impacts of artificial groundwater recharge. In a case in Cyprus at Germasogeia, with artificial recharge of a streambed aquifer, of 5 km length by 0.5 km width and an average depth of 50 m, the active storage is replaced 2.5 times a year by recharge of water released from an upstream dam and by pumping of the water for domestic supply (Simmers 1997). The quality of recharge water is very important in the whole operation so as not to deposit fine sediments on the recharge area and thus reduce the infiltration capacity.

7.5.7 Environmental Impacts of Artificial Recharge

Groundwater abstraction lowers the water table that might lead to a number of adverse environmental effects as analysed in Section 7.4. Such a situation becomes worse under conditions of water shortage when excessive groundwater is pumped to meet the demand. Artificial groundwater recharge helps in ameliorating these adverse conditions by blocking sea intrusion, augmenting the groundwater reserves and raising the water table. The same technique may also be used for introducing wastewater of variable degrees of treatment for further quality improvement and for storage in the aquifer.

Over a prolonged period, artificial recharge may result in some degree of clogging of the soil pores. This causes a reduction of the recharge efficiency of that particular area, not only for the artificial recharge operation but also for natural recharge. However this problem is of limited spatial extent, and does not seriously affect the overall recharge capacity of the system. Removal of the deposited silt and use of soil treatments may ameliorate the problem to a large extent. Compaction of the soil surface within the recharge area should be avoided at all times.

Adverse effects could be avoided within the spreading grounds if a regular dry and wet cycle is observed. Any growth of algae, and odours from biological growth in standing water, are controlled through drying the recharge grounds at regular intervals. Proximity of spreading grounds to population centres may cause problems, especially if the water remains stagnant for a long period of time. Operating the spreading grounds with some intermittent schedule tends to discourage breeding of mosquitoes and development of odours. Fencing and landscaping of the grounds could help suppress hazards and environmental concerns.

7.6 Conjunctive Use of Surface and Groundwater

Sound water resources development and management seeks to maximize the water resources availability at the least cost. This is even more important in arid and semi-arid climates and where droughts are quite frequent. In those regions and in areas with great fluctuation of demand, conjunctive use of surface and groundwater storage is often relied upon to offset deficits in the dry season and accommodate storage and recharge of excess water in the wet season (Margeta et al. 1997). This is also the case for small to medium size islands, where the economic dimensions of dams is usually small, whilst the water supply demand in tourist areas increases disproportionately due to the seasonal character of tourism.

In many such regions, the water managers may have a choice of sources between surface water and groundwater. Often it is found expedient, especially with regard to increasing the reliability of yields and the economic viability, to develop both sources and use them jointly for improved performance. Such combined use would be called a conjunctive use scheme. The term conjunctive use of surface and ground-water is defined as the coordinated use of both these sources for the supply of water, as distinct from the separate development of each source without regard to the other. Conjunctive use may be enlarged to include water from desalination plants and treated wastewater. It should be emphasized that in a conjunctive use scheme, none of the water sources could meet a prescribed demand on its own but the joint use of them could satisfy it at an acceptable reliability level. These qualities point to the differences between the concepts of conjunctive use and of multiple resource utilization.

The ability to incorporate use of groundwater storage with surface water storage, or with water made available from other non-conventional sources (see Chapter 8) has significant importance. A small groundwater storage that can be relied upon on the occasion of a drought or during the season of increased water demand could result in a considerable increase in the reliability of the supply in satisfying a defined demand. Multiplicity in the type of sources of supply has the advantage of using the different characteristics of availability of water from each source, whether a surface water or groundwater source. Optimal switching from one source to the other may result in an increased, steadier and more reliable supply.

Multiple resource systems can rely on components with different hydrological and economic characteristics such as surface reservoirs, aquifers and/or desalina-tion units etc. Inherent to the advantages of these systems when conjunctive use is considered is the selection of a set of component sizes and operation rules that may lead to acceptable performance at minimal capital and operation costs.

The optimum development and management of scarce water resources is of great importance. Finding an optimal system often requires mathematical descrip-tions of the behaviour of complex water resources systems and of their conjunctive operation. As the scarcity of water increases, so does the necessity for more wide-ranging and comprehensive methods for conjunctive use of several sources. The optimisation of water usage through operating policies that not only maximize the water availability but also minimize the operating costs has far reaching effects for national development. As development moves forward, water is often one of the major constraints.

The physical constraints pertaining to a water resources system consisting of a surface reservoir and an aquifer and the advantages that make the conjunctive use of its components desirable and useful are as follows:

- The demand is greater than the surface reservoir yield.
- The maximum borehole abstraction rate is less than the demand.
- The combined yield of the reservoir and the aquifer is greater than the demand.

In such a system, the aquifer is a source of water and a storage facility for which the best use and pumping must be coordinated with releases of water from the surface reservoir. For this, an operating policy consisting of a set of control rules is required indicating volumes of water to be kept in storage or released at given points in time. In such a system, neither of the two sources can meet the demand on its own but the joint use of them could satisfy the demand at an acceptable reliability level. In addition, optimisation of the use of the two sources needs to be made so that the cheapest source could be used to its maximum. This is in essence the basic objective of conjunctive use.

Control rules for the management of the two sources are usually based on computer modelling of the supply and demand to predict future water shortages and to assess economic aspects of operating the scheme at the highest possible reliability of supply. Application of techniques such as dynamic programming allows the optimum yield of water for the system to be evaluated for any chosen combination of sources.

In the case of a conjunctive use scheme between a surface water source and a groundwater source, the advantage is that during the dry season groundwater is a more reliable source when available surface runoff is at its minimum. The surface water source can be drawn upon first during the wet season allowing groundwater to recover. The same approach could practically be employed in the conjunctive use of any two sources, be they surface reservoir, aquifer or desalination plant. In the latter case where the desalination costs are very high, this non-conventional source of water could be used only when the other source (surface or groundwater) cannot meet the demand. Thus, only the capital investment of setting up such a plan should be considered since this plant will be switched on and off according to the need for supplementing the demand. The conventional source could be relied upon to its maximum with the expensive sources called upon only to maintain the high reliability normally required for water supply. Such an approach could make desalination an attractive proposition.

Wastewater reuse could also be considered as an extra source of water that could be conjunctively operated with other conventional sources so as to maximize the yield of water from a combination of sources. Treated wastewater from urban centres has a high reliability of supply and during the irrigation season could be used directly for the irrigation of selected crops. During the wet season, when the demand for it is reduced, this could either be stored in surface reservoirs or used to recharge an aquifer (see Section 7.2.6) from which it could be re-pumped during the irrigation season. Conjunctive use of this source of water with other conventional sources could help alleviate water scarcity problems.

The quest for a better use of the available water, for which demand exceeds supply under water scarcity conditions, calls for well planned monitoring programs of hydrological information, the application of descriptive and optimising models and conjunctive use of more than one source.

7.7 The Use of Groundwater in Coping with Water Scarcity

The physical and storage characteristics of aquifers, large or small, make the use of groundwater essential in alleviating a major part of water scarcity in a region. The major issues pertaining to large aquifers systems are quite different from those of minor ones. Note that minor aquifers can be of great local importance. A good understanding of an aquifer system is required and its operation and management demands greater attention if groundwater is to play a key role in reducing water scarcity.

Managing groundwater reservoirs under the stress conditions of meeting the demand in an environment without sufficient water resources is a very difficult target, particularly when over-exploitation of the aquifers is the norm. Appropriate monitoring of the aquifers and their exploration is essential to support management. Such over-use of groundwater has serious effects on the aquifer and considerable environmental impacts. The fall of water levels increases energy costs for pumping, causes water quality deterioration, seawater intrusion, land subsidence and collapse, reduction of low flows in streams and drying of wetlands. High economic and social impacts also result from these exploitation conditions. Therefore, the main considerations for exploiting and managing an aquifer under scarcity conditions generally aim at combating and overcoming these problems.

Emphasis should be placed on artificial groundwater recharge as a management intervention that could help to partly restore diminished groundwater reserves during periods of surface or other water availability. The same technique could also help to control seawater or other inferior quality water intrusion into the aquifer by establishing favourable controlling hydraulic conditions. Recharge could also be used to improve and further the reuse of treated wastewater. The various techniques available for artificial groundwater recharge, the benefits provided, and the generally positive environmental impacts produced, make artificial recharge an attractive issue for water scarcity regions.

Finally, adopting the conjunctive use of surface- and groundwater, and other non-conventional water sources, provides for sound water resources management seeking to maximize water availability. In this respect, the ability to integrate supply from groundwater storage and surface water storage or with treated wastewater or water from desalination plants attains significant importance in regions with limited availability of water resources.

A good evaluation of the available water resources and of the demands placed upon them, and of the impacts of any exploitation scenario, as well as measures to counteract adverse effects on the aquifers are very important for the sustainable use of groundwater resources. This is even more important if irreparable damages are to be avoided, and groundwater is to continue to play its important role in coping with meeting the demand in regions under water scarcity.

Chapter 8
Using Non-conventional Water Resources

Abstract Taking account of its importance in water scarce regions, the use of non-conventional water resources to complement or replace the use of usual sources of fresh water is discussed. The use of wastewater is considered for irrigation purposes, for aquaculture and for aquifers recharge. Attention is paid to the control of possible associated health risks. Another resource considered in this chapter is saline water and assessment of its suitability for irrigation and related uses. Also dealt with are the use of desalinated water, fog water capture, water harvesting, groundwater harvesting, cloud seeding, and water transfers.

8.1 Introduction

Whenever good quality water is scarce, water of inferior quality will have to be considered for use in agriculture, irrigation of lawns and gardens, washing of pavements, and other uses not requiring high quality water. Inferior quality water is also designated as non-conventional water or marginal quality water. Non-conventional water can be defined as water that possesses certain characteristics which have the potential to cause problems when it is used for an intended purpose (Pescod 1992). Thus, the use of non-conventional water requires adoption of more complex management practices and more stringent monitoring procedures than when good quality water is used.

Non-conventional waters most commonly include saline water, brackish water, agricultural drainage water, water containing toxic elements and sediments, as well as treated or untreated wastewater effluents. All these are waters of inferior or marginal quality. Also included under the designation of non-conventional waters are the desalinated water and water obtained by fog capturing, weather modification, and rainwater harvesting.

The expansion of urban populations and the increased population served by domestic water supply and sewerage give rise to greater quantities of municipal wastewater. With the current emphasis on environmental health and water pollution issues, there is an increasing awareness of the need to dispose of these wastewaters safely and beneficially. The use of wastewater in agriculture is already expanding, particularly in water scarce regions. However, the quantity of wastewater available

L.S. Pereira et al., *Coping with Water Scarcity*, DOI 10.1007/978-1-4020-9579-5_8,
© Springer Science+Business Media B.V. 2009

in most countries will account for only a small fraction of the total irrigation water requirements. Nevertheless, wastewater use will result in the conservation of higher quality water and its use for purposes other than irrigation. The nitrogen and phosphorus content of sewage might reduce the requirements for commercial fertilisers. As the marginal cost of alternative supplies of good quality water will usually be higher in water scarce areas, it is important to incorporate agricultural reuse into water resources and land use planning.

Reuse of treated water for non-agricultural uses is quite small relative to irrigation of agricultural crops. These applications include the reuse of treated industrial effluents for low quality uses in the same industrial plant, the reuse of treated municipal wastewater in aquaculture, for the irrigation of lawns and recreational areas, and for low quality domestic water uses when separated (dual) municipal distribution systems are available (Asano et al. 2006).

The increase in food production for the continuously growing world population will have to be met in large proportion by expansion of irrigation. In water scarce regions, irrigation has the ability not only to increase production per unit area of land but also to stabilise production. In arid and semi-arid areas, irrigation is the only reliable means of increasing agricultural production on a sustainable basis. However, good quality water for irrigation is increasingly short since a number of countries are approaching full utilisation of their water resources and priority is being given to uses requiring higher quality water. Therefore water of inferior quality produced from saline aquifers or resulting from drainage waters has to be used/reused for irrigation of agricultural crops along with wastewaters. In case of reuse of drainage waters, agriculture acts both as a user of the water and as a disposal site, so contributing temporarily to control of potential environmental impacts from salts carried by those waters. Salts leaching is discussed in Section 8.3.4.

As for treated effluents, saline water may also be used for non-agricultural purposes such as washing, low quality domestic uses or irrigation of recreational areas. However, the use of saline waters in irrigation has positive impacts for non-agricultural uses because it decreases the irrigation demand for good quality water, and the good quality water then becomes available for uses requiring more stringent water quality standards.

Desalinated waters are commonly added to the fresh waters for domestic uses. They are free from toxic substances or pathogens and can, therefore be used to satisfy most human requirements. By contrast, because of its low level of salts and the high costs associated with treatment, desalinised water is less appropriate for agricultural uses.

Fog capturing is used in isolated arid areas in mountains and islands where the occurrence of fog is common but rainfall is rare. Water production by this process is small but may be essential for assuring living conditions for small populations in these areas. An alternative for isolated households and villages in remote areas where surface- or groundwater is not easily available for domestic uses or animal drinking is the use of roof-water which can be captured and stored in cisterns. Water harvesting systems provide for irrigation of vegetables, fruit trees and subsistence crops. Water harvesting is analysed in Section 6.4.

Cloud seeding is a process of augmenting rainfall by adding substances to the cold clouds that act as nucleii for the formation of large water drops that otherwise would not fall to the ground. Its interest is however limited by the lack of cold wet air masses travelling over the arid low-lying areas where populations live. Water importation mainly consists of transferring water from a basin where it may be in excess to the demand into another basin where water demand is much above the natural supply. Generally this concerns long distance water transfer from basins in sub-humid to humid regions into semi-arid areas. These transfers may be quite effective from a water management perspective but they may create large environmental and developmental impacts in both the region from where the water is transferred and the region where it becomes available unless appropriate management and monitoring is applied.

8.2 Wastewater Use

8.2.1 Wastewater and Effluent Characteristics

8.2.1.1 General Wastewater Characteristics

Municipal wastewater is mainly comprised of water with relatively small concentrations of suspended and dissolved organic and inorganic solids. Organic substances include carbohydrates, lignin, fats, soaps, synthetic detergents, proteins and their decomposition products, as well as various natural and synthetic organic chemicals from the process industries. In arid and semi-arid countries, water use is often fairly low and sewage tends to be very strong, as indicated in Table 8.1 for Amman, Jordan, where per capita water use is only 90 l/day while 200 l/day to 400 l/day (USA) is common in water abundant areas.

8.2.1.2 Pathogens in Wastewaters

Municipal wastewater also contains a variety of inorganic substances from domestic and industrial sources, including a number of potentially toxic elements, including

Table 8.1 Major constituents of typical domestic wastewater (Pescod 1992)

Constituent	Concentration, mg/l			
	Strong	Medium	Weak	Amman
Total dissolved solids (TDS)	850	500	250	1170
Suspended solids	350	200	100	900
Nitrogen (as N)	85	40	20	150
Phosphorous (as P)	20	10	6	25
Alkalinity (as $CaCO_3$)	200	100	50	850
BOD_5[a]	300	200	100	770

[a] Biochemical oxygen demand at 20 °C over 5 days, which is an indicator of the biodegradable organic matter in the water.

heavy metals, such as arsenic, cadmium, chromium, copper, lead, mercury and zinc. This water might also be at phytotoxic levels, which would limit its use in agriculture. From the point of view of health, the contaminants of greatest concern are the pathogenic micro- and macro-organisms. Pathogenic viruses, bacteria, protozoa and helminths may be present in raw municipal wastewater and will survive in the environment for long periods. Pathogenic bacteria are generally present in wastewater at much lower levels than the coliform group of bacteria, which are easy to identify and enumerate as total coliforms/100 ml. The *Escherichia coli* are the most widely adopted indicator of faecal pollution and they can also be isolated and identified fairly simply, with their numbers usually being given in the form of faecal coliforms (FC)/100 ml of wastewater.

The principal health hazards associated with the chemical constituents of wastewaters arise from the contamination of crops or groundwaters. Particular concern is attached to the cumulative poisons, which comprise mainly the heavy metals, and carcinogens, which are mainly organic chemicals. The World Health Organization guidelines for drinking water quality (WHO 2006a,b) include limit values for the organic and toxic substances based on acceptable daily intakes. These can be adopted directly for groundwater protection purposes. However, considering the possible accumulation of certain toxic elements in plants (for example, cadmium and selenium), the intake of toxic materials through eating the crops irrigated with contaminated wastewater must be carefully assessed.

Pathogenic organisms give rise to the greatest health concern in the use of wastewaters. In areas of the World where helminthic diseases caused by *Ascaris* and *Trichuris* spp. are endemic in the population and where raw untreated sewage is used to irrigate salad crops and/or vegetables eaten uncooked, transmission of these infections is likely to occur through the consumption of such crops. Further evidence was provided to show that cholera can be transmitted through the same channel. There is also evidence that cattle grazing on fields freshly irrigated with raw wastewater, or drinking from raw wastewater canals or ponds, can become heavily infected with the cysticerosis disease. Indian studies have shown that sewage farm workers exposed to raw wastewater in areas where *Ancylostoma* (hookworm) and *Ascaris* (nematode) infections are endemic have significantly higher levels of infection than other agricultural workers. More detailed information on health risks and vector-borne diseases are given by WHO (2006a,b,c,d,e).

In respect of the health impact of use of wastewater in irrigation, pathogenic agents are ranked as shown in Table 8.2. However, negative health effects were only detected in association with the use of raw or poorly treated wastewater, while inconclusive evidence suggested that appropriate wastewater treatment could provide a high level of health protection (Pescod 1992). Risks from diseases such as schistosomiasis, clonorchiasis, and taeniasis vary from high to nil depending on local circumstances (WHO 2006a,e).

The main microbiological parameters that are particularly important from the health point of view when characterizing wastewaters are summarized in Table 8.3.

Table 8.2 Relative health impact of pathogenic agents (Pescod 1992)

Risk	Agents
High risk	Helminths
(high incidence of excess infection)	(*Ancylostoma, Ascaris, Trichuris* and *Taenia*)
Medium Risk	Enteric bacteria
(low incidence of excess infection)	(*Vibrio cholera, Salmonella typhosa, Shigella*)
Low Risk	Enteric viruses
(low incidence of excess infection)	(Viral diarrhoeas, hepatitis A)

Table 8.3 Main pathogenic parameters relative to wastewaters (adapted from Mara and Caimcross 1989, Pescod 1992)

Pathogens	Survival time	Presence in wastewater
Coliforms & faecal coliforms (Citrobacter, Enterobacter, Klebsiella, Escherichia coli)	up to 60 days in water and 70 days in the soil	E. coli count is a main indicator
Faecal Streptococci (S. bovis, S. equines, S. faecalis)	–	occur both in man and in other animals
Clostridium perfringens	survival characteristics similar to viruses or even helminth eggs	useful in wastewater quality reuse studies
Salmonella spp (S. typhi agent for typhoid)	if removal of Salmonellae is achieved Shigellae and Vibrio cholera are probably also removed	typical in a tropical urban sewage
Enteroviruses Poliomyelitis and Meningitis, and respiratory infections	may attain 120 days in water	especially under tropical conditions
Rotaviruses gastro-intestinal problems	more persistent than enteroviruses	removal in parallel with that of SS, virus are solids-associated
Intestinal Nematodes Ascaris lumbricoides	several months	infections can be spread by effluent reuse practices

8.2.2 Wastewaters Characteristics Relative to Agricultural Use

The quality of irrigation water is of particular importance in arid zones where high rates of evaporation occur, with consequent salt accumulation in the soil profile. The physical and mechanical properties of the soil, such as dispersion of particles, stability of aggregates, and permeability, are very sensitive to the type of exchangeable ions present in irrigation water. Thus, when effluent use is being planned, several factors related to soil properties must be considered. Ayers and Westcot (1985) provide appropriate recommendations on these water quality aspects. Questions relative to the use of saline waters in agriculture are dealt with in Section 8.3.

Another aspect of agricultural concern is the effect of dissolved solids (TDS) in irrigation water on the growth of plants. Dissolved salts increase the osmotic potential of soil water and therefore increase the amount of energy which plants must expend to extract water from the soil. As a result, growth and yield of most plants decline progressively as osmotic pressure increases due to the presence of salts in the soil and the soil water (see Section 8.3).

Many of the ions carried in the wastewaters are harmless or beneficial at relatively low concentrations, but may become phytotoxic at high concentrations, or may negatively affect several metabolic processes.

Important agricultural water quality parameters include a number of specific properties of water that are relevant in relation to the yield and quality of crops, maintenance of soil productivity and protection of the environment. These are analysed in the next section. These parameters mainly consist of certain physical and chemical characteristics of the water (Pescod 1992, Kandiah 1990, Rhoades et al. 1992, 1999), such as:

- Total salt concentration or the total dissolved solids, because the salinity of the soil is directly affected by the salinity of the irrigation water.
- Electrical conductivity, which is used to indicate the total ionised constituents of water and is closely correlated with the total salt concentration.
- Sodium adsorption ratio (SAR), because when sodium is present in the soil in exchangeable form, it causes adverse physico-chemical changes, particularly to soil structure. Due to the ability of sodium to disperse the soil aggregates, a crust is formed on the soil surface reducing the infiltration rates and affecting germination and seedling emergence. The SAR is defined by the ionic ratio

$$SAR = \frac{Na}{\sqrt{\frac{Ca+Mg}{2}}} \qquad (8.1)$$

where the ionic concentrations of sodium (Na), calcium (Ca) and magnesium (Mg) are expressed in meq/l. If significant precipitation or dissolution of calcium due to the effect of carbon dioxide (CO_2), bicarbonate (HCO_3-) and total salinity of the water (EC_w) is suspected, the adjusted sodium adsorption ratio (SAR_{adj}) can be used as reported by Ayers and Westcot (1985).

- Toxic ions. When at concentrations above threshold values, they can cause plant toxicity problems, which affect growth and yield of crops. The degree of damage depends on the crop, its stage of growth, and the concentration of the ions. The most common phytotoxic ions in municipal sewage and treated effluents are boron (B), chloride (C1), and sodium (Na).
- Trace elements. Attention should be paid to trace elements in sewage effluents if industrial wastewater is included. The main ones are Aluminium (Al), Beryllium (Be), Cobalt (Co), Fluoride (F), Iron (Fe), Lithium (Li), Manganese (Mn), Molybdenum (Mo), Selenium (Se), Tin (Sn), Titanium (Ti), Tungsten (W) and Vanadium (V).
- Heavy metals, a special group of trace elements, which have been shown to cause health hazards when taken up by plants: Arsenic (As), Cadmium (Cd), Chromium (Cr), Copper (Cu), Lead (Pb), Mercury (Hg) and Zinc (Zn).
- pH. The normal pH range for irrigation water is from 6.5 to 8.4. pH values outside this range indicate water is abnormal in quality.

8.2.3 Wastewater Treatment

Wastewater treatment aims at safe disposal of human and industrial effluents, without danger to human health or damage to the natural environment. Irrigation with wastewater is both disposal and utilisation. Some degree of treatment needs to be provided to raw municipal wastewater before it can be used for agricultural and landscape irrigation, aquaculture or other uses. The required quality of effluent will depend on the proposed water uses, crops to be irrigated, soil conditions and the irrigation system. In aquaculture, more reliance will have to be placed on control through wastewater treatment.

The most appropriate wastewater treatment for agricultural uses is that which will produce an effluent meeting the recommended microbiological and chemical quality guidelines both at low cost and with minimal operational and maintenance requirements (Arar 1988). Adopting as low a level of treatment as possible while achieving the desired results is important, especially in developing countries.

The design of wastewater treatment plants is usually based on the need to reduce organic and suspended solids loads to limit pollution of the environment. Pathogen removal has rarely been considered an objective but, for reuse of effluents in agriculture, this must be of primary concern (Asano et al. 2006, WHO 2006d). Treatment to remove wastewater constituents that may be toxic or harmful to crops, aquatic plants and fish is normally not economically feasible. However, the removal of toxic elements and pathogens that may affect human health needs to be considered.

The daily variations in flow from a municipal treatment plant make it generally not feasible to irrigate with effluent directly from the treatment plant. Some form of short-term storage of treated effluent is necessary to provide a relatively constant supply of reclaimed water for efficient irrigation, and additional benefits can result from storage in reservoirs.

8.2.3.1 Conventional Wastewater Treatment

Conventional wastewater treatment consists of a combination of physical, chemical, and biological processes and operations to remove solids, organic matter and, sometimes, nutrients from wastewater (Fig. 8.1).

The different degrees of treatment considered are (Pescod 1992):

a) *Preliminary treatment*, where the objective is the removal of coarse solids and other large materials from the raw wastewater. Treatment operations include coarse screening, grit removal in most small treatment plants and, in some cases, comminution (trituration) to reduce the size of large particles so as to remove them in the form of sludge in subsequent treatment processes. In addition, equalization (homogenization) may be considered.

b) *Primary treatment*, Its objective is the removal of settable organic and inorganic solids by sedimentation, and the removal of materials that float by skimming. Large fractions of the total suspended solids, oil and grease are then removed; however, only a relative small fraction of biochemical oxygen demand (BOD_5)

is removed. Some organic nitrogen, organic phosphorous and heavy metals associated with those solids are also removed, but not colloidal and dissolved constituents. It may be considered sufficient treatment if the wastewater is to be used to irrigate crops that are not consumed by humans or to irrigate orchards, vineyards, and some processed food crops. However, to prevent potential nuisance conditions in reservoirs, which may affect nearby populations and workers, some form of secondary treatment may be required even in the case of non-food crop irrigation. It is important to remember that part of the pathogenic organisms are removed from the wastewater but they will be present in the primary sludge, thus, it is important to promote some hygienization, e.g., addition of lime or solar disinfection.

Primary sedimentation tanks or clarifiers are used. The sludge of settled solids is removed from the bottom of tanks and floating solids are swept across the tank surface by water jets or mechanical means. In large sewage treatment plants, primary sludge is commonly processed biologically by anaerobic digestion. Gas (Biogas) containing methane is then produced and can be used as an energy source. In small sewage treatment plants, sludge is processed by methods such as aerobic digestion, storage in sludge lagoons, land application, and others.

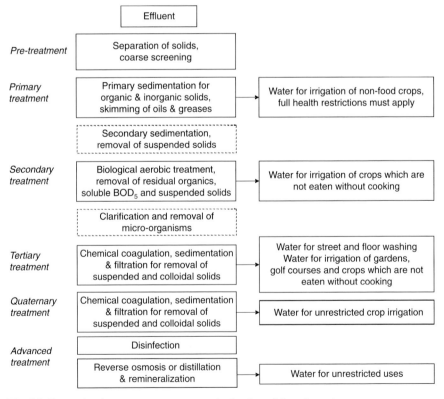

Fig. 8.1 Conventional wastewater treatment and related conditions for water reuse

c) *Secondary treatment*, where the objective is the further treatment of the primary treated effluent to remove the residual organics (as indicated by soluble BOD_5) and suspended solids. In most cases, secondary treatment involves the removal of biodegradable dissolved and colloidal organic matter using aerobic biological treatment processes. Several aerobic biological processes are used for secondary treatment, differing primarily in the manner in which oxygen is supplied to the microorganisms and in the rate at which microorganisms metabolise the organic matter.

High-rate biological processes are characterised by relatively small reactor volumes and high concentrations of microorganisms compared with low rate processes. The microorganisms must be separated from the treated wastewater by sedimentation to produce clarified secondary effluent. The biological solids removed during secondary sedimentation are normally combined with primary sludge for sludge processing. Common high-rate processes include the activated sludge processes, trickling filters or biofilters, oxidation ditches and lagoons, and rotating biological contactors (RBC). A combination of two of these processes in series (e.g., biofilter followed by activated sludge) can be used to treat municipal wastewater containing a high concentration of organic material from industrial sources.

High-rate biological treatment processes remove nearly 85% of the BOD_5 and suspended solids as well as some of the heavy metals. Activated sludge generally produces an effluent of slightly higher quality than biofilters or RBCs. When coupled with a disinfection step such as chlorination, these processes can provide substantial but not complete removal of bacteria and viruses. However, they remove very little of the phosphorous, nitrogen, non-biodegradable organics, or dissolved minerals.

d) *Tertiary and/or advanced treatment*, which is employed when specific undesirable wastewater constituents cannot be removed by secondary treatment. This may be the case for nitrogen, phosphorous, additional suspended solids, refractory organics, heavy metals and dissolved solids.

Where the risk of public exposure to the treated water is high, the intent of the advanced treatment is to minimise the probability of human exposure to enteric viruses and other pathogens. Because effective disinfection is believed to be inhibited by suspended and colloidal solids in the water, these solids must be removed by advanced treatment before the disinfection step. Therefore, the sequence of treatment often is secondary treatment followed by chemical coagulation, sedimentation, filtration, and disinfection. This level of treatment is assumed to produce an effluent free from detectable viruses.

e) *Disinfection*, Disinfection normally involves the injection of a chlorine solution (5–15 mg/l) at the head end of a chlorine contact basin. Ozone and ultra violet irradiation can also be used to meet advanced wastewater treatment requirements. A chlorine contact time as long as 120 min is sometimes required. In Near East countries adopting tertiary treatment, the tendency has been to introduce pre-chlorination before rapid-gravity sand filtration and post-chlorination afterwards. A final ozonation treatment after this sequence has seldom been considered

Table 8.4 Qualitative comparison of various disinfection technologies (adapted from Lazarova 2000)

Characteristics	Chlorine	Ozone	UV	Membrane filtration	Ultra/nano-filtration
Bactericidal action	++	+++	++	+++	+++
Virucidal action	++	+++	+	none	+++
Bacterial regrowth potential	+	++	+	none	none
Residual toxicity	+++	+	none	none	none
By-products	+++	+	none	none	none
Operation costs	+	++	++	+++	++++
Investment costs	+	+++	++	+++	+++

+ low; ++, medium; + + +, high; + + ++, very high.

(Al-Nakshabandi et al. 1997). A simplified analysis of disinfection processes are given in Table 8.4.

f) *Advanced treatment*, for unrestricted water use. It includes membrane technology or ultra/nanofiltration or reverse osmosis as well as water remineralization. Because the price of the technology is now much less than in the past, these advanced technologies gain interest for unrestricted uses of water. However, it is important to make an economical/social evaluation including the realistic cost of water to allow full consideration of consumers (non)acceptance of the price they will have to pay for their water.

8.2.3.2 Natural Biological Treatment Systems

Natural low-rate biological treatment systems for the treatment of municipal sewage tend to be less costly, but are also less sophisticated in operation and maintenance than the high-rate biological processes mentioned above. They may be more effective in removing pathogens if properly designed and not overloaded. Natural biological treatment systems consist of (Pescod 1992):

a) *Stabilisation ponds.* Stabilisation ponds are the preferred wastewater treatment process for effluent use in agriculture in developing countries. These systems are designed to achieve different forms of treatment in a series of three stages. The number of stages used will depend on the organic strength of the input waste and the effluent quality objectives. Strong wastewaters, having $BOD_5 > 300\,mg/l$, are introduced into first-stage anaerobic ponds, which achieve a high rate of removal. Where anaerobic ponds are environmentally unacceptable (mainly because of odour and flies) or are not required (e.g. for weaker wastes, with low BOD_5), wastewater is discharged directly into primary facultative ponds. Effluent from first-stage anaerobic ponds will overflow into secondary facultative ponds which comprise the second-stage of biological treatment. Maturation ponds to provide tertiary treatment are introduced following the primary or secondary facultative ponds, if further pathogen reduction is necessary.

Solids in the inflows to a facultative pond and excess biomass produced in the pond will settle out forming a sludge layer at the bottom. The benthic layer will be anaerobic and, as a result of anaerobic breakdown of organics, will release soluble organic products to the water column above. Organic matter dissolved or suspended in that water is metabolised by heterotrophic bacteria, with uptake of oxygen, as in conventional aerobic biological treatment processes. Unlike in conventional processes, the dissolved oxygen utilised by the bacteria in facultative ponds is replaced through photosynthetic oxygen produced by microalgae rather than by aeration equipment. High temperature and sunlight create conditions which encourage algae to utilise the carbon dioxide (CO_2) released by bacteria in breaking down the organic components of the wastewater and to take up nutrients, mainly nitrogen and phosphorous. This contributes to the overall removal of BOD_5 in facultative ponds. However, the organic loading must be strictly limited otherwise insufficient oxygen will be produced. Wind is important to the satisfactory operation of facultative ponds for mixing the contents and to prevent thermal stratification that would cause anaerobiosis and subsequent failure of the processes.

The effluent from facultative ponds normally contains at least 50 mg/l BOD_5. If lower BOD_5 concentration is required it will be necessary to use maturation ponds. A more important function of maturation ponds, however, is the removal of excreted pathogens. Longer retention in anaerobic and facultative pond systems will make them more efficient than conventional treatment processes in removing pathogens. Effluents from a facultative pond treating municipal sewage generally require further treatment. Maturation ponds should then be designed to achieve a given reduction of faecal coliforms (FC). Protozoan cysts and helminth eggs are removed by sedimentation in stabilisation ponds. A series of ponds with overall retention of 20 days or more will produce an effluent totally free of cysts and ova (Pescod 1992). Pathogen die-off, is linked to algal activity. The coliform and faecal coliform die-off coefficients vary with retention time, water temperature, organic loading, total BOD_5 concentration, pH and pond depth (Pescod 1992).

b) *Overland treatment of wastewater*. In overland flow treatment, the effluent is distributed over gently sloping grassland on fairly impermeable soils. Ideally, the wastewater moves evenly down the slope to collecting ditches at the bottom edge of the area. Suspended and colloidal organic materials are then removed by sedimentation and filtration through the surface grass and organic layers. Overland flow systems also remove pathogens from sewage effluent at levels comparable with conventional secondary treatment systems, without chlorination. This form of land treatment requires intermittent applications of effluent (usually treated) alternating with resting of the land, to allow the soil to react with the sediments and for grass cutting. Basic site characteristics and design features for overland flow treatment have been suggested (Pescod 1992). The impact on groundwater should be considered in the case of highly permeable soils. The application rate for wastewaters will depend principally on the type of soil, the quality of wastewater effluent and the physical and biochemical activity in the near-surface environment (Pescod 1992).

The cover crop is an important component of the overland flow system since it should prevent soil erosion, provide nutrient uptake and serve as a fixed-film medium for biological treatment. Crops best suited to overland flow treatment are grasses with a long growing season, high moisture tolerance and extensive rooting. Reed canary grass, rye grass and tall fescue are among the suitable species.

c) *Macrophyte treatment*. This occurs in maturation ponds that incorporate floating, submerged or emergent aquatic species (macrophytes). They can be used for upgrading effluents from stabilisation ponds, thus acting as maturation ponds. Macrophytes take up large amounts of inorganic nutrients (N and P) and heavy metals (Cd, Cu, Hg and Zn), but of course the macrophytes must be harvested occasionally otherwise the nutrients and metals remain in the pond water due to recycling as the macrophytes mature and die.

Among the floating macrophytes, having large root systems and very efficient nutrient extraction, are water hyacinth, *Eichornia crassipes*, able to double in mass about every 6 days, and the coontail, *Ceratophyllum demersum*. The aquatic vascular plants also serve as living substrates for microorganisms that remove BOD and nitrogen, and achieve reductions in phosphorus, heavy metals and some organics through plant uptake. The basic function of the macrophytes in the latter mechanism is to assimilate, concentrate and store contaminants on a short-term basis. Subsequent harvest of the plant biomass results in permanent removal of stored contaminants from the pond treatment system. Fly and mosquito breeding are a problem in floating macrophyte ponds. This nuisance can be partially controlled by introducing into the ponds of fish species such as *Gambusia* and *Peocelia* that eat these larvae. Pathogen die-off is poor in macrophyte ponds as a result of light shading and the lower dissolved oxygen and pH.

Natural and artificial wetlands and marshes having emergent macrophytes can also be used to treat raw sewage and partially-treated effluents. The main species of emergent macrophytes (reeds) are *Phragmites communis* and *Scirpus lacustris*. These macrophytes not only take up the inorganic nutrients but also create a favourable environment in the root zone for microorganisms, through pathways created by their highly developed root systems. BOD and nitrogen are then removed by bacterial activity. Aerobic treatment takes place in the rhizosphere, since oxygen passes to it from the atmosphere through leaves, stems, and roots, while an anoxic and anaerobic treatment takes place in the surrounding soil. Suspended solids in the sewage are aerobically composted in the above ground layer of decaying vegetation formed from dead leaves and stems. Nutrients and heavy metals are then removed by plant uptake.

The growth rate and pollutant assimilative capacity of emergent macrophytes such as *Phragmites communis* and *Scirpus lacustris* are limited by the culture system, wastewater loading rate, plant density, climate and management factors.

d) *Nutrient film technique*. The nutrient film technique (NFT) is a modification of the hydroponics plant growth system, in which plants are grown directly on an impermeable surface to which a thin film of wastewater is continuously applied.

Root production above the impermeable surface is high and the large root surface area traps and accumulates matter. Plant growth produces nutrient uptake, and provides shading for protection against the development of algae, while the large mass of roots and accumulated material serve as living filters.

8.2.4 Minimising Health Hazards in Wastewater Use in Irrigation

A potential for disease transmission exists when wastewater is used for irrigation, because pathogens brought with the wastewater can survive for many days in the soil or on the crop. Factors influencing transmission of disease include the degree of wastewater treatment, the crops grown, the irrigation method used to apply the wastewater, and the cultural and harvesting practices used (Westcot 1997). These define the degree of public health risk (Fig. 8.2).

The possible infection of field workers results from direct contact with the crop or soil in the area where wastewater is used. This path is directly related to the level of protection needed for field workers. The only feasible means of dealing with the worker safety problem is to adopt preventive measures against infection. The following *risk situations for field workers* are often identified (Westcot 1997):

a) Low risk of infection

- Mechanised cropping practices
- Mechanised harvesting practices
- Irrigation ceasing long before harvesting
- Long dry periods between irrigations

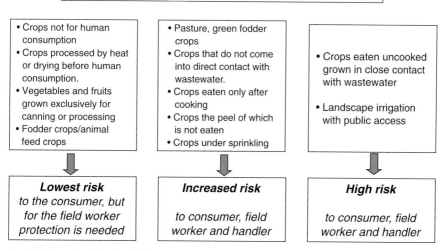

Fig. 8.2 Levels of health risk for consumers and workers due to wastewater use in irrigation (adapted from Westcot 1997)

b) High risk of infection

- High wind and dust areas
- Hand cultivation and hand harvesting
- Moving of sprinkler or other irrigation equipment
- Direct contact with irrigation water

To minimise health hazards for field workers preventive measures are required. These include wearing protective clothing, including impermeable boots that prevent any direct skin contact with the wastewater, maintenance of high levels of hygiene, and immunisation against infections likely to occur.

International guidelines or standards for the microbiological quality of irrigation water used on a particular crop do not exist. Because there is a lack of direct epidemiological data, the standards and guidelines for the quality of wastewater used for irrigation are focused on effluent standards at the wastewater treatment plant, rather than at the point of use. These standards are most often used for process control at wastewater treatment plants. Based on an epidemiological review WHO adopted the water quality guidelines for wastewater use in agriculture, shown in Table 8.5.

Table 8.5 Health-based targets for wastewater use in agriculture (adapted from WHO 2006d,f)

Exposure scenario	Health-based target (DALY* per person (per year)[a]	Log_{10} pathogen reduction needed[a]	Number of helminth eggs per litre[b,c]	E. coli (number per 100 ml)
Unrestricted irrigation	$\leq 10^{-6}$			$< 10^3$
Lettuce		6	≤ 1	Relaxed to $< 10^4$ for high-growing leaf crops or drip irrigation
Onion		7	≤ 1	
Restricted irrigation	$\leq 10^{-6}$			$< 10^5$
Highly mechanized		3	≤ 1	Relaxed to $< 10^6$
Labour intensive		4	≤ 1	when exposure is
Localized (drip) irrigation	$\leq 10^{-6}$			limited or regrowth is likely
High-growing crops		2	No recommendation[d]	
Low-growing crops		4	≤ 1[c]	

* Disability adjusted life years
[a]The health-based target can be achieved, for unrestricted and localized irrigation, by a 6–7 log unit pathogen reduction (obtained by a combination of wastewater treatment and other health protection measures); for restricted irrigation, it is achieved by a 2–3 log unit pathogen reduction.
[b]When children under 15 are exposed, additional health protection measures should be used (e.g. treatment to ≤ 0.1 egg per litre).
[c]The mean value of ≤ 1 egg per litre should be obtained for at least 90% of samples
[d]When no crops to be picked up from the soil.

It seems prudent to utilise the WHO (2006c,d,f) guidelines for controlling the quality of water used in irrigation (Table 8.5). These guidelines should be a performance goal for those water supplies which presently exceed this level, and for cropping areas that would present a high risk of infection. Using the guidelines as irrigation standards would help to:

- Identify the areas currently being contaminated
- Reduce the disease infection risk until suitable wastewater treatment is adopted
- Improve the basic health level in rural areas; and
- Provide indication and data of unsatisfactory areas to assist in planning for wastewater management

The purpose of applying the helminth standard throughout all cropping systems was to increase the level of protection for agricultural workers, who are at high risk from intestinal nematode infection (Mara and Cairncross 1989, WHO 2006d). Meanwhile, it was implied that if the recommended helminth egg limit could be reached, that equally high removals of all protozoa would be achieved. Then no bacterial guideline was needed for protection of the agricultural worker since there was little evidence indicating a risk to such workers from bacteria, and some degree of reduction in bacterial concentration would be achieved with efforts to meet the helminth levels.

The guidelines in Table 8.5 are for the microbiological quality of treated effluent from a wastewater plant when that water is intended for irrigation. They could be used as design goals in planning wastewater treatment plants, but they are not intended as standards for quality monitoring of irrigation water. However, because urban populations grow enormously, the degree of river and irrigation water supply contamination in developing countries will likely increase. Pressure will also increase to use partially treated wastewater for irrigation until adequate treatment facilities can be constructed. Thus, there is an immediate need to control wastewater use in high risk cropping systems such as vegetable crop production. New guidelines have been proposed by WHO (WHO 2006c,d,f). These guidelines could be applied in areas where wastewater is utilised directly for irrigation or where use is indirect, for example where contaminated river water, including when a treatment plant exists, is diverted for irrigation (Fig. 8.3).

Indicative expected removal levels of pathogens are shown in Table 8.6 for various treatment processes. Results depend not only upon the treatment as indicated in Table 8.7, but also on time of detention (Mara and Cairncross 1989, Westcot 1997).

8.2.5 Crop Restrictions and Irrigation Practices

To minimise the health risk from using wastewater in irrigation, the prime approach is to treat the wastewater to the level recommended above. However the reality is that untreated or insufficiently treated wastewaters are still used for irrigation, for example when uncontrolled flows are used by irrigators (Fig. 8.3). Then, the application of crop restrictions can be the most effective measure to protect the consumer.

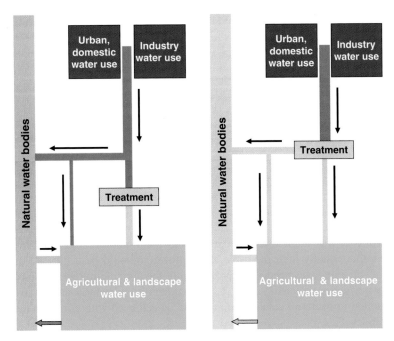

Fig. 8.3 Uncontrolled (*left*) vs. controlled (*right*) wastewater reuse

Table 8.6 Qualitative comparison of various treatment systems (adapted from Westcot 1997)

Criteria and factors considered	Package plant	Activated sludge plant	Extended aeration activated sludge	Biological filter	Oxidation ditch	Aerated lagoon	Waste stabilisation pond system
Plant performance							
BOD removal	F	F	F	F	G	G	G
FC removal	P	P	F	P	F	G	G
SS removal	F	G	G	G	G	F	F
Helminth removal	P	F	P	P	F	F	G
Virus removal	P	F	P	P	F	G	G
Economic factors							
Simple & cheap	L	L	L	L	M	M	H
Simple operation	L	L	L	M	M	L	H
Land requirement	H	H	H	H	H	M	L
Maintenance cost	L	L	L	M	L	L	H
Energy demand	L	L	L	M	L	L	H
Sludge removal cost	L	M	M	M	L	M	H

BOD – biochemical oxygen demand; FC – faecal coliform; SS – total suspended solids. G – good; F – fair; P – poor. H – high; M – medium; L – low (e.g. low demanding or low cost).

Table 8.7 Indicative log unit reductions or inactivation of pathogens achieved by selected wastewater treatment processes (adapted from WHO 2006d)

Treatment process	Log unit pathogens removals[a]			
	Viruses	Bacteria	Cysts	Helminth eggs
Low-rate biological processes				
Waste stabilization ponds	1–4	1–6	1–4	1–3
Wastewater storages and treatment reservoirs	1–4	1–6	1–4	1–3
Constructed wetlands	1–2	0.5–3	0.5–2	1–3
High-rate processes				
Primary treatment				
Primary sedimentation	0–1	0–1	0–1	0–<1
Chemically enhanced primary treatment	1–2	1–2	1–2	1–3
Secondary treatment				
Activated sludge + secondary sedimentation	0–2	1–2	0–1	1–<2
Trickling filters + secondary sedimentation	0–2	1–2	0–1	1–2
Aerated lagoon + settling pond	1–2	1–2	0–1	1–3
Tertiary treatment				
Coagulation/flocculation	1–3	0–1	1–3	2
High-rate granular or slow-rate sand filtration	1–3	0–3	0–3	1–3
Dual-media filtration	1–3	0–1	1–3	2–3
Membranes	2.5–>6	3.5–>6	>6	>3
Disinfection				
Chlorination	1–3	2–6	0–1.5	0– <1
Ozonation	3–6	2–6	1–2	0–2
Ultraviolet radiation	1–>3	2–>4	>3	0

[a]The log unit reductions are $\log_{10}$ unit reductions. A 1 log unit reduction = 90% reduction; a 2 log unit reduction = 99% reduction; a 3 log unit reduction = 99.9% reduction; and so on. For data sources see WHO (2006d).

Crop restrictions constitute the most widely used measure to protect public health. Crop restrictions focus on salad or vegetable crops that are normally eaten raw as indicated before. Unfortunately in most developing countries, institutional arrangements and enforcement of standards are usually not strong enough to ensure unsafe food products are not presented to an unsuspecting consuming public.

Crop restrictions need a strong institutional framework and the capacity to monitor and control compliance with the regulations. The following factors favour the adoption of crop restrictions (Mara and Cairncross 1989):

- A law-abiding society or strong law enforcement;
- Allocation of wastewater is controlled by a public body that has legal authority to enforce crop restrictions;
- The irrigation water conveyance and distribution system is controlled by strong central management;
- There is high demand and price advantage for the unrestricted crops;
- There is little market pressure in favour of the excluded crops, and
- Wastewater is used by a small number of large farms.

On the contrary, very large, dispersed irrigation schemes and those having poor or weak management make it difficult to enforce crop restrictions. Difficulties also occur when producers are mainly small farmers and the market prices do not favour the adoption of lower risk crops. In many developing countries, wastewater, including untreated effluent, is discharged directly to surface waters and these are diverted downstream for irrigation purposes. This leads to widespread distribution of the wastewater and makes crop restriction extremely difficult.

Irrigation practices should be designed according to the quality of wastewater being used. Questions concerning salinity hazards that may be associated with wastewater are described in Section 8.3. Several aspects referring to the control of health hazards have been dealt with above. However, the selection of methods of irrigation with wastewater to comply with guidelines for controlling health risks deserves particular attention.

Irrigation methods are briefly described in Section 10.6. Readers may get more appropriate details on irrigation methods and their respective characteristics and performance in specialised literature (e.g. van Lier et al. 1999, Tiercelin and Vidal 2006, Hoffman et al. 2007). The irrigation methods differ on several aspects as summarised in Table 8.8. They mainly concern:

- The probability of direct contact of workers with the irrigation water; which refers to the need for adopting more stringent preventive measures,
- The direct contact of the water with the harvestable yield that implies the need for more care on consumer protection measures,
- The foliar contact with the water that may cause phytotoxic problems to the crop,
- The capability for avoiding salt concentration in the crop root zone, which would cause soil degradation.

Aspects relative to health risk reduction have been discussed above and may be found in several references such as WHO (2006d), Pescod (1992) and Westcot (1997). Toxicity hazards to the crops are well covered by Ayers and Westcot (1985). Their suggested controls mainly rely on avoiding direct contact between the water charged with toxic ions and the crop leaves and/or other sensitive parts of the crop. Dilution of ion concentrations by mixing charged waters with fresh water may be considered but is generally more costly than to select an irrigation method where such foliar contact is minimised.

The salt accumulation in the root zone is generally controlled by leaching of the salts from the root zone naturally when rainfall is abundant, or by applying a leaching fraction with the irrigation water. Leaching with the irrigation water is carefully dealt with by Rhoades et al. (1992), and is briefly discussed in Section 8.3.4. The appropriate application of a leaching fraction depends on the irrigation method and the performance of the irrigation system. Practising over-irrigation to be sure that salts are leached down from the root zone is common but this produces excessive percolation to the groundwater. The groundwater then rises close to the soil surface, and the groundwater quality can be degraded. Therefore, to avoid problems due to deep percolation, drainage has to be considered. However, problems may be

Table 8.8 Evaluation of the irrigation methods for irrigation with wastewater

Irrigation methods	Basin irrigation and border irrigation	Corrugated basin irrigation	Furrow irrigation	Sprinkler irrigation	Micro irrigation: Drip and subsurface irrigation	Micro irrigation: Micro-sprinkling and microspray
Human contact (health hazard)	Likely to occur, mainly when water is controlled manually. Preventive measures including clothing requirements	Likely to occur when water is controlled manually, less when automation is adopted. Preventive measures including clothing are required	Likely to occur, when water is controlled manually, less when automation is adopted	Generally workers are not in the field when irrigating but they may have contact with wetted equipment. Small droplets will inevitably be spread by wind, so workers will always be at risk"	Not likely to occur except contact with wetted irrigation equipment	Generally workers are not in the field when irrigating but they may have contact with wetted equipment. Small drops will inevitably be spread by wind, so workers will always be at risk
Contact with fruits and harvestable yield (contamination hazard)	Not occurring for tree crops and vines, and most horticultural and field crops. May occur for low vegetable crops such as lettuce and melon	Not likely to occur because crops are grown on ridges	Not likely to occur because crops are grown on ridges	Fruits and harvestable yield are contaminated	Not likely to occur	Fruits and harvestable yield of vegetable crops may be contaminated. Less likely for under-tree irrigation with no wind
Salt accumulation in the root zone (salinity hazard)	Not likely to occur except for the under-irrigated parts of the field when uniformity of water application is very poor	Salts tend to accumulate on the top of the ridge. Leaching prior to seeding or planting is required for assuring germination and plant establishment	Salts tend to accumulate on the top of the ridge. Leaching is required prior to seeding/planting	Not likely to occur except for the under irrigated parts of the field resulting from low uniformity of water application	Not likely to occur except for the under-irrigated parts of the field resulting from low uniformity of water application	Not likely to occur except for the under-irrigated parts of the field resulting from low uniformity of water application
Foliar contact (toxicity hazard)	Possible for bottom leaves in low crops (e.g. lettuce, melon) and fodder crops. Possible during first stage of growth of annual crops	Exceptionally because crops are grown on ridges and water flows in furrows between them	Exceptionally because crops are grown on ridges	Severe leaf damage can occur affecting yields. Edible leaves would be contaminated	Not likely to occur	Severe leaf damage can occur definitely affecting yields of annual crops but not for irrigation of trees and vines if drops are large enough and jets are oriented to the soil. Hazardous for edible leaves

controlled easily if the irrigation system is designed carefully allowing for control of volumes applied and for an even distribution of water over the field.

Summarising, the safe use of wastewater in irrigation requires not only compliance with guidelines for the control of health risks, but also well designed and efficient irrigation systems. Case study examples are provided in the literature (e.g. Oron et al. 1999, Ragab et al. 2001, Hamdy et al. 2005).

8.2.6 Monitoring and Control for Safe Wastewater Use in Irrigation

Developing a program to promote safe crop production areas should occur alongside and as an alternative to crop restrictions. This can be achieved with a phased process (Westcot 1997) as depicted in Fig. 8.4.

The first phase is to develop a sound water quality monitoring program that is used to evaluate the existing levels of contamination in the water being used. This includes selection of contamination indicators, establishing field-sampling methods, defining laboratory methods and participating laboratories, selecting field monitoring sites, and then conducting a field water quality monitoring program. The second phase consists of evaluating the water quality data and developing procedures to assess the levels of contamination. The resulting database can be used to define safe production areas and as a basis to control or regulate contaminated water use in vegetable or other high-risk areas. From combining field monitoring and exploration

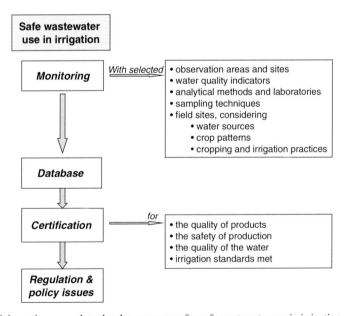

Fig. 8.4 Schematic approach to develop a program for safe wastewater use in irrigation

of a database it may lead to certification programs that assure consumers on the quality and safety of production methods applied in a given area. The last phase is developing mechanisms to regulate the use of potentially contaminated water on high-risks crops, thus also supporting certification programs.

8.2.7 Wastewater for Aquifer Recharge

When soil and groundwater conditions are favourable for artificial recharge of groundwater through infiltration basins (see Section 7.5 referring to recharge) partially-treated wastewater may be used to recharge the groundwater. The unsaturated or "vadose" zone then acts as a natural filter and can remove mainly suspended solids, biodegradable materials, bacteria, viruses, and other microorganisms (Table 8.9).

When this water has reached the aquifer, it usually flows within it before it is collected. This movement through the aquifer can produce further purification. The soil and aquifer are then used as natural treatment systems, so they are often called soil-aquifer treatment systems (SAT systems). They constitute a low-technology but advanced wastewater treatment system having the advantage over conventional sewage treatment systems that the water is not only well purified but is also stored for later use.

Various types of SAT systems exist. The simplest is when the wastewater is applied to infiltration basins from where it moves down to the groundwater and drains naturally through the aquifer. Another type is recharge with the help of pumping wells. The infiltration basins are arranged in parallel strips and the wells are located midway between the strips. Alternatively, the infiltration basins may be located close together and the wells placed in a circle around them. Pumping creates a water table depression in the area where the water infiltrates, so accelerating the recharge process.

Despite the fact that SAT systems greatly improve the quality of the sewage effluent, generally the quality of the resulting treated water is not as good as that of the native groundwater. In that case, appropriate management has to be adopted for exploiting the aquifer in such a manner that native groundwater may continue to be pumped for uses requiring higher water quality without being compromised by

Table 8.9 Water quality of secondary effluent and percolated water (from Aertgeerts and Angelakis 2003)

	Secondary effluent	Percolated water
Suspended solids (mg/l)	18	1.2
COD (mg/l)	97	51
NH_3-N (mg/l)	28	0.5
NO_3-N (mg/l)		47
Fecal coliform (CFU/100 ml)	$6.1 \times 10^5 \sim 7.3 \times 10^6$	Variable, in the range of 100–500

COD = chemical oxygen demand, CFU = colony forming units.

the recharge effluent. This requires groundwater level and quality monitoring using observation wells located carefully throughout the aquifer system.

Sewage water should travel sufficient distance through the soil and aquifer, and residence times in the SAT system should be long enough to produce water of the desired quality. Travel distance and residence time depend on the quality of sewage effluent, the soil types in the vadose zone and aquifer, the depth to the groundwater, and the desired quality of the water to be taken from the aquifer.

The infiltration basins for SAT systems should be located on soils that are permeable enough to give high infiltration rates. The soils should also be fine enough to provide good filtration and quality improvement of the effluent, ideally fine sand, loamy sand, or sandy loam soils. Aquifers should be sufficiently deep and transmissive to prevent excessive rise of the groundwater table due to percolation infiltration. Bare soils often provide the best conditions for the infiltration basins in SAT systems. Dense vegetation can hamper the soil drying process and the recovery of infiltration rates, contribute to evaporation losses, and aggravate mosquito and other insect problems. Suspended algae are also detrimental to infiltration rates and tend to increase the pH of the water.

During flooding, organic and other suspended solids in the sewage effluent accumulate on the bottom of the basins, producing a clogging layer. Drying of the basins causes the clogging layer to dry, crack, and therefore re-establishes the infiltration rates. Besides intermittent drying, periodic removal of this material is also necessary.

The main constituent that must be removed from raw sewage before it is applied to any SAT system is suspended solids. Reductions in BOD and bacteria are also desirable, but less essential. In the USA, the sewage used for recharge typically receives conventional primary and secondary treatment. The latter removes biodegradable material, as expressed by the BOD, but bacteria in the soil can also degrade organic material and reduce the BOD. Primary treatment could be sufficient when aquifers are non potable while more advanced treatments are required when recharging potable aquifers. Criteria relative to increased water quality requirements are summarized in Table 8.10. A detailed description on the process involved in removing the sewage constituents in SAT systems, public health considerations, risk assessment and other considerations relative to groundwater recharge with wastewaters are given by Aertgeerts and Angelakis (2003).

8.2.8 Non-agricultural Uses of Wastewater

Non-agricultural uses of wastewater are less common then those for irrigation. They include uses in aquaculture, industry, low quality municipal uses, irrigation of recreational areas including golf courses, and for wetlands. The use of wastewater for irrigation of recreational areas has been dealt with above, being included among the higher risk cases that require more stringent water treatment and monitoring. Lower degrees of treatment may be required when wastewater is used in wetlands since these may act as tertiary treatment facilities.

Table 8.10 Guidelines for criteria applied to recharge into potable aquifers (from Aertgeerts and Angelakis 2003)

Recharge by	Indicative criteria to be considered according to local conditions and requirements
Spreading	• *Primary treatment and disinfection*, plus soil aquifer treatment, handling of dry and wet cycles as well as hydraulic and mass charges to avoid soil sealing if suspended solids are mostly mineral. • *Primary advanced treatment and disinfection*, plus handling of dry and wet cycles as well as hydraulic and mass charges so as to avoid soil sealing, if suspended solids are mostly mineral. • *Secondary treatment and disinfection* with a well operated soil aquifer treatment. • *Possibly advanced treatment* under site-specific considerations. • *Meet drinking-water standards after percolation.* • *Monitoring* for coliforms. pH, chlorine residual, drinking-water standards plus site-specific others. • *Distance to point of extraction* (600 m, or dependent on site-specific factors)
Injection	• *Secondary treatment, filtration, disinfection, advanced wastewater treatment.* • *Meet drinking-water standards*.* • *Monitoring* for turbidity, coliforms, chlorine residual, pH, others. • *Distance to point of extraction* as above

*no detectable faecal coliforms in 100 ml, turbidity limits, 1 mg/l chlorine residual, $6.5 < pH < 8.5$

Aquaculture uses of wastewater are quite common in many parts of the World, including use of non-treated effluents (WHO 2006e). As defined by these authors, aquaculture means water farming for fish production, mainly carp and tilapia, and growing selected water crops such as water spinach, water hyacinth, water caltrop and lotus. Human excreta and animal manure are utilised as fertilisers. The use of untreated materials is fortunately decreasing. However, the risk of infection from fish raised in these waters is small since fish is usually consumed after cooking.

Risks associated with wastewater use in aquaculture (Mara and Cairncross 1989) consist of:

- Passive transfer of pathogens by fish and cultivated aquatic macrophytes,
- Transmission of trematodes such as *Clonorchis sinensis* and *Fasciolopsis buski*, whose life cycle includes passing through fish and aquatic macrophytes
- Transmission of schistosomiasis, and
- Mainly, field workers exposure to infections.

Techniques for minimisation of risks include the cultivation of species that are not eaten raw, to hold shell-fish in clean water to remove excreted organisms (depuration), keeping fish in clean water for 2–3 weeks before harvest, the control of schistosomiasis through the removal of all snails, and the adoption of preventive measures that enable workers to avoid direct contact with wastewater. The use of wastewater resulting from pre-treatment and first stage treatment processes is also to be considered. Health-based targets defined by WHO are presented in Table 8.11.

Treated municipal wastewater may be used in domestic water supply when separate distribution systems are available as discussed in Section 10.1.6. Uses refer

Table 8.11 Health-based targets for waste-fed aquaculture (adapted from WHO 2006e)

Exposed group	Hazard	Health-based target	Health protection measure
Consumers, workers, and local communities	• Excreta-related diseases	10^{-6} DALY*	• Wastewater treatment • Excreta treatment • Health and hygiene promotion • Chemotherapy and immunization
Consumers	• Excreta-related diseases • Foodborne trematodes • Chemicals	10^{-6} DALY and • Absence of trematode infections • Tolerable daily intakes (as per the Codex Alimentarius Commission)	• Produce restriction • Waste application/timing • Depuration • Food handling and preparation • Produce washing/disinfection • Cooking foods
Workers and local communities	• Excreta-related pathogens • Skin irritants • Schistosomes • Vector-borne pathogens	10^{-6} DALY and • Absence of skin disease • Absence of schistosomiasis • Absence of vector-borne disease	• Access control • Use of personal protective equipment • Disease vector control • Intermediate host control • Access to safe drinking-water and sanitation at aquacultural facilities and in local communities • Reduced vector contact (insecticide-treated nets, repellents)

*DALY = Disability adjusted life years

to demand for low quality water such as for toilet flushing and outdoor washing. Treatments usually follow the more stringent standards to prevent infections by direct contact, since direct contact with the water is always possible. However, costs associated with these dual systems are very high and may be justified only for tourist areas and in regions where water is extremely scarce.

New approaches to on-site wastewater treatment and reuse of wastewater generated at individual households and small communities are reported by several authors, e.g. Manel (2001). The main uses are for gardening, green spaces and wetland conservation, but also include domestic uses not requiring very stringent water quality. A good example comes from Brazil, where a small plant consisting of a collection system - home and roof waters -, septic collection tanks, bioreactors and solar radiation (UV) disinfection units is becoming operational in Jaiba, Northeast Brazil (Fig. 8.5). The resulting treated water is intended to be used for irrigation.

Industrial wastewaters generally require treatment specific to the industrial processes from which they are produced. They are determined by the chemical composition of the effluents and the reuse objectives. Commonly, the treated effluents are

Fig. 8.5 Small wastewater treatment plant for households and small communities developed in Brazil (system STAR) (courtesy by INTEC, Viçosa, Brazil)

reused in the same industrial plant for uses that do not require high water quality such as cooling, washing and outdoor uses as referred in Section 10.2. Generally health risks are not related to pathogens but to toxic ions such as heavy metals. The degree of treatment is site specific.

8.3 Use of Brackish, Saline and Drainage Waters

8.3.1 Characteristics and Impacts of Saline Waters

Saline water includes water commonly called brackish, saline or hyper-saline from different sources, including aquifers which are naturally saline or became saline due to human activities, and drainage effluents from agricultural land. It also includes water that contains one or more specific elements in concentrations above those found in good quality water.

Fresh water is considered to have a total dissolved solids (TDS) concentration of less than 500 mg/l (EC < 0.7 dS/m). Saline and brackish water have 500–30,000 mg/l (0.7–42 dS/m) TDS, while sea water has TDS averaging 35,000 mg/l (49 dS/m).

The annual use of saline drainage water in Egypt is above 5 billion m^3 for irrigating nearly 500,000 ha of land. Drainage water is used in many parts of the world including California, USA. Saline groundwater is used extensively in many countries such as Tunisia, India, and Israel. Reports on use of saline water are abundant in the literature (e.g. Kandiah 1990, Rhoades et al. 1992, Minhas 1996, Tyagi and Minhas 1998, Gupta et al. 1998). The use of highly saline waters, unusable by common agricultural crops, may be feasible for halophytes, which could be explored for human and animal consumption (e.g. Lieth and Al Masoom 1993, Choukr-Allah et al. 1996).

Waterlogging, salinity and related problems have arisen in many irrigation areas where fresh water is used for irrigation. Such problems could arise even more quickly and more severely when saline water is used. The major potential hazards associated with the use of saline water in agriculture are summarized in Table 8.12.

The negative impacts which result from the use of saline waters can be overcome when water management is performed appropriately, that is having the control of the negative impacts as a main goal. However, this implies good knowledge of the mechanisms influencing plant stress and soil degradation, and appropriate information on the quality characteristics of the waters being used and of the toxic ions, nutrients or disease vectors likely to be present. Only when such information is available will it be possible to select the crops, cultivation techniques, irrigation methods and water management practices that facilitate saline water use.

Table 8.12 Major potential hazards associated with the use of saline and poor quality waters in agriculture (adapted from Ayers and Westcot 1985, Kandiah 1990)

Hazards	Causes and/or effects
Yields decrease	• Reduced soil water availability to the crop due to augmentation of the soil water osmotic potential • Reduced soil infiltration rates, thus reducing soil water availability • Soil crusting, affecting infiltration of water and crop emergence • Toxicity to the crop when the concentration of certain ions is above crop tolerance • Imbalance of nutrients available to crops
Soil degradation	• *Salinisation*, particularly in the absence of adequate leaching and drainage, when salts accumulate in the root zone and in the groundwater • *Sodification* when there is a high sodium content as compared to other cations, then the soil complex accumulates this excess sodium. Sodification causes soils to lose their structure and tilth, become dispersive, and have reduced infiltration and permeability • *Loss of soil productivity* when the processes of salinisation and/or sodification are continued because conditions for plants to extract water and nutrients become progressively worse
Effects on the environment	• Soil degradation, which contributes to desertification. • Damages to the soil environment, causing negative changes in plant and animal communities, and soil biodiversity. • Nutrients in reused drainage effluents give rise to uncontrolled algal blooms, development of aquatic weeds in water bodies, which lead to clogging problems in hydraulic structures and equipment, in waterways and canals, and reduce the wild life which inhabit farm ponds, lakes and reservoirs. • When nitrogen in reused drainage waters is excessive or when it is not taken into consideration in the fertilisers' balance, the resulting excess nitrates add to groundwater pollution and to eutrophication. • The presence of particular ions to levels exceeding specific health and safety thresholds affect either plants sensitive to those ion concentrations, or animals that drink those waters, e.g. selenium in drainage waters in California
Risk for public health	• Toxic ions such as heavy metals that, although they may be present in minute concentrations, are cancerous when accumulated in humans • Vectors of disease, such as mosquitoes and snails, which develop better in saline waters, mainly those rich in nutrients, so creating or increasing health hazards for the populations living in areas using saline water or reusing drainage waters.

An FAO expert consultation (Kandiah 1990) proposed a set of criteria for use of saline water for irrigation. These include the need for

- Integrated management of water of different qualities at the levels of the farm, the irrigation system and the drainage basin when sustaining long-term production potential of land and water resources is a main goal.
- Adopting irrigation methods with high performance, using minimal leaching fractions to reduce drainage volumes, implementing reuse of drainage water for progressively more tolerant crops, and reusing otherwise unusable saline water for halophyte production.
- Monitoring of soil and water quality, providing feedback to management to provide for optimal operation and control of the irrigation systems.
- Further development in understanding the effects of saline irrigation water on soil-plant-water relationships, including cumulative effects on perennials.
- Promotion of tools to predict long and short-term effects of irrigation water quality on crop yields, soil properties and quality of the environment.
- Establishment of pilot areas to test and assess irrigation methods and complementary soil and crop management practices for using saline water.
- Training of irrigation and agricultural officers.

8.3.2 Criteria and Standards for Assessing the Suitability of Water for Irrigation

A great deal of research on the water quality requirements for irrigation has been developed over a long time. Consolidated standards have been made available in FAO publications (Ayers and Westcot 1985, Kandiah 1990, Rhoades et al. 1992). It is then possible to define the main water quality parameters which need to be known to allow saline waters to be used safely in irrigation. Recommendations by FAO include:

1) Water quality characteristics to be considered for irrigation to assess the suitability of saline water concerning, particularly:

   - Salinity hazards (total dissolved salts, TDS and electrical conductivity, EC)
   - Crusting and permeability hazards (SAR or adjusted SAR, EC, and pH)
   - Specific ion toxicity hazard (Na, Cl, B, and Se, among others)
   - Nutrient imbalance hazard (excess NO_3, limited Ca, phosphate, etc.)

2) Parameters required to evaluate the quality of saline water on a routine basis including: TDS, EC, concentration of cations and anions (mainly Ca, Mg, Na, CO_3, HCO_3, Cl, SO_4), SAR or the adjusted SAR under certain conditions, trace elements (such as Se, As, B, Mo, Cd, Cr, Cu), as well as other potentially toxic substances of agricultural origin.

3) Water quality standards already available should be evaluated through research and adjusted for the specific conditions of saline water use, including soils,

climate, crops and crop sequences, and irrigation methods. More emphasis should be given to the development of appropriate models, criteria and standards applicable under non-steady conditions.
4) Guidelines on use of saline water should also include the possible hazardous effects of trace elements on people or livestock which consume crops produced using such water (As, Se, etc.).

The main water quality characteristics for saline waters are summarised in Table 8.13, with indications of the restrictions on their use in relation to the overall quality of the water. Restrictions mainly concern crop tolerance to salts, need for leaching, specific requirements for the irrigation methods and systems, and soil management practices.

Agricultural drainage waters are also a resource for irrigation and other uses (Table 8.14). When treated, for example by desalination it may be reused for several processes as outlined by Tanji and Kielen (2002) and van der Molen et al. (2007).

8.3.3 Crop Irrigation Management Using Saline Water

Water for agricultural use is normally considered to be in one of five salinity classes. These classes are outlined in Table 8.15, their boundaries being defined by the total dissolved solids and electrical conductivity of the water.

Table 8.13 Water quality for irrigation and required restrictions in use (adapted from Ayers and Westcot 1985)

Problems	Water characteristic	No restrictions	Slight to moderate restrictions	Severe restrictions
Salinity effects on water availability	EC (dS/m)	< 0.7	0.7–3.0	> 3.0
	TDS (mg/l)	< 450	450–2000	> 2000
Salinity effects on soil infiltration	SAR < 3	EC > 0.7 dS/m	EC: 0.7–0.2 dS/m	EC < 0.2 dS/m
	3–6	> 1.2	1.2–0.3	< 0.3
	6–12	> 1.9	1.9–0.5	< 0.5
	12–20	> 2.9	2.9–1.3	< 1.3
	20–40	> 5.0	5.0–2.9	< 2.9
Toxicity	*Sodium*			
	Surface irrigation: SAR	< 3	3–9	> 9
	Sprinkle/spray (me/l)	< 3	> 3	
	Chloride concentration			
	Surface irrigation (me/l)	< 4	4–10	> 10
	Sprinkle/spray (me/l)	< 3	> 3	
	Boron (mg/l)	< 0.7	0.7–3.0	> 3.0
	Trace elements	Variable		
	Bicarbonate (me/l)	< 1.5	1.5–8.5	> 8.5
	(Sprinkle/spray irrigation)			
Plant nutrition	pH	6.5–8.5		

Table 8.14 Quality of drainage water for use in irrigation and other uses (adapted from van der Molen et al. 2007)

Quality	EC (dS/m)	Application
Very good	< 1	all crops
Good	1–2	most crops
Moderate	2–3	tolerant crops
Poor	3–6	tolerant crops, with appropriate leaching
Very poor	> 6	not recommended for irrigation. Other uses when not polluted: • irrigate halophytes • maintain water levels in fish ponds • secure water levels in brackish coastal lakes • leaching salt-affected soils during the initial stage

Table 8.15 Salinity classification of water (adapted from Rhoades et al. 1992)

	Non saline	Slightly saline	Medium saline	Highly saline	Very highly saline
Total dissolved solids, TDS (mg/l)	< 500	500–2000	2000–4000	4000–9000	> 9000
Electrical conductivity, EC (dS/m)	< 0.7	0.7–3.0	3.0–6.0	6.0–14.0	> 14.0

Table 8.16 Summary guidelines for use of saline waters (adapted from Rhoades et al. 1992)

Crop tolerance to salinity		Non saline	Slightly saline	Medium saline	Highly saline	Very highly saline
I sensitive	Limitations to use	None	Slight to medium	Restricted	Not usable	Not usable
	Yield reduction	None	Up to 50%	> 50%	–	–
II moderately sensitive	Limitations to use	None	Slight	Medium	Restricted	Not usable
	Yield reduction	None	Up to 20%	Up to 50%	> 50%	–
III moderately tolerant	Limitations to use	None	None	Slight to medium	Medium	Very restricted
	Yield reduction	None	None	20–40%	40–50%	> 50%
IV tolerant	Limitations to use	None	None	Very slight	Slight to medium	Restricted
	Yield reduction	None	None	Practically none	20–40%	> 50%

Crop responses to salinity vary with species and, to a lesser degree, with the crop variety. The tolerance of crops to salinity is generally classified into four to six groups (Table 8.16), from the sensitive (or non tolerant), where most horticultural and fruit crops are included, to the tolerant, which includes barley, cotton, jojoba, sugarbeet, several grass crops, asparagus and date palm. Full lists of crop tolerance classes are given by Hoffman and Shalhevet (2007). In addition, halophytes may be commercially explored and they use very highly saline water.

The behaviour of various crops under irrigation with water of different degrees of salinity varies with species, varieties and the crop growth stages. As irrigation water salinity increases, germination is delayed. Germination is adversely affected for most field crops when the EC of the irrigation water or the soil saturation extract reaches a threshold of 2.4 dS/m. Adverse effects occur at lower values ($<$ 1 dS/m) for non-tolerant crops, and at higher values for tolerant crops (generally not exceeding 4 dS/m).

The germination and seedling stages are the most sensitive to saline water irrigation. Any adverse effects at such stages will lead to a reduction in crop production proportional to the degree of plant loss during germination and plant establishment. At this stage, water of good quality should be used, especially if plants are sensitive. If fresh water is lacking at this stage, irrigation during seedling development, after germination, must be carried out with fresh water to avoid dramatic effects on yields. Besides germination and crop establishment, another growth stage where most crops are more sensitive to salinity is the reproductive phase. Other critical stages vary from crop to crop. Several case studies are reported in the literature (e.g. Hamdy and Karajeh 1999, Hamdy et al. 2005).

The suitability of water for irrigation based on salinity, leaching and drainage requirements, and crop tolerance to salinity must be related to irrigation management. Several approaches can be adopted in water and crop management to minimise the accumulation of salts in the active root zone and to eliminate salt stress, especially during the critical growing stages of the plants. These include:

- Appropriate selection of irrigation methods.
- Efficient leaching management, including volumes and frequency, and respective drainage of the salty water away from the cropped land.
- Proper irrigation scheduling, in agreement with the available irrigation system
- Crop rotations adapted to the prevailing conditions. Considerations must include irrigation water quality, soil salinity levels, chemical and physical properties of the soils, and climatic conditions.

The selection of the irrigation method must consider the quality of the water and the potential for both the water and the irrigation method to produce negative impacts. A summary of the more relevant considerations is presented in Table 8.17. These refer to the capabilities for controlling:

- Soil salinity hazards due to salt accumulation in the root zone,
- Toxicity hazards caused by direct contact of the salty water with the plant leaves and fruits,
- Soil infiltration and permeability hazards caused by the modification of the soil physical properties, mainly due to the Na ion, and
- Yield hazards which may occur when the irrigation system does not allow adoption of appropriate irrigation management, i.e. frequency and volumes of irrigation.

Table 8.17 Evaluation of the irrigation methods for use with saline water

Irrigation method*	Flat basin irrigation	Corrugated basin irrigation	Border irrigation[1]	Furrow irrigation[2]	Sprinkler irrigation	Micro irrigation: drip and subsurface irrigation	Micro irrigation: micro-sprinkling and microspray
Salt accumulation in the root zone	Not likely to occur except for the under-irrigated parts of the field when uniformity of water application is very poor. Leaching fraction difficult to control in traditional systems	Salts tend to accumulate on the tops of the ridges. Leaching prior to seeding/planting is required for germination and crop establishment	As for basin irrigation but infiltration control is more difficult, as is the control of the leaching fraction	Salts tend to accumulate on the tops of the ridges. Leaching is required prior to seeding/planting	Not likely to occur when set systems are used except for the under-irrigated parts of the field due to low uniformity. Instead, problems occur with equipment designed for light and frequent irrigation	Not likely to occur except for the under-irrigated parts of the field due to low uniformity, including that resulting from nozzle clogging when water filtration is poor	Not likely to occur except for the under-irrigated parts of the field due to low uniformity and clogging; leaching could then be more difficult
Foliar contact, avoiding toxicity	It is possible only for bottom leaves in low crops and fodder crops, and during the first stage of growth of annual crops	Unlikely because crops are grown on ridges.	As for flat basins	Unlikely because crops are grown on ridges	Severe leaf damage can occur affecting yields. Damage greater with more frequent irrigation.	Not likely to occur	Leaf damage can occur, affecting yields of annual crops. Less damage for tree crops
Ability to infiltrate water and refill the root zone	Adequate because large volumes of water are generally applied at each irrigation and water remains in the basin until infiltration is complete	As for flat basins, above	Because water infiltrates while flowing on the soil surface, runoff losses increase when infiltration decreases	Salinity induced infiltration problems cause very high runoff losses	Salinity induced infiltration problems including soil crusting may cause very high runoff losses	Problems generally do not occur except when there are not enough emitters and under-irrigation is practised	Problems are similar to those for set sprinklers, so runoff losses may be important
Control of crop stress and yield reduction	Adequate because toxicity is mostly avoided, salts are moved down through the root zone, infiltration is completed and irrigation can be scheduled to avoid crop stress	As for flat basins but depending on avoiding salt stress at plant emergence and crop establishment	Crop stress is likely to occur due to reduced infiltration, so inducing relatively high yield losses	Crop stress is very likely to occur due to reduced infiltration, so inducing significant yield losses	Crop stress is very likely to occur due to toxicity by contact of water with the leaves and fruits, and due to reduced infiltration, thus significant crop stress and yield losses may occur	These systems are able to provide for crop stress and toxicity control, so yield losses are minimised	Toxicity due to direct contact with the leaves. Non uniform application may produce crop stress and runoff. Yield losses likely

* See a brief description of the methods in Section 10.6

One of the most important factors in crop management when using saline irrigation is the irrigation frequency. Saline water requires more frequent irrigation than
for fresh water because salts in the water and the soil increase the osmotic potential
of the soil water, which makes water uptake by the crop roots more difficult. However, increasing the frequency implies reducing the depth of water applied at each
irrigation to avoid gross accumulation of salts in the soil. The irrigation application
depths that can be used depend on the irrigation method and the off-farm system
delivering water to the fields.

Surface irrigation methods make it extremely difficult to apply small irrigation
depths, as outlined in Section 10.6. When water is delivered to the farms through
surface canal systems, the delivery schedules are generally of the rotation type, and
are rigid, delivering large irrigation volumes at long intervals. These systems are
inappropriate for irrigation of less tolerant crops.

8.3.4 Leaching Requirements and Control of Impacts on Soil Salinity

The leaching requirement is usually computed from (Ayers and Westcot 1985)

$$LR = \frac{EC_{iw}}{5EC_e - EC_{iw}} \tag{8.2}$$

where EC_{iw} is the electrical conductivity of the irrigation water and EC_e is the electrical conductivity of the saturated extract of the soil. EC_e should be the average soil
salinity tolerated by the crop. This value should not be that for achieving maximum
yield but that which will provide an acceptable yield decrease, i.e. for attaining
70–90% of the potential yield.

A methodology for estimating salinity impacts on crop yields is proposed by
Rhoades et al. (1992), and Allen et al. (1998) propose a method for estimating crop
water requirements under salinity conditions. These methods rely on knowledge of
the threshold EC_e at which crops are affected, and the rate of yield decrease when
the salinity of the irrigation water increases by 1 dS/m. These values are tabulated
in both quoted publications and Hoffman and Shalhevet (2007). The EC_e threshold
ranges from 1.0 dS/m for very sensitive crops such as carrots and beans up to more
than 7.5 dS/m for barley, cotton, and tolerant grasses. The rate of decrease in yield
per unit increase in EC varies from more than 15% per dS/m for the sensitive crops
down to 5% per dS/m for tolerant crops.

When a leaching fraction is applied with the irrigation water the salinity built up
of the soil is reduced non-linearly with the size of that fraction. In general this may
be expressed by

$$EC_e = \frac{1 + LF}{LF} \times \frac{EC_{iw}}{5} \tag{8.3}$$

where LF, the actual leaching fraction, is used in place of the leaching requirement, LR. This equation shows that the soil salinity EC_e increases proportionally to the salinity of the irrigation water, EC_{iw}. The salinity built up can not be prevented by drastically increasing the leaching fraction, because this would dramatically increase the percolation of water, causing the water table to rise, waterlogging and degradation of the quality of the groundwater. Under conditions of water scarcity, this would be a very poor use of the available water. Moreover, the degrading of the quality of local water bodies, surface or groundwater, by adding water of inferior quality would further contribute to the problems of water scarcity. Therefore, depending on the crop, and the salinity of the water and soil, a 15–20% leaching fraction is commonly recommended to do not be exceeded. Equation (8.3) above predicts that $EC_e = 1.5\ EC_{iw}$, when the salinity of the soil and the irrigation water are in equilibrium. Lower EC_e can be produced if good quality water, including rainwater is employed for leaching. Therefore, the conjunctive use for irrigation of saline water and good quality water is advocated, and similarly for the case of using wastewater for irrigation. An updated discussion on LF practices is given by Hoffman and Shalhevet (2007).

Several strategies are usually adopted to facilitate irrigation with saline water (e.g. Kandiah 1990, Hamdy and Karajeh 1999, Hamdy et al. 2005). Mainly the strategies consist of:

a) The *dual rotation* management strategy (Rhoades 1990), in which sensitive crops (e.g. lettuce, alfalfa) in the rotation are irrigated with low salinity river water, and salt-tolerant crops (e.g. cotton, sugarbeet, wheat, barley) are irrigated with saline drainage water. For the tolerant crops, the switch to drainage water is usually made after seedling establishment, i.e. irrigations at pre-planting and at the initial crop stages are made with low salinity water. Benefits from this strategy include (Rhoades 1990): (i) harmful levels of soil salinity in the root zone do not occur because saline water is used only for a fraction of the time; (ii) substantial alleviation of salt build-up in the soil occurs during the time when salt-sensitive crops are irrigated with fresh water; (iii) proper pre-planting irrigation and careful irrigation management during germination and seedling establishment leach salts out of the seed layer and from shallow soil depths. The main difficulties with this strategy are the complexity of water management for farmers and system managers and the possible lack of fresh water when it is required.

b) *Blending*, which is a drainage reuse strategy where water supplies of different salinity levels are mixed in variable proportions before or during irrigation (e.g. Shalhevet 1994, Tyagi 1996). Irrigation water having a quality superior to that of the saline water and able to satisfy the allowed salinity threshold for the crops to be irrigated is obtained. Blending may be more practical and appropriate on large farms because it is practised by the farmer himself, taking into consideration the crops grown and the respective growth stages. On the other hand, in large irrigation systems supplying many small farms it would be difficult to properly satisfy the requirements of all the crops to be grown, unless a cropping pattern was imposed for all farms.

c) *Cyclic application* of saline and fresh water (Tyagi 1996). Salinity must not be above acceptable thresholds for the crops grown. Cycles of application of fresh water should coincide with the more sensitive growth stages, particularly for planting and seedling development, and for the leaching of the upper soil layers. This strategy has more potential and flexibility than the blending strategy and may be easier to implement.

8.3.5 Long-Term Impacts: Monitoring and Evaluation

Most studies indicated that using irrigation water containing salts in excess of conventional suitability standards, can be successful on numerous crops for at least seven years without a loss in yield (Rhoades 1990). However, uncertainty exists concerning the long-term effects of these irrigation practices on the physical quality of the soil. These effects largely depend on the soil chemical and physical characteristics, on the climate and on the possibility of leaching with natural rain or using higher quality water for leaching as mentioned above.

The greatest concern with regard to long-term reuse of drainage water for irrigation is its effects on the soil physical quality, particularly the reduction of water infiltration capacity. According to Rhoades (1990), this is especially important where reuse is practised on poorly structured soils and the drainage water has SAR $> 15\,(\text{mmol/l})^{1/2}$. The capability to predict changes in soil infiltration and permeability is still beyond current knowledge of soil physics.

Long-term effects on soil salinisation are already considered when using simulation models. However existing knowledge is quite limited. Soil variability in space and irrigation variability both in time and space produce large uncertainty in predictions. For instance, soil salinity under a cyclic strategy of application will fluctuate more, both spatially and temporally, than if using a blending strategy. Therefore, predicting or anticipating plant response would be more difficult under the former. Nevertheless, management schemes must be practised that keep the average root zone salinity levels within acceptable limits in both strategies.

Drainage water often contains certain elements, such as boron and chloride, that can accumulate in plants to levels that cause foliar injury and a subsequent reduction in yield. This is another cause of uncertainty and, in many cases, may produce more long-term detrimental effects than salinity. Another long-term consideration with regard to reuse of drainage water is the potential for accumulation of heavy metals in plants and soils. These metals can be toxic to human and animal consumers of the crops. For these cases also, there is only limited potential for using prediction models to estimate when long term impacts would be important.

Monitoring soil salinity, leaching and drainage adequacy is therefore required to evaluate the long term impacts from using saline waters, irrespective of whether the source of the saline water is groundwater or drainage water. Monitoring and evaluation should be concerned with:

• The status of salts throughout the soil profile on a continuous basis for detecting changes in salinity levels and to identify when salt build up is steadily increasing.

- The functioning and performance of the drainage system, including observations of the hydraulic head, drainage outflows and salts transported with the drainage water.
- The irrigation performance, mainly relative to the uniformity of water distribution and the leaching fractions actually applied.
- The irrigation schedule in respect to the satisfaction of irrigation and leaching requirements, and the constraints on delivery or other restrictions that may hamper the appropriate irrigation management
- Sampling EC throughout the irrigated area to identify the occurrence of problem areas requiring specific water management.
- Sampling for specific ions that may be present in the irrigation water and that could have toxicity effects or, as for heavy metals, could create health risks.
- Follow-up non agricultural impacts from using saline water, such as changes in groundwater quality, in plant ecosystems and in wetland or river-bed fauna and flora.

Traditional laboratory techniques of salinity measurement using soil samples are impractical for the inventory and monitoring requirements of large areas, so new approaches have recently been developed, particularly those described by Rhoades et al. (1999) that were developed especially for application in large areas. Environmental impact assessment methodologies for irrigation and drainage projects may also be adapted for surveying and monitoring areas where water of inferior quality, i.e. saline and wastewater, is used for irrigation.

8.3.6 Non-agricultural Use of Saline Waters

Non-agricultural uses of saline waters are quite limited. Domestic water uses for small villages and isolated households are traditional in areas where fresh water is not available. Low salinity brackish water may then play a major role. However, these waters are not appropriate for drinking nor for animal consumption, and thus conjunctive use of this water with high quality water, for example from roof-top collection, is required.

For large populations, when a separated municipal distribution system is available, brackish saline water may be used for low quality demanding uses such as toilet flushing and washing, similar to what was discussed for wastewater reuse (8.2.7). However, saline waters have an advantage over treated wastewaters in that they tend to be free of pathogens. However the presence of salt and its corrosive potential may affect the piping system and the equipment where it would be used. Therefore, as for treated wastewater, its use is only recommended when water of good quality is definitely insufficient to satisfy the demand, such as during peak periods. The cyclic use of good quality water for flushing salts accumulated in the pipe systems is then recommended. For the above reasons, the use of saline waters in industry is very limited.

Saline and brackish water can also be used in outdoor facilities since it is free
of pathogens. However, it may be less appropriate for irrigation because lawns
and most garden plants are not salt tolerant. Advice provided above for irrigation
of agricultural crops could be followed for ornamentals but little is known of the
response to salts of most of the commonly used flowers, shrubs and trees. Nevertheless, landscapes may be planned with plants known to have some degree of tolerance
to salinity.

8.4 Desalinated Water

8.4.1 General Aspects and Treatment Processes

Seawater and brackish water desalination is progressively being used by more countries in their effort to cope with water scarcity. This is especially accentuated by the
growth of urban areas, and the development of arid areas and tourism. The latter
creates a very large water demand in coastal areas, which is often resolved through
seawater desalination.

Desalination is a water treatment process that removes salts from saline water to
produce water that is low in total dissolved solids (TDS). The use of this process in
areas of water scarcity has obvious benefits, whether ocean water or brackish inland
water is used. Quite a number of different desalination processes can be applied but
in practice, only a limited number of them are economically viable. The various
processes considered here are distillation; electrodialysis; reverse osmosis and solar
desalination (Semiat 2000).

A desalination plant can be pictured as a black box into which feed water and
high-grade energy are fed and from which low grade energy, brine and desalinated
water are produced. The TDS of the feed water may range from 500 to 50 000 mg/l.
The feed water must be treated to the level for which the plant was designed before
it is fed into the plant itself. This treatment may include physical filtration, chemical
conditioning and other processes. The desalinated water produced by the plant is
usually in the range of 50–500 mg/l TDS and to be used for human consumption
must at least comply with WHO limits (WHO 2006a,b).

The energy input into desalination plants is usually thermal, electrical or mechanical and very often is a combination of all these. Energy rejection from the plants
is usually low-grade (thermal) energy. To be able to assess the performance of a
desalination plant two important parameters are used – the recovery ratio (R_c) and
the performance ratio (R). R_c is defined as the ratio of the product water to the
feed water while R is reciprocal of the energy consumption. New developments
consist in using alternative energy sources, including solar and wind energy (e.g.
Lindemann 2004, Mathioulakis et al. 2007, Trieb and Müller-Steinhagen 2008).

The main desalination processes are:

a) *Thermal Distillation (TD).* When a saline solution is boiled, the vapour that
   comes off is pure water and when this is cooled (condensed) the resulting water

is found to contain no salt. This separation is perfect and the fact that the vapour separates very easily makes this very old method of desalination popular. The main drawback is that the energy required to evaporate a saline solution is quite high because of the high latent heat of vaporisation of water. For this to be done economically in a desalination plant, the boiling point is altered by adjusting the atmospheric pressure on the water being boiled to produce the maximum amount of water vapour under controlled conditions. The temperature required for boiling water decreases as the pressure above the water decreases. The reduction of the boiling point is important in the desalination process for two major reasons: multiple boiling and scale control. These two concepts, boiling temperature reduction and multiple boiling, have made various forms of distillation successful in locations around the world. Three types of thermal distillation units are used commercially; namely, Multistage Flash (MSF); Multiple Effect Distillation (MED); and, Vapour Compression (VC).

b) *Electrodialysis (ED)*. If a current is passed through a saline solution the different ions (cations and anions) in the solution will carry the current from one electrode to the other by drifting in opposite directions. If an ion selective membrane is placed in this flow, say an anion permeable membrane, only the anions will manage to pass through. If an alternating series of cation/anion selective membranes are placed in the path of the ion flow, the channels formed between the membranes will alternately become concentrated and diluted. The overall effect is that salt is being removed from every alternate channel to its neighbouring channels. The basic ED unit consists of several hundred-cell pairs bound together with electrodes on the outside and referred to as a membrane stack. Feedwater passes simultaneously through the cells to provide a continuous, parallel flow of desalted product water and brine that emerge from the stack. ED is only an economical process when used on brackish water, and tends to be most economical at TDS levels of up to 4,000–5.000 mg/l. ED units have a waste discharge of brackish water ranging in volume from 10% to 50% of its output of freshwater. The feedwater must be pre-treated to prevent materials from entering the membrane stack that could harm the membranes or clog the narrow channels in the cells. Post-treatment consists of stabilising the water and preparing it for distribution by removing gases such as hydrogen sulphide and adjusting the pH.

c) *Reverse Osmosis (RO)*. This is a membrane separation process in which the pressure of the water is raised above the osmotic pressure of the membrane. No heating or phase change is necessary for this separation, and the major energy requirement is for pressurising the feedwater. In practice, the saline feedwater is pumped into a closed vessel where it is pressurised against the membrane. As a portion of the water passes through the membrane, the salt content of the remaining feedwater increases. A portion of this saltier feedwater is discharged without passing through the membrane. RO units have a waste discharge of brackish water or brine which could range from 35% to 100% of its output of fresh water, depending on the feedwater being treated. Two improvements have helped reduce the operating costs of RO plants during the past decade. These are the development of membranes that can operate efficiently at lower pressures and the use of energy recovery devices.

d) *Solar Desalination (SD).* There are three basic ways in which solar energy is used to desalinate saltwater. These are humidification, distillation, and photovoltaic separation. Solar humidification imitates a part of the natural hydrologic cycle by using the Sun's rays to heat a saline water source to produce water vapour. This vapour, or humidity, is then condensed on a cooler surface and the condensate collected as product water. In the solar distillation process, a solar collector is used to concentrate solar energy to heat the feedwater so that it can be used in the high temperature end of a standard thermal desalination process. This is usually a multiple effect or multistage flash process. These units tend to be very capital intensive and require specialised staff to operate them over a long period of time. In addition, they require additional energy inputs to pump the water through the process. Desalination with photovoltaics use photovoltaics to provide electrical energy to operate standard desalting processes like reverse osmosis or electrodialysis. Batteries are used to store energy and inverters are needed to supply alternating current when necessary. The availability of solar energy for only part of the day requires commercial units to be oversized to produce the quantity of water required. Solar desalination is not used extensively and remains largely experimental. There are no large-scale installations, generally because of the large solar collection area requirements, high capital cost, vulnerability to weather-related damage and complexity of operation.

Water desalination has evolved from traditional water distillation, with high energy consumption, to the modern membrane technologies, mainly reverse osmosis (RO), which is more energy efficient and requires lower investment costs (Ma et al. 2007, Khawaji et al. 2008). Distillation technologies were predominant in the past. The appearance of RO membranes in the 1970s for brackish water and in the 1980s also for seawater has changed the technological panorama of desalination. Water desalination involves high energy consumption, which is the main cost in desalinating water. Distillation technologies consume considerable energy even for treating brackish water. Energy consumption for membrane technologies depends on the salt content of the feed water and of the product water. RO can be adapted to different water salinity contents which makes RO attractive for applications in agriculture and in particular for treating brackish water for rural populations. Electro-dialysis reversal (EDR) is another membrane technology but it is less flexible than RO and can be used only for special brackish water applications in agriculture (Martínez Beltrán and Koo-Oshima 2006).

In general, desalinated seawater has not been considered as an alternative source of water for irrigation, including in areas bordering the sea, except in highly profitable cash-crops, produced out-of-season in green-house environments, and where the tourist demand for fresh horticultural products is high, e.g. in southeast Spain.

Reverse osmosis is considered to be the most promising for the treatment of agricultural drainage water and other brackish waters mainly due to its comparatively low cost and because it is a process capable of removing different contaminants including dissolved salts and organics (Tanji and Kielen 2002, Martínez Beltrán and Koo-Oshima 2006). The energy consumption of the process depends on the salt concentration of the feed-water and the salt concentration of the effluent. Depending on the quality of the water to be treated, pre-treatment might be crucial to prevent fouling of the membrane. For brackish water and oilfield produced water systems, pre-treatment is critical because of the impurities that the water may contain. Even the most sophisticated RO facility can experience poor flux and high maintenance costs where pre-treatment is inappropriate for the feed water being treated (Martínez Beltrán and Koo-Oshima 2006).

8.4.2 Extent of Use, Costs and Environmental Impacts of Desalination

Distillation used to have the largest share of the worlds installed desalination capacity, with the MSF process making up the highest proportion of distillation units. The MSF and MED processes are often used as part of a dual purpose facility where the steam to run the desalination unit is taken from the low pressure end of a steam turbine that is used to generate electricity. The remaining steam and condensate is then returned to the boiler to be reheated and reused. Individual MSF or MED units generally have a capacity of $1,000–20,000\,\mathrm{m^3/day}$ but several of these units can be grouped around an electrical generating plant. Facilities with a total water output of $200,000\,\mathrm{m^3/day}$ or more are not uncommon in the Middle East, while smaller facilities, consisting of several $5,000\,\mathrm{m^3/day}$ units, are common. VC units are also widely used but, individually, these have much smaller capacities, and, hence, a lower overall total capacity than that of the MSF and MED plants.

Electrodialysis makes up about 5% of the world's installed desalination capacity. Electrodialysis units are used in applications requiring smaller volumes of water and can be purchased in units with individual capacities ranging from 10 to $4,000\,\mathrm{m^3/day}$.

Reverse osmosis is becoming the most used technology, and is the most popular for small units and for treating brackish water. RO units are small relative to thermal distillation plants, and can be purchased in packages with individual capacities from 10 to $4,000\,\mathrm{m^3/day}$. The largest plants are in the range of $40,000\,\mathrm{m^3/day}$.

The capital cost of the MSF and MED distillation units tends to be in the range of $1000–$2000/m^3/d$ of installed capacity, exclusive of the steam supply and site preparation. Recent economic analyses of desalination are produced by El-Sayed (2007) and Tian et al. (2007). Improved technologies, pre-treatment and increased energy efficiency (Ma et al. 2007, Khawaji et al. 2008) are leading to reduced costs both for seawater and brackish water desalination (Tables 8.18 and 8.19, respectively).

Table 8.18 Indicative seawater desalination costs (Euro/m^3) (from Martínez Beltrán and KooOshima 2006)

	Multistage flash	Multiple-effect distillation	Vapour compression	Reverse osmosis
Energy, Fuel	0.52	0.46	0	0
Electricity	0.15–0.16	0.07–0.08	0.50–0.55	0.22–0.27
Labour	0.032–0.036	0.03–0.04	0.054–0.08	0.018–0.081
Chemicals	0.032–0.045	0.027–0.036	0.021–0.036	0.018–0.054
Membrane replacement	0	0	0	0.001–0.036
Chemicals cleaning	0.001–0.002	0.001–0.002	0.001–0.002	0.001–0.002
Maintenance	0.018–0.032	0.018–0.032	0.016–0.027	0.018–0.032
Total O&M costs	0.76–0.79	0.61–0.65	0.59–0.70	0.31–0.49
Payback costs	0.34–0.35	0.34–0.35	0.36–0.38	0.15–0.22
Total costs	1.10–1.15	0.96–1.01	0.96–1.08	0.45–0.71

Table 8.19 Indicative costs for desalination of brackish water with membrane technologies (Euro/m^3) (from Martínez Beltrán and Koo-Oshima 2006)

	Reverse osmosis, RO	Electro-dialysis reversal, EDR
Energy	0.08–0.12	0.10–0.17
Labour	0.02–0.07	0.02–0.07
Chemicals	0.02–0.03	0.006–0.01
Membrane replacement	0.015–0.022	0.006–0.013
Chemical cleaning	0.0013–0.0025	0.0006–0.0012
Maintenance and others	0.012–0.018	0.006–0.013
Total O&M costs	0.15–0.27	0.14–0.29
Payback	0.07–0.09	0.08–0.11
Total costs	0.21–0.36	0.22–0.38

Desalination has evident advantages for areas of very high water scarcity. However, in addition to high costs it also has some disadvantageous environmental impacts:

- It requires land occupation for the facilities, although this is not much larger than any other treatment facility, which may cause particular impacts on landscapes;
- It is energy intensive, with energy consumption greater than for any other water treatment process. This high consumption has implications for the primary energy associated with carbon dioxide (CO_2) production
- It may be noisy, which impacts the nearby populations.
- Major impacts in desalination are usually related to brine disposal, particularly when desalination plants are located inland. In case of very large units, provisions must be made either for transporting the brine to the sea through brine pipelines, or for drying in evaporation ponds. For small units this could be the solution or the brine could be used to raise fish as in some communities in Northeast Brazil. Where the facilities are coastal, because the very high brine salinities mean that discharges must be directed into water currents to provide dilution and where marine ecosystems, fauna and flora, will not be detrimentally affected. Otherwise

brine would need to be diluted before disposal into the sea. Temperature (more relevant in distillation technologies) and pH should be considered in addition to salt concentration.

- Contamination of the brine by chemicals used in pre-treatment that are rejected by the membranes, with by-products of membrane and tube cleaning, and by coagulants and aids in filters, should also be considered.

8.5 Fog-Capturing, Water Harvesting, Cloud Seeding, and Water Transfers

Dew, mist and fog generally make only a small contribution to precipitation in water scarce regions. However, they can be responsible for peculiar vegetation ecosystems in arid or semi-arid areas. Several physical processes influence the occurrence of these ecosystems. Agnew and Anderson (1992) refer to the importance of fogs in several coastal areas in Namibia, Oman and Peru.

Fog collection is used in isolated arid areas, in mountains and islands where the occurrence of fog is common but rainfall is rare. Ancient examples of fog capturing have been found in the Negev, the so-called "dew mounds", and in Crimea, these ones constituted by piles of rocks named "aerial wells" (NAS 1974). Modern examples are reported by Agnew and Anderson (1992) for Peru, Oman and the Negev. Several case studies are described by NGOs devoted to fog collection such as FogQuest and OFUR. Documentation is abundant relative to conferences (e.g. FogQuest 2004) and manuals or guidelines. Case studies refer to the use of fog collectors in the Arabian Peninsula, Central and South America and in the islands of Cape Verde (Fig. 8.6), located offshore of the Sahel zone of Africa. Particular attention is given to the chemical composition of fog water in view of its use by humans, animals and ecosystems in dry areas (Schemenauer and Bridgman 1998).

Fog capturing is achieved by means of appropriate screens that favour the coalescence of the small fog droplets to create larger drops with enough dimensions to flow down by gravity into collectors. Fog screen collector design is described

Fig. 8.6 Fog collectors on Santiago Island, Cape Verde

by Schemenauer and Cereceda (1994a,b) and in documentation available from the FogQuest and OFUR websites. The amount of water collected every day is small and depends on the properties of the air masses passing over the collection sites.

The reliability of fog capturing is associated with the frequency of fog occurrence. When fog occurs nearly everyday, a water yield of $3\,l/day/m^2$ of collector net may occur as in Mexico. In general, fog capturing is only feasible for human consumption for small populations living in such areas (e.g. Gandhidasan and Abualhamayel 2007), and for drinking of animals, mainly those grazing on high pasture lands where frequent fogs favour vegetation despite the lack of rains.

Rainwater harvesting, or simply water harvesting (WH), has been practised for millennia in the Negev and in Petra, Jordan, by the Nabateans. Prinz (1996) refers to water harvesting structures in Jordan from 9000 years ago and in southern Mesopotamia as early as 4500 BC. Rainwater collection was a common practice in the Roman cities of North Africa and the Near East. Water harvesting for agriculture is an ancient practice common in Mexico, Tunisia, the Arabian peninsula, Afghanistan, Pakistan, India and the Sahel countries, among others. New developments are being introduced in areas where WH was not traditional such as the arid Northeast Brazil, or where social changes require new approaches, such as the Sub-Saharan Africa (Reij and Critchley 1996, UNEP 2002c).

Roof-top rainwater harvesting, commonly associated with water collection from impermeable areas surrounding the households, plays a major role for making water available for domestic uses and animal drinking in a large variety of rural environments throughout the World, e.g. in the arid environments of Cape Verde islands (Fig. 8.7) However, care is required to keep stored water in cisterns free from contamination by animals and sewage water. Drinking water has to be boiled before consumption and filtering is commonly also required.

WH for animal drinking is commonly achieved by collecting runoff into small reservoirs created by excavation or using small earth dams. Animals should not have direct access to the stored water to avoid contamination by excreta and silting due to stirring up of the soils at the edge of the storage. Appropriate drinking facilities should be used such as drinking fountains connected to the reservoir by a pipe.

Fig. 8.7 Roof-top rainwater harvesting in Cape Verde (courtesy by Angela Moreno)

WH for crops varies from one region to another according to the indigenous knowledge, the land form, the soil type, the runoff intensity and the crops to be irrigated. Generally, distinction is made among: inter-row WH, where water collection is made in the location where it is used; micro-catchments WH, where rain is collected in part of the area and is infiltrated in the downstream cropped zone; macro-catchment WH, where runoff water is collected in large areas to be stored in earth dam reservoirs or tanks to irrigate the downstream fields; and floodwater diversion systems, also called spate irrigation systems, where appropriate structures in the river bed divert the flood runoff in a controlled manner to the nearby cropped fields. Runoff WH is discussed in Section 6.4.

Another traditional process for capturing water in arid zones is groundwater harvesting. This term is used to make a distinction from modern groundwater exploitation, which is mainly with tube wells (see Chapter 7). It includes dug-wells, which capture quite shallow groundwater and continue to be the basic water supply system in many arid or semi-arid areas of Africa and West Asia, horizontal hand made wells, which are commonly in use in many parts of the world, the Nasca aqueducts or *puquios* (see Fig. 5.5) and the *qanats*. The latter are horizontal tunnels that tap the water in an alluvial fan and transport it to the surface by gravity, without any pump or other lifting equipment. The *qanats* are a heritage of the Persians, who developed them about 3000 years ago. *Qanats* are used for irrigation in arid zones of Iran, Pakistan, Afghanistan, Xinjiang in West China, the Arabian Peninsula, and North Africa. Water may be transported underground for several kilometres, in many cases 10–20 km. *Qanats* are composed of dug or horizontal wells for capturing the water, a tunnel having a series of vertical shafts used for digging out the excavation debris and for respiration, and the downstream diversion structures that provide for the distribution of waters into different canals. *Qanat* systems are of great importance. Modernisation of these systems is desirable but careful approaches have to be used, regarding both the structural and management aspects. A recent analysis of *qanats* in Oman, locally called *aflaj*, is provided by Al-Marshudi (2001), who pays particular attention to the traditional organisational aspects as a key for maintenance and operation.

Weather modification has been an objective of research for a long time but successes reported in the literature are limited. Essentially it consists in using aircrafts to spray into the clouds passing over an arid area substances such as ice, frozen carbon dioxide and silver iodide, which are intended to serve as nucleii for condensation and coalescence of rain drops. When large enough, these drops could fall to the ground. This method for rainfall augmentation is called cloud seeding.

Results show that wet air masses must be cold enough that the formed water particles are in the form of ice crystals that coalescence to be large enough to create drops which can reach the ground as rain. In most low-lying arid areas, wet air masses are not as large, cold or frequent as in mountainous areas. There, cloud seeding could be effective, but these are often unpopulated regions and benefits to the populations can result only from augmentation of runoff due to that increased rainfall. Nevertheless, as reported by Khouri et al. (1995), during the 1980 drought in Morocco, cloud seeding over the Atlas mountains increased the rainfall by 10–15%, which caused runoff that could be stored in small dams downstream, and contributed to alleviation of the drought.

Water transfers from one basin to another have not been discussed much until recently. The idea is mostly concerned with long distance transfer of water from basins in a humid or semi-humid zone to another basin in a semiarid environment. A typical example is the Grand Canal in China, for which construction started in the Qin Dynasty, around 400 B.C. Its main purpose was navigation but it also served to transfer water from the Yangtse River to the north across the lower reaches of the Yellow River and the Huai River. New water transfers are currently planned to import water from the Yangtse basin to the municipal areas of Beijing and Tianjin and for irrigation and municipal uses in the lower Yellow River and Huai River basins. One of these projects is essentially an update of the Grand Canal system but focuses on the South to North Water transfer.

Several water transfer projects are operating in various parts of the world. This is the case for the Tajo – Segura transfer from Central Spain to the semi-arid south, the Basento – Bradano system in south Italy, the transfer of water from the Colorado river to irrigate the arid southern California and to supply the Los Angeles and San Diego areas, and in Western Australia the 700 km pipeline that transports water from south of Perth to Kalgoorlie. Another example is the Snowy Mountains Scheme in Australia, where the coastal flowing Snowy river water is turned inland to provide for irrigation and generate hydropower. In this case power generation pays for the works and irrigation water is put into the inland rivers at no net cost.

Although water transfer is a rational way for increasing water availability in areas where water scarcity is evident or demand largely exceeds the available supply, several factors contribute to a decline in interest in water transfers. They may contribute to an unbalanced regional development since the excess demand in the water importing area may not be fully in balance with its overall ecological potential. Similarly the water exporting region may do not receive all the investments and policy measures that could help it to develop better. These issues are of course questionable, but they give rise to political, cultural and social attitudes around which pressure groups develop to oppose inter-basin water transfers. More important are the recognisable environmental impacts on the basin from where water is transferred due to the consequent decrease in river flow, particularly during the low-flow period, with detrimental effects on the river ecosystems and biodiversity. The environmental impacts in the water importing areas due to quick growth of irrigation areas, urbanisation, and other activities that affect the fragile ecosystems in the water scarce regions may be less evident, but they occur. Finally, costs associated with inter-basin transfers are very high because distances to be covered are large, compensatory measures are costly and environmental control measures are quite demanding. Water transfers are usually only feasible when the water transferred has high value, such as for large population survival and high value industry. Appropriate evaluation and implementation of measures that provide for the sustainable development of both the areas exporting and importing water are required. Yevjevich (2001) presents an updated discussion of these matters. Water transfers should go together with the implementation of water conservation and saving measures since the corresponding increase in availability does not turn the importing area into one of water abundance.

In many areas, the reallocation of water rights may be an alternative or a complementary measure to water transfer. This reallocation generally relates to the water value and may be easily performed when a water market can be implemented.

Chapter 9
Water Conservation and Saving: Concepts and Performance

Abstract Water use concepts and performance that may be useful in analysing water conservation and saving are dealt with in some detail in this chapter. New indicators are proposed include the consideration of water reuse and assist in identifying beneficial and non-beneficial water uses. An analysis of water productivity concepts useful in irrigation and for other uses is also included. These approaches are complemented with a review of water conservation and saving measures relative to the various water scarcity regimes.

9.1 Concepts

9.1.1 Water Conservation and Water Saving

The terms water conservation and water saving are generally associated with the management of water resources under scarcity. However, these terms are often used with different meanings in accordance with the scientific and technical disciplines or the water user sector considered. Very often, both terms are used as synonymous.

The term *water conservation* is used herein referring to every policy, managerial measure, or user practice that aims at conserving or preserving the water resources, as well as combating the degradation of the water resource, including its quality. Differently, *water saving* aims at limiting or controlling the water demand and use for any specific purpose, including the avoidance of water wastes and the misuse of water. In practice both perspectives are complementary and inter-related. Despite not being easy to distinguish between them, these terms should not be used synonymously. In particular, questions relative to preservation and upgrading of water quality are essential in water conservation.

This chapter analyses the measures and practices relative to water saving and water conservation that can be applied by the different user sectors – urban systems, domestic water uses, landscape and recreational uses, industrial and energy, and agriculture, both dryland and irrigated agriculture – to cope with the various water scarcity regimes: aridity, droughts and man-made desertication and water shortage. These suggested measures and practices are the base for establishing local,

L.S. Pereira et al., *Coping with Water Scarcity*, DOI 10.1007/978-1-4020-9579-5_9,
© Springer Science+Business Media B.V. 2009

regional or national programs for water conservation and saving when focusing on the aspects and issues that are more relevant to the areas under consideration.

9.2 Water Use, Consumptive Use, Water Losses, and Performance

9.2.1 Water Systems, Efficiency, and Water Use Performance

The performance of water supply systems and water use activities are often expressed with terms relative to efficiency. However, there are no widely accepted definitions, and the efficiency terms are used with different meanings, mainly relative to the various water use sectors. In certain cases, both water conservation and water saving are used as synonymous with water use efficiency. For a better understanding of terminology utilised in relation to water use performances, a more consistent conceptual approach is required.

The term efficiency is often used in the case of irrigation systems and it is commonly applied to each irrigation sub-system: storage, conveyance, off- and on-farm distribution, and on-farm application sub-systems. It can be defined by the ratio between the water depth delivered by the sub-system under consideration and the water depth supplied to that sub-system, usually being expressed in percentage. In case of on-farm application efficiency, the numerator is replaced by the amount of water added to the root zone storage and the denominator is the total water applied to that field. However, in reality an efficiency indicator refers to a single event and should not be applied to a full irrigation season without adopting an appropriate up-scaling approach. These delivery/demand ratios relate to individual processes and their use as a bulk term does not provide information on the processes. A scheme on processes involved in irrigation water use is given in Fig. 9.1. For non-irrigation water systems, the term efficiency is less used but could be similarly applied referring to the various processes involved.

The term efficiency, often leads to misconceptions and misunderstandings (e.g., Jensen 1996, 2007, Allen et al. 1997, Burt et al. 1997, Pereira et al. 2002a,b). A common misconception is that of considering that increasing irrigation efficiencies is almost synonymous with creating more available water. In fact, there is the need to quantify the fraction of water used (diverted for a given use) that is beneficially consumed, and the fraction that is not consumptive use and is available for reuse or becomes degraded after use. For the latter case, improving efficiencies would represent a reduction in water losses and contribute to the conservation of the available resource. In many cases, the non-consumed fraction is not degraded and is used by other systems downstream; then, improving efficiencies would not be advantageous to the total system.

The present trend is to abandon the term efficiency for irrigation water conveyance and distribution and to adopt service performance indicators. In fact, it is recognized that impacts on agricultural yields, farmers' incomes, and farm water

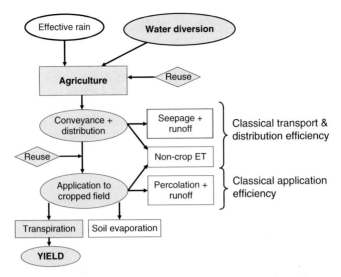

Fig. 9.1 Processes influencing the irrigation efficiency off- and on-farm: in grey boxes, the processes leading to the crop yield; in white boxes, those leading to water wastes and losses

management largely result from the quality of the water delivery service. Indicators referring to the reliability, dependability, adequacy, or equity of deliveries may be used for that purpose. These and other indicators are measures of the capability of collective water systems for timely water delivery with appropriate discharges, pressure head, time intervals and duration to satisfy the farm requirements throughout the irrigation season and independently of the location of the gate or hydrant. Similar water service indicators are also used for other non-irrigation networks.

The term application efficiency is still used to characterize the management relative to a given event. However, it must be adopted together with an indicator of the uniformity of water application to the field (Burt et al. 1997, Pereira 1999). In fact, if a system does not provide for uniform water application, efficiency is necessarily low and percolation through the bottom of the root zone is high.

Another term commonly used is water use efficiency (WUE), but again no common definition has been adopted. Some authors refer to it as a non-dimensional output/input ratio as for the single term efficiency noted above. Others adopt it to express the productivity of the water, as a yield to water used ratio. In crop production, the term WUE is applied with precise meanings, such the yield WUE, which is the ratio of the harvested biomass to the water consumed to achieve that yield. In plant physiology and eco-physiology WUE expresses the ratio between assimilates produced during a certain period of time and the corresponding plant transpiration. In this case, WUE expresses the performance of a given plant or variety in using water (Steduto 1996). To avoid misunderstandings, the term "water use efficiency" should be only used to measure the performance of plants and crops, irrigated or non-irrigated. The term "water productivity" (WP) should be adopted to express the quantity of product or service produced by a given amount of water used. Given its importance in water conservation and saving, water productivity is discussed in a specific section hereafter.

9.2.2 Water Use, Consumption, Wastes and Losses

New concepts to clearly distinguish between consumptive and non-consumptive uses, beneficial and non-beneficial uses are being developed. Similarly the differences between reusable and non-reusable fractions of the non-consumed water diverted into an irrigation system or subsystem are being clarified (Allen et al. 1997, Pereira et al. 2002b). These consist of alternative performance indicators that are much more relevant than "irrigation efficiency" when adopted in regional water management for the formulation of water conservation and water saving policies and measures. These concepts and indicators refer to irrigation and non-irrigation water uses.

When water is diverted for any use only a fraction is consumptive use. The non-consumed fraction is returned after use with its quality preserved or degraded. Quality is preserved when the primary use does not degrade its quality to a level that does not allow further reuse, or when water is treated after that primary use, or when water is not added to poor quality, saline water bodies. Otherwise, water quality is considered degraded and water is not reusable (Fig. 9.2).

Both consumed and non-consumed fractions concern beneficial and non-beneficial water uses. These are beneficial when they are fully oriented to achieve the desirable yield, product, or service. Alternatively, when that use is inappropriate or unnecessary, it is called non-beneficial. Reusable water fractions are not lost because they return to the water cycle and may be reused later by the same or by other users. They are not losses; but are wastes since they correspond to water unnecessarily mobilized. Contrarily, the non-beneficial water consumed or returned to poor quality, saline water bodies, or that contribute to degradation of any water body are effectively water losses (Fig. 9.2).

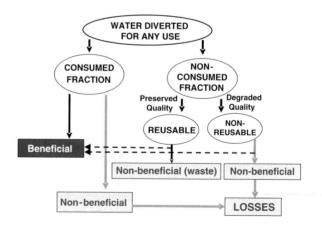

Fig. 9.2 Water use, consumptive and non-consumptive use, beneficial and non-beneficial uses, water wastes and losses

Since a water use is purposeful, it is important to recognize in the water economy perspective both the beneficial and non-beneficial water uses (Fig. 9.3). In crop and landscape irrigation, the beneficial uses are those directly contributing to an agricultural product or an agreeable garden, lawn or golf course. Non beneficial are those uses that result from excess irrigation, poor management of the supply system, or from misuse of the water.

These concepts may also be applied to the use of water in industry, urban regions, energy production and other activities. Then beneficial uses include all the activities and processes leading to achievement of some production or service which results in some good or benefit, such as washing, cooking, heating, cooling, or generating hydropower energy. The uses are not beneficial when water is used in non-necessary processes, is misused or is used in excess of the requirements (Fig. 9.4).

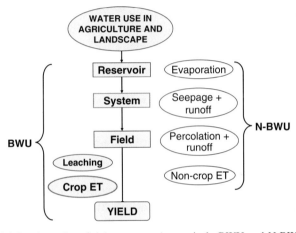

Fig. 9.3 Beneficial and non-beneficial water use (respectively BWU and N-BWU) in crop and landscape irrigation

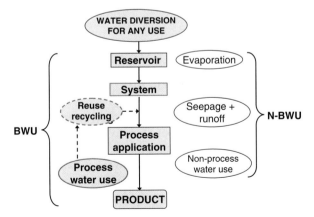

Fig. 9.4 Beneficial and non-beneficial water use (respectively BWU and N-BWU) in agriculture, industry, urban, energy and landscape, with reference to reuse or recycling

Pathways to improve water use

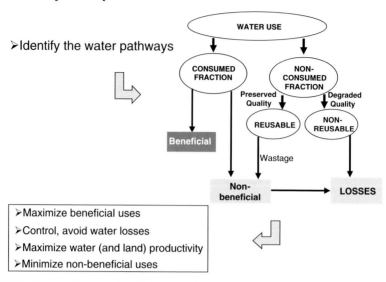

Fig. 9.5 Pathways to improve the efficient use of water

Assuming the concepts above, it is important to recognize what is meant by "efficient water use". To support this concept, a few main ideas are developed in Fig. 9.5.

First, it is required to identify the water pathways in any water use (Fig. 9.5), to distinguish what is consumptive and non-consumptive water use, what is a beneficial or a non-beneficial water use, and which fractions are really losses or just wastes. This requires that productive and non-productive processes, i.e., oriented to achieve the water use goal, be recognized. Then, a water use is more efficient when beneficial water uses are maximized, the water productivity is increased, and water losses and wastes are minimized.

9.3 Water Use Performance Indicators

9.3.1 Consumptive Use and Beneficial Use

Assuming the concepts above, it is possible to define water use indicators adapted to any water use or system and to adopt them to make more efficient the use of water, i.e., aiming at improved performances from the perspective of water resources conservation. These indicators may be useful for water resources planning and management under scarcity. They may be combined with process indicators, including those which relate to the quality of service of water systems.

The indicators refer to the three water use fractions (Fig. 9.2) and to the respective beneficial and non-beneficial water use components. These indicators can be characterised in equations such as those shown below:

(a) *The consumptive use fraction* (CF), consisting of the fraction of diverted water which is evaporated, transpired or incorporated in the product, or consumed in drinking and food, which is no longer available after the end use:

$$CF = \frac{E + ET_{process} + ET_{weeds} + IN_{food} + IN_{product}}{TWU} \tag{9.1}$$

where the numerator refers to process evaporation (E) and evapotranspiration (ET) and incorporation in products (IN), and the denominator is the total water use (TWU). Subscripts identify the main sinks of water consumption.

The CF beneficial and non-beneficial components are:

$$BCF = \frac{E_{process} + ET_{process} + IN_{food} + IN_{product}}{TWU} \tag{9.2}$$

and

$$N-BCF = \frac{E_{non-process} + ET_{weeds}}{TWU} \tag{9.3}$$

(b) *The reusable fraction* (RF), consisting of the fraction of diverted water which is not consumed when used for a given production process or service but that returns with appropriate quality to non degraded surface waters or ground-water and, therefore, can be used again:

$$RF = \frac{(Seep + Perc + Run)_{non-degraded} + (Ret\ flow + Effl)_{treated}}{TWU} \tag{9.4}$$

where the numerator consists of non-consumptive use processes that did not degrade the water quality, thus allowing further uses, including when the return flows (Ret flow) and effluents (Effl) are treated. The RF components are:

$$BRF = \frac{(LF + Runoff_{process})_{non-degraded} + Effl_{treated}}{TWU} \tag{9.5}$$

which includes water used for salt leaching (LF), runoff necessary to the processes (such as furrow and border irrigation), and controlled effluents (Effl) required by non agricultural uses, as for many domestic uses. The non-beneficial reusable fraction (N-BRF) is then

$$N-BRF = \frac{(Seep + Perc + Exc\ Runoff)_{non-degraded} + Exc\,Effl_{treated}}{TWU} \tag{9.6}$$

and refers to excess water use in the processes involved such as seepage and leaks from canals and conduits, spills from canals, excess percolation in irrigation uses or excess runoff that is non-degraded, and effluents due to water wastes when treated.

(c) *The non-reusable fraction* (NRF), consisting of the fraction of diverted water which is not consumed in a given production process or service but which returns with poor quality or returns to degraded surface waters or saline ground-water and, therefore, cannot be used again

$$NRF = \frac{(Seep + Perc + Run)_{degraded} + (Ret\ flow + Effl)_{non-treated}}{TWU} \quad (9.7)$$

which refers to the same process as the RF but where the water has lost quality and is not treated or is added to water bodies which are not usable for normal processes, such as saline groundwater, saline lakes and the oceans. The NRF shall also be divided into a beneficial and a non-beneficial component

$$BNRF = \frac{(LF + Runoff_{process})_{degraded} + Effl_{non-treated}}{TWU} \quad (9.8)$$

and

$$N - BNRF = \frac{(Seep + Perc + ExcRunoff)_{degraded} + ExcEffl_{non-treated}}{TWU} \quad (9.9)$$

In addition to the indicators defined above, it is also worthwhile to define the beneficial and the non-beneficial water use fractions (BWUF and N-BWUF), which are obtained from the various components defined above, respectively through Equations (9.2, 9.5, and 9.8) for the first, and Equations (9.3, 9.6 and 9.9) for the second. These indicators are further used in the next chapter.

These indicators have various advantages in view of water conservation and saving in water scarce areas: they allow understanding of what is or is not consumed, they focus on the water use processes, and they direct attention to which uses are beneficial under the perspective of the product or service achieved with the water. Moreover, they focus not only on the quantities but also on the water quality when identifying the degradation of the water after use and when it is treated or not treated.

Illustrations of the main processes of water use referring to the above described water use fractions are presented in Tables 9.1 and 9.2 for agricultural and non-agricultural uses, respectively.

9.3.2 Water Productivity: Irrigation

Nowadays, there is a trend to call for increasing water productivity (WP) as a main issue in irrigation (Molden et al. 2003, Clemmens and Molden 2007). The attention formerly given to irrigation efficiency is now transferred to water productivity. However, this term is used with different meanings in relation to various scales (Fig. 9.6). The analysis herein is oriented only to the total WP with limited references to irrigation process WP. WUE was referred in Section 9.2.1.

Table 9.1 Beneficial and non-beneficial water use and its relation to consumptive and non-consumptive uses in irrigation

	Consumptive	Non-consumptive but reusable	Non-consumptive and non-reusable
Beneficial uses	• ET from irrigated crops • Evaporation for climate control • Water incorporated in product	• Leaching water added to reusable water	• Leaching added to saline water
Non-beneficial uses	• Excess soil water evaporation • ET from weeds and phreatophytes • Sprinkler evaporation • Canal and reservoir evaporation	• Deep percolation added to good quality aquifers • Reusable runoff • Reusable canal seepage and spills	• Deep percolation added to saline groundwater • Drainage water added to saline water bodies
	Consumed fraction	**Reusable fraction**	**Non-reusable fraction**

Table 9.2 Beneficial and non-beneficial water use and its relation to consumptive and non-consumptive uses in non-irrigation user sectors

	Consumptive	Non-consumptive but reusable	Non-consumptive and non-reusable
Beneficial uses	• Human and animal drinking water • Water in food and process drinks • Water incorporated in industrial products • Evaporation for temperature control • ET from vegetation in recreational and leisure areas • Evaporation from swimming pools and recreational lakes	• Treated effluents from households and urban uses • Treated effluents from industry • Return flows from power generators • Return flows from temperature control • Non-degraded effluents from washing and industrial processes	• Degraded effluents from households and urban uses • Degraded effluents from industry • Degraded effluents from washing and process waters • Every non degraded effluent added to saline and low quality water
Non-beneficial uses	• ET from non beneficial vegetation • Evaporation from water wastes • Evaporation from reservoirs	• Non-degraded deep percolation from recreational and urban areas added to good quality aquifers • Leakage of non-degraded water from urban, industrial and domestic systems added to good quality waters	• Deep percolation from recreational and urban areas added to saline aquifers • Leakage from urban, industrial and domestic systems added to low quality waters and saline water bodies
	Consumed fraction	**Reusable fraction**	**Non-reusable fraction**

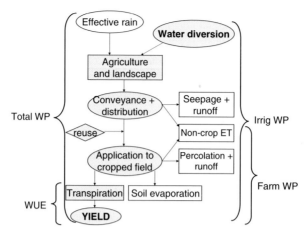

Fig. 9.6 Water productivity in agriculture at various scales: (a) the plant, through the water use efficiency WUE; (b) the irrigated crop at farm scale (Farm WP); (c) the irrigated crop, at system level (Irrig WP); and the crop including rainfall and irrigation water (Total WP)

Water productivity in agriculture and landscape irrigation may be generically defined as the ratio between the actual crop yield achieved (Y_a) and the water use, expressed in kg/m³. For landscape, a convenient definition of Y_a has to be selected because irrigating gardens, lawns or golf courses produces qualitative yields. The denominator may refer to the total water use (TWU), including rainfall, or just to the irrigation water use (IWU). This results in two different indicators:

$$WP = \frac{Y_a}{TWU} \tag{9.10}$$

and

$$WP_{Irrig} = \frac{Y_a}{IWU} \tag{9.11}$$

The meaning of these indicators is necessarily different. The same amount of grain yield depends not only on the amount of irrigation water used but also on the amount of rainfall water that the crop could use, which relates to rainfall distribution during the crop season. Moreover, the pathways to improve crop yields are often not so much related to water management as to agronomic practices and the adaptation of the crop variety to the cropping environment. However, a crop variety for which WUE is higher than that of another variety has the potential for using less water than the second when achieving the same yield. Therefore, discussing how improving WP could lead to water saving in irrigation requires the consideration of various different factors: (a) the contribution of rainfall to satisfy crop water requirements, (b) the management and technologies of irrigation, (c) the agronomic practices, (d) the adaptability of the crop variety to the environment, and (e) the water use efficiency of the crop and variety under consideration.

Equation (9.10) may take a different form:

$$WP = \frac{Y_a}{P + CR + \Delta SW + I} \tag{9.12}$$

where Y_a is the actual harvestable yield (kg), P is the season amount of rainfall, CR is the amount of water originated from capillary rise, ΔSW is the difference in soil water storage between planting and harvesting, and I is the season amount of irrigation, all expressed in m^3. When appropriate soil water conservation practices are adopted (cf. Section 10.5), the proportion of total P that is available to the crop is increased and soil water storage at planting may also be increased. When irrigation practices are oriented for conservation (cf. Section 10.6), crop roots may be better developed and the amount of water from CR and ΔSW may become higher. The results may then be a lower demand for irrigation water.

The same Equation (9.10) may be written in another form:

$$WP = \frac{Y_a}{ET_a + LF + N - BWU} \tag{9.13}$$

where Y_a is the actual harvestable yield (kg), ET_a is the actual season evapotranspiration, LF is the water used for leaching when controlling soil salinity is required, and N-BWU is the non-beneficial water use, i.e., the water in excess to the beneficial ET_a and LF water uses. N-BWU consists of percolation through the bottom of the root zone, runoff out of the irrigated fields, and losses by evaporation and wind drift in sprinkling. WP may be increased by minimizing these N-BWU components. However, a higher WP should also be attained through higher yields; achieving this may require an increase in ET_a to its optimum level, ET_c. It could be that attaining the maximal value for WP in irrigation requires that yields are maximized, ET and LF are optimized and N-BWU are minimized:

$$max(WP) = \frac{Y_{max}}{ET_c + LF + min(N - BWU)} \tag{9.14}$$

A high WP may also be obtained when a crop is water stressed, but then the yield is reduced. It is also observed that the resulting increases in WP are often small. Under these conditions the economic results of production may be detrimental, particularly for small farms. This implies that in addition to WP also economic water productivity should be considered.

Replacing the numerator of equations above by the monetary value (€) of the achieved yield Y_a, the economic water productivity (EWP) is expressed as €/m^3 and defined by:

$$EWP = \frac{Value(Y_a)}{TWU} \tag{9.15}$$

However, the economics of production are less visible in this form. It may be better to express both the numerator and the denominator in monetary (€) terms, respectively the yield value and the TWU cost, thus yielding the following ratio:

$$EWPR = \frac{Value\,(Y_a)}{Cost\,(TWU)} \tag{9.16}$$

Alternatively, considering the Equation (9.12), we have:

$$EWPR = \frac{Value\,(Y_a)}{Cost(soil\ water\ conservation) + Cost\,(I)} \tag{9.17}$$

This Equation (9.17) shows both the costs for improving rainfall and capillary rise water uses and the costs of irrigation. Improving this ratio implies finding a balance between production and yield costs, as well as appropriate soil and water conservation and irrigation practices. This is not easy to achieve and explains why farmers may retain low irrigation performances and poor conservation practices if related costs for improvement are out of their economic capacity.

Alternatively, considering Equation (9.13), the following ratio is obtained:

$$EWPR = \frac{Value\,(Y_a)}{Costs\,(ET_a + LF + N - BWU)} \tag{9.18}$$

This suggests that the costs for reducing the N-BWU may be the bottleneck in improving water productivities: to reduce N-BWU implies investment in improving the irrigation system that may be beyond the farmers' capacity, particularly for small farmers. This calls attention to the need for support and incentives for farmers when a society requires they decrease the demand for water and increase the water productivity. In collective and cooperative irrigation systems a part of the difficulties results from poor system management and inadequate delivery services, which are often outside the control of the farmers.

Maximizing EWPR, when all costs other than for water use are kept constant, means finding the limit to the ratio between the yield value and the water use costs, which corresponds to maximizing crop revenue in the form:

$$\max\,(EWPR) = \max \frac{Value\,(Y_{opt})}{Water\ Costs\ for\ Y_{opt}} \tag{9.19}$$

This maximal EWPR generally relates to the maximal farm income. The optimal yield, Y_{opt}, is often different from the maximum yield, depending upon the structure of the production costs.

9.3.3 Water Productivity for any Water Use Sector

The concept of water productivity is also applied in other water user sectors. It must be adapted to the specificities of each sector and activity. The term water productivity probably needs to be used or defined separately for each production or service process (Fig. 9.7). Similar to WP being expressed in kg of grain per m^3 of water used in the case of irrigation, it is also possible to express WP in meters of fabric per m^3 of water in the textile industry; kWh produced per m^3 of water in energy generation; m^2 of lawns irrigated per m^3 of water in recreational areas; or m^2 of area washed per m^3 of water in commercial areas.

Extending the water productivity concepts used in agriculture to other user sectors yields:

$$WP = \frac{End\ Product\ or\ Service}{TWU} \qquad (9.20)$$

where the numerator is expressed in units appropriate to the activity under consideration, and the denominator consists of the total water used to yield that product or service.

Equation (9.20) may take a different form to distinguish the beneficial and nonbeneficial water uses contributing to yield of product or service

$$WP = \frac{End\ Product\ or\ Service}{BWU + N-BWU} \qquad (9.21)$$

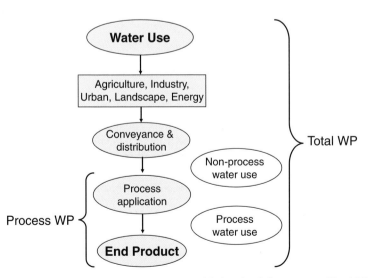

Fig. 9.7 Water productivity in any user sector considering the full water use (Total WP) and a single process WP

This equation shows that WP may be increased through improved production processes that may require less BWU or when N-BWU (non-process water uses) is minimized. In the industrial applications, BWU is commonly reduced by recycling and reuse for less stringent processes and applications. In urban and domestic uses, BWU may also be decreased when treated wastewater is used for processes not requiring the highest water quality.

Replacing the numerator of equations above by the monetary value (€) of the achieved product or service produces the economic water productivity (EWP) expressed as €/m^3

$$EWP = \frac{Value\,(Product\ or\ Service)}{TWU} \qquad (9.22)$$

and the economic water productivity ratio

$$EWPR = \frac{Value\,(Product\ or\ Service)}{Cost\,(TWU)} \qquad (9.23)$$

This Equation (9.23) shows that when this ratio is very large, which is common to most non-agricultural water uses, there is no incentive to reduce TWU unless water policies relative to the quality of effluents and respective treatment induce a reduction of the amounts to be treated and, therefore, used. Water scarcity may be another reason for reducing TWU, mainly when competition among users is high.

Maximizing EWPR is a question of minimizing the costs of water used to yield the desired product or service:

$$\max\,(EWPR) = \max \frac{Value\,(Product\ or\ Service)}{Water\ Costs} \qquad (9.24)$$

Differently from agriculture, where water use and related costs (including equipment, labour, and energy) may constitute a large percentage of production costs, water costs in other user sectors are often a small fraction of the production costs, but often include wastewater treatment and water recycling. Therefore, the rationale behind water productivity for most sectors and activities is very different from that in agriculture.

In urban supply systems consumption data are usually available in terms of litres/person/day, and there needs to be an aim to have these numbers continually falling. In all the above cases, for farms, factories and domestic supply operations there need to be policies and incentives aimed at bringing water consumption to the lowest possible level for each unit of production or activity in all areas, i.e. increasing the water productivity in all uses.

9.4 Water Conservation and Saving to Cope with the Various Water Scarcity Regimes

In other chapters of this book, different facets of water conservation and water saving are analysed, described and proposed. In this section, some of most important measures and policies are summarised (Tables 9.3 and 9.4) with the aim of giving some perspective to their relative importance for coping with the various natural and man-made water scarcity regimes identified in Chapter 2.

When *aridity* is the cause of water scarcity, the stress on the natural resources, the vulnerability and fragility of the ecosystems, and the strong inter-sectorial competition for water require that specific resource conservation policies and measures need to be enforced. Among those referred in Table 9.3, water conservation for coping with aridity requires particular attention relative to water quality issues because quality degradation makes water unavailable for other users, produces enormous health problems, easily destroys the riparian ecosystems, and may be a cause for desertification and water shortage. In case of agriculture, the problem is also relevant under the soil resource perspective because inappropriate use of water of poor quality also degrades the soil and land resources.

A problem whose importance is rising relates to the overuse and often mining of groundwater. In arid zones recharge is small and uncontrolled withdrawals may destroy an essential resource. The same happens with careless contamination of aquifers. As analysed in Chapter 7, the sustainable use of groundwater in arid zones is a major challenge for sustainable development. Rediscovering ancient management practices that favoured the control of withdrawals and recharge would be welcome. The traditional know-how may also play an important role in agriculture, particularly in relation to practices that favoured water infiltration and storage in the soil, or controlling evaporation of soil water. A few millimetres of water conserved in the soil may make the difference between success and failure of a crop.

Another aspect deserving attention is the use of water for nature, or environmental flows in rivers as discussed in Chapter 6, and the use of water by natural vegetation, with large consequences on biodiversity. Because water scarcity in arid zones is generally acute and competition among users is very high, these aspects, apparently without economic returns in the short term, and without spokesmen to put their case, are often neglected.

Water saving programs (Table 9.4) are essential in arid zones. However, that saving must be economically feasible. Recycling in industry and in energy generation is spreading in more developed economies, but reuse of treated water in agriculture is increasing only slowly. This relates to the economics of water use and water productivity as analysed in the preceding section. Reduced irrigation water use is often advocated but farmers often do not willingly follow these practices because they believe the economic consequences will be negative for them.

Water saving, as well as conservation, are also affected by the way how water is valued and water services are priced. Nevertheless, these economic issues must be in agreement with the social and cultural values given to the water by the users

Table 9.3 Water conservation measures and policies and their relative importance to cope with different water scarcity regimes

Water conservation measures and practices	Aridity	Drought	Desertification and water shortages
(a) Meteorological and hydrological information networks to support planning and real time operation and management of reservoirs and water supply and irrigation systems	H	H	H
(b) Storage and regulation reservoirs for improved availability of the water resources	H	H	M
(c) Land and water use planning and management	H	L	H
(d) Enforcing water quality management measures and practices	H/M	L	H
(e) Improved conditions for operation, maintenance, and management of water supply systems	H/M	H	M
(f) Maintenance of required discharges for ecological purposes in natural streams and water bodies	H	L	H
(g) Controlling ground-water withdrawals, recharge and contamination	H/M	L	H
(h) Enforcement of water allocation policies focusing on the prevalent water scarcity problems	H	H	H
(i) Augmentation of available water resources through the re-use of treated waste-waters, drainage and low quality waters, for specified uses	H	H	M/L
(j) Exploring non-conventional water sources by households, farmers, industry and water supply bodies	H/M	H	L
(k) Adoption of water technologies and practices by the end users for resource conservation	H	H	H
(l) Development of soil and water conservation practices in rainfed and irrigated agriculture	H	H	H
(m) Erosion control and soil conservation	M	L	H
(n) Development of soil and crop management practices for restoring the soil quality	M	L	H
(o) Combating soil and water salinisation	H	L	H
(p) Development of participative institutions to water management, including legal and regulatory measures	H	H	H
(q) Application of water pricing and financial incentives that favour efficient water uses, and treatment, re-use and recycling of the water	H	H	H
(r) Adoption of penalties for water wasting and misuse, and degradation of the resource	H	H	H
(s) Enhancing public awareness of the economic, social and environmental value of water, including nature conservation	H	H	H

H = high to very high importance; M = important but not having first priority; L = important but having low priority

Table 9.4 Water saving measures and policies and their relative importance to cope with different water scarcity regimes

Water saving measures and practices	Aridity	Drought	Desertification and water shortages
(a) Control of evaporation losses from reservoirs	H/M	H	L
(b) Control of leaks from canals and conduits	H/M	H	L
(c) Control of spills from canals and other hydraulic infrastructures	H/M	H	L
(d) Exploring information systems for decision makers, water system management and improved water uses	H	H	H
(e) Implementing reservoir and ground-water management rules	H	H	H
(f) Enforcing water scarcity oriented water allocation and delivery policies	H	H	H
(g) Adoption of reduced demand crops, cropping patterns, cultivation practices, and irrigation techniques	H	H	H
(h) Cropping and irrigation practices oriented to control non-point source pollution by agro-chemicals, fertilisers and erosion sediments	M	L	H
(i) Adoption of deficit irrigation practices	M	H	M
(j) Adoption of irrigation and drainage practices favouring salinity management	H	L	H
(k) Adoption of farm water storage and soil water conservation practices	H/M	H	H/M
(l) Use of inferior quality water for irrigation	H/M	H	L/M
(m) Water recycling in industry and in energy generation	H	H	L/M
(n) Adopting water saving tools and practices for reducing domestic, urban, and recreational water uses	H/M	H	L/M
(o) Water price policies in relation to the used water volumes, the specific uses, and the productivity of water use	H	H	H
(p) Incentives for reducing water demand and consumption	H	L/M	H
(q) Penalties for excessive water uses as well as for low quality effluents and return flows	H	H	H
(r) Education and campaigns for adoption by end-users of water saving tools and practices	H	H	H

H = high to very high importance; M = important but not having first priority; L = important but having low priority

societies. An effort to understand the users responses to water values and prices is required because related policies often fail when they do not fit in with the social and cultural reality of the communities where such policies are to be applied.

Educational programs and specific campaigns for adoption of water conservation and saving tools and practices by end-users, and the population in general, are essential for coping with water scarcity in arid zones. However, these programs and campaigns must be well focused on recognized problems and related solutions, and must be in agreement with the local culture and social values, as discussed further in Chapter 11.

When water scarcity is due to ***drought***, water conservation and saving requires some policies and practices that are common with aridity (Tables 9.3 and 9.4) while others are specific to drought risk management, as analysed in Chapter 4. Coping with droughts requires a distinction between preparedness and reactive or mitigation measures, the first essentially consisting in preparing for the application of the mitigation measures during drought.

Among water conservation measures for drought preparedness, the development and effective implementation of drought watch systems plays a main role for risk management: These systems may provide information about the incidence of a drought, its evolution with time, and its ending. That information may produce prediction of the development of drought severity evolution in the short time or, when combined with global atmospheric data, some short term forecast may be feasible. The resulting information is essential for warning users, and for decision and policy makers to timely implement the mitigation measures.

Drought requires that reduced demand be enforced by all users sectors. However, reductions in supply generally affect the irrigators first. To minimize related impacts due to reduced water availability it is required that improved soil water conservation practices be adopted in addition to appropriate irrigation methods and management, as discussed later in Chapter 10.6. Often, it is said that crop patterns must include drought tolerant crops. However, a main issue is to provide information and incentives for farmers that help them to be prepared to face droughts and thus to timely decide on crop patterns that help in coping with drought. Nevertheless, other users also need to adopt reduced demand practices, and also need information and incentives. The development of water technologies and practices to be adopted by the end users that help in temporarily reducing the demand and controlling the water wastes under conditions of diminished water availability is therefore required and should receive high priority. It is too late to search for water saving strategies once a drought has arrived. Drought coping strategies need to be prepared before the onset of drought.

Another aspect requiring attention is the need for planning for the augmentation of available water resources during drought, including wastewater reuse and the use of non conventional water resources. However, as stated above it is important that possible alternative water sources and storages with water in them must be identified before the drought begins, because it is unlikely any more water will become available for capture until the drought ends.

Enforcing preparedness measures, and later, mitigation strategies during a drought, needs an appropriate institutional framework, including widespread for public involvement. Those measures also refer to establishing water pricing and financial incentives and penalties aimed at reducing water consumption and use, and avoiding water wastage and misuse, including the control of water quality degradation by effluents and return flows. Of course, encouraging and enhancing public awareness on the economic, social and environmental value of the water, particularly oriented to produce a favourable attitude in regard to the adoption of drought mitigation measures, is essential.

When a drought occurs, *water conservation for drought mitigation* should be implemented. Then, measures and practices include:

a) Exploring the drought watch system to monitor the drought onset, development and termination, as well as to produce information for decision makers and water users,
b) Implementing changes in reservoir and ground-water management rules,
c) Enforcing drought oriented water allocation and delivery policies, and
d) Adoption of farm water storage and soil water conservation practices, which cannot be effective if only implemented during drought but must be in place before the drought starts.

Water conservation is particularly important in the preparedness for droughts but it must be complemented with water-saving programs, which are essentially reactive. *Water saving for drought mitigation* concerns measures and practices common to those for coping with aridity (Table 9.4), and others that might be specific for drought conditions such as:

a) Adoption of drought tolerant crops and drought oriented cropping patterns,
b) Reduction of the irrigated areas and/or adoption of deficit irrigation practices,
c) Extended use of inferior quality water for irrigation for saving high quality water for more stringent uses,
d) Adoption of water saving tools and practices for reducing domestic, urban, and recreational water uses, including the use of inferior quality water for landscape irrigation,
e) Under extreme conditions, ceasing supply by pipe and implementing tanker delivery. This is a drastic move but in a desperate situation it greatly reduces water demand,
f) Enforcing specific water price policies in relation to the used water volumes, the type of uses, and the efficiency of use,
g) Adopting incentives for reducing water demand and consumption and penalties for excessive water uses, for non authorised uses, as well as for degrading the available waters with low quality effluents and return flows, and
h) Developing campaigns for end-users to adopt drought oriented water saving tools and practices.

Desertification and water shortage (see definitions in Section 2.1.2) are man-made water scarcity regimes, and refer to problems such as land degradation by soil erosion and salinisation, over exploitation of soil and water resources, and water quality degradation. Therefore, they require that water conservation and water saving policies and measures be combined with others oriented to solve the existing problems (Table 9.3). Here the aim is to adopt land and water use planning and management which will facilitate re-establishment the environmental balance of the natural and man-made ecosystems, as well as in the use of natural resources.

Appropriate reservoir and groundwater management rules, including artificial recharge and water quality preservation measures, water allocation policies including criteria relative to water quality, measures and practices for water quality management and focusing on the control of related impacts on the environment, are among the main water conservation measures and policies. In addition, referring to agriculture, other important measures refer to soil and water conservation practices in rainfed and irrigated agriculture, soil and crop management practices for restoring the soil quality, and combating soil and water salinization.

To enforce these policies, it is required to strengthen the related institutional framework, to implement education campaigns aimed at managers, farmers, factory managers and households, on how to save water, and to develop the public awareness for the combating of desertification and water shortage.

Water saving policies and practices also play a fundamental role in combating water scarcity due to desertification and water shortage. As for droughts, several measures and practices are generally common with those for coping with aridity (Table 9.4). Others are specifically oriented to combating desertification and water shortage and include:

a) Changes in crops and crop patterns for reducing the irrigation demand and taking advantage of soil moisture conservation,
b) Adoption of irrigation and drainage practices favouring salinity management and reduced water demand,
c) Control of effluent quality to avoid pollution of surface- and ground-water,
d) Enforcing specific water price policies in relation to water use performances and the quality of effluents to favour adoption of practices that contribute to restoring the environmental balances,
e) Enforcing a policy of incentives and penalties in relation to users adoption of tools and practices that contribute to restoration of the water and soil quality and resource conservation, and
f) Developing campaigns to encourage end-users to adopt water saving and environmentally friendly tools and practices.

Coping with water scarcity requires that measures, practices and policies of water conservation and water saving be effectively applied in relation to the respective causes, as discussed above. Those measures, practices and policies are analysed in the next Chapter in relation to the respective economic and social water user sectors.

9.5 Implementing Efficient Water Use for Water Conservation and Saving

Adopting water conservation and saving is not easy. If it was easy many problems would have been solved. In fact it requires that an efficient water use be implemented

Fig. 9.8 Steps to implement a
plan for efficient water use
(Courtesy of E. Duarte)

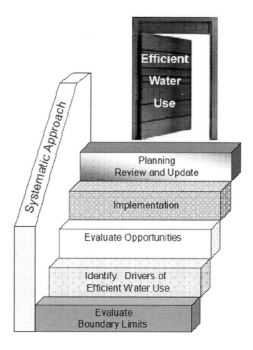

at various time scales and at various sectors and by the multiple users. Decision-makers and stakeholders often have contradictory interests and perspectives. Policy-makers may also be facing conflicts among diverse and opposed interests. Therefore, despite the rationale behind efficient water use, its implementation is difficult. In fact it requires (Fig. 9.8) that:

a) the practical limits and limitations be recognized prior to implementation of policies and measures by policy- and decision-makers, as well as by the users when they wish to adopt new practices. This implies the recognition of the water pathways (Fig. 9.5) in the system under appreciation.

b) When the various boundaries are identified, then it is necessary to recognize the driving forces that may lead to the desired change. Driving forces refer among others to the economic opportunity for adopting new technologies or manage-ment issues, advantages relative to adoption of innovative technologies, existing regulatory compliances that may force to change, or the public perception about the value of water or the need to adopt conservation and saving.

c) The next step is the evaluation of the positive and negative consequences of changes, or no-changes, which leads to the final decision. It is important that the appropriate people are involved in the decision making. The days of "all knowing" decision makers is passing. Everyone involved in water use or affected by the decisions will wish to have some influence on the decisions. If all parties interests are not given a hearing the decisions are unlikely to lead to success.

A dissatisfied user or public will ensure decisions about which they were not consulted will never be fully supported and will operate with difficulty.

d) When a decision is made, then the change in water uses has to be planned, eventually designed when it implies modifications in the processes of production or when it requires adoption of new production technologies, or water technologies such as for treating, reusing or recycling the water. Fundraising is part of this implementation phase.

e) The last step is planning the implementation, monitoring and ongoing revision it when required for successful application.

In this step by step approach there is the opportunity for adopting indicators as such those described above relative to water use and water productivity. However, this process of change to make a water use system more efficient requires the support of extension services and consultants in case of irrigators and small water users, or the support of specialists and consultants when large industry or water supply enterprises are considered. A large number of solutions for water conservation and saving are presented in the preceding section and in the next chapter but it is not possible to provide more than basic information on these topics and to calling attention to the need for their adoption. In many cases the corresponding technologies may not be easily available. However, the adoption of water conservation and saving, despite it being essential to the sustainable development of water scarce regions, results mainly from opportunistic implementation of strategies that clearly provide significant economic benefit.

Chapter 10
Water Conservation and Saving Measures and Practices

Abstract This chapter aims to present a variety of water conservation and saving measures and practices applicable to the diverse water user sectors: urban water supply systems, domestic use, applications to landscape and recreational areas, industry and energy uses, rainfed and irrigated agriculture. Measures and practices are described particularly with respect to applications in local, regional or national action plans for efficient water use.

10.1 Water Conservation and Saving in Urban Systems

10.1.1 General Aspects

The main supply problems faced in urban centres include the over-exploitation and depletion of the water supply sources, their contamination, high costs of operation and maintenance of supply systems, increased competition with other water uses (agriculture, industry, recreation), the maintenance of water quality standards, and the avoidance of water wastes. Cities often face a high incidence of leaks, the use of water-wasteful technologies, very low levels of water reuse, less appropriate water pricing and billing systems, water revenues that often do not cover costs, and a lack of public awareness of water scarcity. Unfortunately, these problems are more acute in water scarcity regions and less developed countries than in temperate climates and developed countries.

Urban water supplies are used in households, industry, and commercial areas, as well as for urban services such as city washing, combating fire, maintenance of recreational lakes and swimming pools and irrigation of recreational areas. Water conservation and water saving are required in all these domains, both by the users themselves and directly by the municipal authorities. Numerous sites provide information on water conservation and saving in all those domains. Literature is also abundant (e.g. Goosen and Shayya 2000, Versteeg and Tolboom 2003).

Municipal water services may be public or private. Despite differences in institutional arrangements, funding sources, water pricing and billing systems, and relationships between water management authorities and their customers, water conservation and water saving should be approached using a common perspective.

L.S. Pereira et al., *Coping with Water Scarcity*, DOI 10.1007/978-1-4020-9579-5_10,   243
© Springer Science+Business Media B.V. 2009

Water conservation and water saving in an urban water system includes water metering, leak detection and repair, water pricing and billing, assessment of service performance, proper use of water in municipal and public areas and services, and development of public awareness on water saving. These items are developed below and the respective main issues, benefits and limitations are summarised in Table 10.1.

10.1.2 Monitoring and Metering

Metering is required at both the supply and distribution systems and at household connections. At the supply level, it concerns monitoring and measuring the water stored, being conveyed, and circulating in the distribution system. The resulting data produces information on the state of the system, and the respective variables. This information is vital for planning system developments and modernisation, for operation, maintenance and management, and for planning and implementation of water conservation and water saving programs.

Metering and monitoring the supply system provides for:

- Updated knowledge on the actual volumes stored and the discharges flowing in the water conveyance systems and in the distribution sectors, as well as water pressure and water levels at key nodes of the networks.
- Information on the supply and distribution network's state variables that allows establishing the real time balance between water availability and demand.
- Assessing the water supply balance in problem areas, namely where large differences in pressure may occur, and to evaluate the pressure distribution and uniformity along the distribution network, thus providing for improved management of the system.
- Data to evaluate the actual hydraulic conditions of the system's operation, including the detection of anomalies and malfunctioning of the networks, pumping stations, and regulation and control equipment.
- Information for planning and implementation of both preventive and corrective maintenance programs for supply lines, distribution networks, electromechanical equipment, water treatment plants, and storage facilities.
- Data for accurate planning of system extensions, developments, and modernisation.
- Estimating the volumes of water not being billed, better identification of the factors causing system operation losses, and base data for upgrading of the rating policies.
- Data to evaluate operation, maintenance and management programs and to support a system of information and control.
- Planning of water conservation measures.
- Establishment of drought preparedness plans and real-time drought mitigation operational and managerial plans.

Table 10.1 Water conservation and water saving in urban water systems

Issue	Benefits	Limitations	Effectiveness
System monitoring and metering	• Information on the system state variables • Estimation of system operational water losses, locations and causes • Base for planning water conservation measures • Support for implementation of water saving	• High investment and ongoing operational costs • Needs an information system to be fully effective	High to very high
Metering	• Provides for fair billing • Allows for water conservation planning • Induces adoption of water saving by customers	• Needs capital investment and solving ongoing operational costs • May require system modernisation in case of old systems	High to very high
Water pricing and billing	• Induces water saving when prices increase with used volumes • Helps people to appreciate the water supply, including giving water a social value	• Requires well designed price structure to be socially acceptable	High
Leak detection and repair	• Reduced volume of non beneficial water use • Induces positive water conservation behaviour of customers • More water is available to satisfy the demand	• Requires appropriate technologies	High
Maintenance	• Prevents equipment and conduit failures • Minimises system water losses • Supports good service	• Requires planning • Requires system monitoring • Requires appropriate technical staff	High
Regulation and control devices	• Improves the hydraulic performance of the network • Helps prevent system failures • Prevents contamination	• Requires appropriate technical staff • High investment costs	High
High service performance	• Makes the customer responsive to water conservation policies • Gives confidence to the customer to adopt water saving practices	• Customer oriented service requires high quality operation and management • Higher costs	Moderate to high

Table 10.1 (continued)

Issue	Benefits	Limitations	Effectiveness
Separated distribution of high quality and reused water	• Conserves good quality water to uses requiring such quality • Implies wastewater treatment • Effective water reuse for non-human and outdoor recreational uses	• Very high investment costs • Higher operation costs than for a single line distributor • Requires the involvement of the users	High to very high
Legislation and regulations	• Create a framework for adopting water conservation and saving • Enforce restriction policies when drought increases scarcity • Favours connections with public water authorities	• Needs an appropriate institutional system for enforcement	High when appropriately enforced
Economic incentives and penalties	• Favour the adoption of water saving devices and practices • Discourage water wastage	• Requires appropriate institutional framework • Needs careful planning	High to very high
Public education and information	• Create public awareness of water conservation • Promote the adoption of water saving practices and devices by customers and the population	• Require well planned and co-ordinated efforts of public authorities, schools, water companies and municipalities	High

Despite advantages, there is a limited use of the metering data when appropriate databases are not installed and used. To fully explore the benefits of metering and monitoring urban water supply systems, it is advisable to make meter data collection, handling and analysis part of an information system, which should also include the production of control, operation, maintenance and management reports. Automation of meters enables the data collected to be transmitted in real time to the information centre, thus to be processed and analysed also in real time.

Metering at the household outlets is required for knowing the users consumption, for billing the customers in accordance with the respective water use, and for the support of measures to be enforced when water availability does not allow the supply to match the current demand.

Metering the water delivered to customers is essential when a water saving price policy is enforced to induce customer water saving. It provides for customers to pay for the used volumes, which is far better than to be billed by any other criterion. Arreguín-Cortés (1994) reports that billing for the volumes of water used induces by itself a reduction in use of up to 25% in areas that previously had no metering. However, the upper and lower socio-economic classes are generally not responsive to metering.

When outlet metering is adopted, it is relatively easy to fix the water price to cover the full costs of operation, maintenance and management, and thus to improve both the network and the water service, and to provide for avoidance of health hazards. Information on metered volumes provides an accurate knowledge of consumer use patterns, which are important for planning system developments and modernisation, as well as water conservation and water saving programmes, particularly to establish drought contingency planning.

Implementing a metering system requires appropriate design and follow-up. Meters with the appropriate capacity should be used. The use of undersized meters leads to false readings and a shorter service life as a result of excessive wear on its components. Conversely, using an oversized meter implies a higher initial investment and less accurate readings of low flow rates.

Although the ideal is to meter 100% of all residential outlets, this is not always possible owing to the installation costs of a metering system. Strategies for the installation of meters when capital investment does not allow for metering all the customers can thus be defined as follows (Arreguín-Cortés 1994):

- Selective metering, where major consumers should be metered first, then continuing the metering process until the target level for the locality is reached.
- Metering by sectors, where metering applies to groups of users having similar consumption patterns, and dividing the cost of the water used among them.
- Combined metering, that is a combination of the two systems above, which is applicable when water use is differentiated. A sector meter can be installed at the upstream end of each sector and individual meters are installed for major consumers. The water use by non-metered users can be estimated by difference.

10.1.3 Maintenance, and Leak Detection and Repair

Maintenance of urban water supply systems plays a major role in water conservation. It may be preventive and reactive. The main purpose of preventive maintenance is to ensure the proper functioning of the water supply system, from the upstream water sources to the customers. Consequently, it includes the network reservoirs and conduits and respective equipment, the pumping stations, the water treatment plants, and the metering system. For the latter, meter readings need to fall within a well-defined range of accuracy, not to over- or under-meter the water use. Each utility and sector should have its own program of maintenance. Computer programs may be helpful in establishing and controlling maintenance programmes.

Reactive maintenance takes place in response to information provided by the field personnel, meter readers and users about system failures, equipment disrepair, and inaccurate meter readings.

System losses in urban drinking water supply systems are mainly due to evaporation and seepage in storage and regulation reservoirs, and leaks in water treatment plants, in distribution networks, and in home outlets. The used volumes not metered due to inaccurate or non-existent metering, the unauthorised outlets and the unrecorded volumes used by municipal services, such as for watering public gardens or used from fire hydrants, are often accounted as "losses", despite the fact that they constitute beneficial uses. Leaks in the system and through the outlets are very high, often exceeding 60% of the total inflow volumes in old and poorly maintained systems.

Leaks in the network are generally very high in systems that breakdown very often. Leaks may be visible or not. Water rising through the soil or pavement provides a visual indication, but it is common to have invisible leaks where the water flows into the drainage system or to an aquifer. The causes of leaks vary, depending on the geotechnical soil characteristics, quality of construction, materials used, particularly in joints and at valve locations, flow pressure and pressure variation, age of the network, water chemistry, and operation and maintenance practices. Leaks can result from crosswise or lengthwise cracking of the conduits, being caused by surface vibrations, poor construction, fatigue and manufacturing flaws of the material, or sudden pressure variations due to fast maneuvering of valves. Other causes include rusting, poor pipe joints or valve failure. In home outlets, faults can be due to fissures, perforations, and cuts or loose fittings. These are due to poor quality materials, poor construction and external loads.

Leak detection and repair is beneficial to reduce system water losses, energy consumption, wear of equipment, operational costs, contamination risks of the network, and negative impacts of the percolating water. Moreover, it provides for improved service performance. Leak detection and repair is also beneficial to better utilise the available resource and to facilitate the promotion of citizen awareness in water conservation and saving.

That activity may be performed easier with the help of models and geo-referenced data bases. Descriptive analysis models provide an initial understanding of the system, including the analysis of records of operations, failures, and repairs undertaken, so making it possible to estimate the percentage of operation losses in the network and in home outlets in the different sectors of the system. Predictive models are based on the knowledge of the age of the network, the materials used, rate of corrosion, most frequent kinds of failures and respective repair to predict the behaviour of the network and its operational losses. Models play an important role when planning for droughts.

Physical analysis includes fieldwork to detect leaks in the distribution network, home outlets and any other part of the system. Several leak detection methods exist including: (a) the acoustic method, based on the fact that leaks under pressure produce sounds that can be detected by using appropriate sensors, amplifiers and headphones; (b) the so-called Swiss method, consisting of injecting water under pressure into a reach or a sector of the network and measuring the amount of water required to keep the pressure constant, where the resulting volume equals the amount of water leaked; (c) the correlation analyser method, which is based on recording the noise produced by a leak using sensors placed up- and downstream of it and then processing the two signals by a correlation analyser to give the respective distances to the leak location; and (d) the tracing method, where tracers are injected upstream and then the leak site is detected with support of chemical or radioactive tools. Enforcing field leak detection programmes as part of a drought contingency planning is required for water conservation.

10.1.4 Water Pricing

Historically, in most countries, water costs have largely been or still are subsidised. In others, water is supplied free. It is of importance to establish rate policies that emphasise greater user involvement in water conservation and saving. When users are charged appropriately for water services and these perform well, the water use as well as the water waste tends to decrease.

Water pricing can help to save water if the price structure meets some essential conditions:

- Prices must reflect at least the actual costs of supply and delivery to the customers to ensure the sustainability of the water supply services and the maintenance of conduits and equipment;
- The price rate should increase when the water use increases to induce customers to adopt water saving and conservation;
- Different price rates should be practiced for diverse types of water use in municipal supply, e.g. differentiating domestic indoor uses from gardening and other outdoor water uses; when water is more scarce than usual, prices for less essential uses could be modified accordingly;
- Differential increases in price must be large enough to encourage water saving;
- Prices must reflect the quality of service, i.e. poor and non reliable service cannot be provided at high cost but costs must change as soon as service is improved; and
- Any change in pricing must be accompanied by information and education programs that encourage increased customers awareness of the value of water and water supply services.

Water benefits are different for each type of user. In urban areas there are several types of water use: domestic, public, industrial, tourism, commercial, services, construction, and recreational. Each of these categories reacts differently to the same financial spur in charging for the service, and differentiated price policies may be adopted for them. In general price should be related to consumption. There is often a desire to impose fixed standing charges in order to guarantee income to the water provider, but these generally do not assist in the conservation of water, and in fact they may encourage wasteful use.

Users must be billed correctly and be informed of the amount of their consumption. Increases and adjustments in prices must be clearly linked with the costs of water services. If they are not, prices may become subject to external influences (e.g. the need to raise tax revenue) and fail to be accepted by the customers. In such a case the customers may react negatively to water saving programmes. Therefore, the political authorities and the users must be clearly informed of why changes are needed, through information campaigns that target the different categories and social sectors, explaining the facts simply and honestly.

10.1.5 Regulation and Control Equipment, and Service Performance

Important progress in design and equipment for urban water supply systems has been made during the last decades. However, investment costs for control equipment are high limiting the adoption of developments. This is especially the case in less developed areas, where funds are limited but population growth is often very high. Particular attention must be paid to the adoption of appropriate regulation and control devices in the network because this equipment plays a fundamental role for the quality of service, and therefore in the implementation of water conservation and water saving programmes. This equipment mainly relates to:

- Support appropriate hydraulic functioning of the network, including the avoidance of sudden variations in pressure due to the manoeuvre of gates and pumping station controls in relation to variations in demand, which could damage conduits and equipment,
- Prevent system failures,
- Prevent the contamination of the network,
- Provide for flexibility in operation,
- Maintain adequate service pressure, mainly when the network covers areas with less favourable topography, with large variations in elevation.

The adoption of advanced regulation and control equipment requires appropriate technical staff, and investment and operation costs are high. Thus it is difficult for small networks in rural and remote areas to incorporate such devices.

A water supply company may have good water metering, pricing and billing systems but may fail in inducing water saving from the customers when the service provided is poor. The quality of service may be evaluated using indicators such as reliability and equity.

A water service is reliable when there is a very high probability that the discharge and pressure delivered to a customer corresponds to those contracted between the water company and the customer. The service is not reliable when there are frequent breaks in the delivery, or when only a small discharge and reduced pressure are available. When the service is not reliable, the user tends to consider that the water price is always excessive and not enough water is made available to him and there is no need to save water. When frequent breaks occur, the user may be forced to create his own reserves of water (household reservoirs), which he may waste as soon as supply is restored.

Problems may be aggravated when there is no equity in the service. Equity occurs when all customers receive a similar service in terms of discharge and pressure irrespective of the location of the user, the volume of water used, the social class and the type of water use (e.g. domestic, commerce, services). When equity does not exist and this is recognised by the poorly served customers, these react negatively to the water price and generally are unwilling to adhere to a water saving program.

As discussed earlier, when a monitoring and metering program exists, information becomes available that allows recognition of the quality of service in terms of reliability, and the existence of locations or type of users that are critical concerning equity of service. When problems that affect them are solved, then the users behaviour may become more favourably oriented towards saving water, which in turn may have positive impacts on the water service.

10.1.6 Dual Distribution Networks for High Quality and for Treated Reusable Water

Diverse water uses in urban areas require different water quality. High quality water is definitely required for uses such as drinking, food preparation, or bathing. However the largest fraction of this water is not consumed but returned as effluent with degraded quality and is not reusable for the same purposes. On the other hand, uses such as toilet flushing, heating, floor washing, or irrigation of lawns and gardens do not need such high quality water and could use treated wastewater.

Originating from these different requirements, in urban areas where extreme water scarcity exists, a feasible but expensive solution is to duplicate the distribution network, mainly in the neighbourhoods or sectors where water users can manage with inferior quality water for uses other than human consumption (Okun 2000, Aoki et al. 2006). At the limit, different sewage systems may be built, separating the less contaminated and less charged effluents from the more degraded ones. The two effluents may require different treatments and may attain different quality levels, and therefore may have different uses. Urban drainage rainwater may also be treated and added to the higher quality treated urban effluents.

This solution has several advantages under the water conservation perspective:

- Urban, namely domestic effluents are treated and thus added to the available resource,
- The scarce high quality water is used only for more exigent uses, and
- Uses demanding inferior quality water may more easily be satisfied through treated wastewater and urban drainage rainwater.

Despite the usefulness of this approach, this is difficult to implement because it requires very large public investments, as well as important investments by the customers in their homes or buildings, and it is necessary to create the public awareness required for a wide acceptability of such a new solution. Financial incentives such as subsidies for investment and deductibility on taxes may then be required.

Adopting roofwater harvesting systems for less stringent uses is nowadays spreading not only in water scarce areas such as in India, western USA and Australia, but also in water abundant countries, e.g. in UK, Germany, New Zealand

and Canada, where several dealers propose these individual systems. Roofwater is to be used for household applications requiring less water quality; high quality water is supplied by the municipal systems. A large number of websites provide related information, including for design. A design manual is authored by Thomas and Martinson (2007).

10.1.7 Legislation and Regulations, Incentives and Penalties

Water laws give each country its general framework for water use and conservation, and are complemented by regulations, which establish the practical application of the legal water policies.

Regulations to improve water saving are often of a restrictive nature. In most cases, they are established for long-term application, but they may be applicable only during periods of limited water availability such as droughts. In that case they generally require very strict surveillance and should only be applied when they are really necessary.

Long term regulations concern questions such as the characteristics and standards for indoor plumbing fixtures, or maximum volumes per flush in toilets (e.g. 6 l/flush), or maximum discharge in showers (e.g. 10 l/min). Regulations also include requirements to replace old type toilets, compulsory use of equipment for filtering, treating and recycling the water in swimming pools, etc. Temporary regulations may concern the prohibition of use of hoses to wash motor vehicles or sidewalks. These water saving issues are analysed in Section 10.2.

Incentives and penalties of a financial nature are required as a complement of the water saving campaigns, and as the prime measure to enforce regulations.

Incentives are often required for urban water supply companies, municipal or private, mainly in less developed areas, to implement high cost water conservation technologies, which may require an expensive investment. Examples could be the full coverage of every customer with metering, the modernisation of the network in an area of low income population where reliability and equity are low, or the investments required to install dual supply systems.

Incentives to urban customers are required when the implementation of water saving implies investments in homes and buildings, in particular to low-income populations. Examples are the installation of modified supply and sewerage systems when separate supply and sewerage networks become available in the neighbourhood, the replacement of toilets for reduced flush volumes, or the adoption of pressure control devices. Incentives could then be given by the government, the municipality or the water supply company. Incentives by the government to the water company and the customers generally relate to:

- Credit facilities relative to a specific investment for water conservation and saving,
- Lowering or exempting custom taxes for selected goods or equipment,
- Subsidising a percentage of the investment costs, which could vary with the income level of the customers, and

- Fiscal benefits for the water company or the customers in relation to selected investments.

Penalties essentially consist in discouraging water users from adopting practices and devices opposed to regulations. Penalties and fees are the easiest methods to enforce regulations but may not be the best approach. While incentives are given to those seeking them and proving the corresponding rights, penalties require control by water company personnel, municipal services or the police. Enforcing penalties and fees involves additional costs in personnel and in bureaucracy, including that relative to the Court when the citizen refuses to pay. Therefore, the implementation of a penalties and fees policy needs appropriate planning to ensure that it can be effectively applied and be useful to the community.

10.1.8 Information and Education

Water saving programs need the users participation to be successful. Furthermore, information and education campaigns are required. The same requirements exist for successful implementation of legislation and regulations or to achieve water saving objectives in relation to the adoption of water metering and water pricing.

Information to the users may include leaflets, including sent out with bills, publicity campaigns in the press, radio and television, billboards on streets (Fig. 10.1) and public transport vehicles. Special campaigns for water saving may also include the free distribution of water-saving devices, assistance with the cost of investments required to renovate or upgrade household water systems, or allow tax deductions for specific water saving investments.

Education mainly concerns the introduction in primary and secondary school curricula of the essential aspects of the hydrological cycle and water sources, which are dealt in Chapter 12. It should also cover limitations in water use, causes and other considerations related to water scarcity, the costs to collect and supply water, the main water uses and benefits, and, finally, how to properly use water both indoors and outdoors.

Fig. 10.1 A wall painted advertisement in Cape Verde, written in Creole language: "Here we don't waste our richness"

10.2 Water Saving in Domestic Applications

Domestic water use can be classified under indoor and outdoor uses. In a typical household having outdoor facilities, nearly 50% of the water use is for the garden (Stanger 2000).

Water uses in commercial and services buildings are considered herein together with the domestic ones since they are similar, they are very often located in the same neighbourhood or distribution sector, and water saving measures are common. Also common are water uses for hotels and restaurants. A summary of water saving in domestic uses is given in Table 10.2.

10.2.1 Indoor Water Uses and Saving

Main indoor water uses refer to bathing, toilet flushing and laundry (Stanger 2000). However, the percentage distribution of indoor water uses varies from country to country or region to region according to local habits.

The main attention must be given to the following types of water uses:

a) *Toilets*: Traditional toilets use 16–20 l/flush, which means an average of 80–100 l/day/capita. Leaks from the toilet reservoir are common. Water saving in toilets may be achieved by using several means such as:

  1) Toilets with a reduced reservoir, such as 6 l/flush,
  2) Toilets with double discharge, one producing larger flushes, the other a smaller flush,
  3) Placing containers or bags filled with water in the tank of traditional large volume toilets to reduce its capacity; however, the toilet's siphoning effect may be reduced.
  4) Changing to longer discharge siphons in traditional toilets, which reduces water flushes,
  5) Toilets having flushes controlled by the pressure in the mains or a pressure regulation device, so that volumes of water discharged may be near 6 l/flush.
  6) Special toilets such as with pressurised tanks having the feed line connected to a hermetically sealed tank, or biological toilets or incinerators, which degrade faecal matter in deposits located below and turn it into fertiliser.
  7) Toilets where leaks are easy to detect and repair, or toilets less susceptible to leaks such as pressurised tanks without the conventional fittings, and models that substitute siphons for new design fittings.

b) *Showers and bathing*: Water saving include:

  1) Using shower-head designs that limit the discharge to not exceed fixed values such as 6–10 l/min,
  2) Using flow or pressure limiting devices to reduce shower discharges to target values,

Table 10.2 Water saving in domestic water uses

Water uses	Techniques	Applicability
Indoors		
Toilets	• Use reduced flush • Use leak proof fixtures	• Easy, affordable costs
Showers	• Use reduced discharges • Shortening shower duration	• Easy but needs willingness
Basin taps	• Reduce pressure and discharge • Use valves with automatic closing • Keep closed when not directly in use • Dish wash in a basin, not in flowing water	• Easy but needs care
Cloth washers	• Use only when full load • Use washers with less water requirement	• Easy
Dish washers	• Use only when full • Use washers with less water requirement	• Easy
Outdoors		
Gardens and lawns	• Irrigate to refill the top 15 cm of soil • Adopt information or tools to decide the irrigation frequency • Select an irrigation method that helps to adopt reduced and localised watering, such as drippers and microsprayers • Irrigate by night, out of the high demand hours and when evaporation is minimal • Under restrictions reduce irrigation frequency, reduce volumes to annuals and lawns but keep irrigating trees and shrubs • Use mulches to avoid soil evaporation • Control weeds • Avoid plants from humid climates • Select native and xerophyte plants	• Costs are affordable • Needs public awareness • Requires information
Side-walks and car washing	• Washing with a hose is water wasteful • Dry clean side-walks • Use recycling in car wash services	• Needs public awareness
Swimming pools	• Use water purification tools • Cover for evaporation control when not being used	• Affordable costs
General issues		
Leak control	• Apply good quality materials and construction techniques • Detect and repair	• Expensive but effective
Pressure control	• Use pressure regulators to avoid excessive discharges	• When pressure is excessive

   3) Reducing the time for showering,
   4) Preferring shower to immersion bathing,

c) *Basin taps*: several water saving measures may be applied to water use from taps in bath rooms and kitchens:

   1) Flow reduction by using aerators, which add air to the stream and disperse it, increasing the spread and the washing efficiency, reducing flow as much as 6%,
   2) Flow reduction by using pressure control devices when pressure is too high,
   3) In hand washing basins, use of valves with sensors that activate the water flow only when the hands are placed beneath them, or use of valves for limited flow time in hand washing basins, caring about keeping taps closed when not being required, such as keeping them closed while soaping the hands and opening again to wash, or wash teeth using a glass of water, not flowing water,
   4) Use of tap valves easy to close and re-open that make it easier to have water flowing only when it is necessary,
   5) Dish washing in a basin, not with flowing water,

d) *Washing machines*: Water saving in these machines can be achieved by:

   1) Loading the washers with the appropriate weight of clothes, and using the water levels necessary for efficient operation; some modern machines may require less water when half charged,
   2) Using washers that require less water; front loading washers can use up to half the amount of water (up to 50% of the hot water and 33% of the detergent) required by tub or top loading washers; modern washing machines have reduced water use, by up to 25% in comparison with traditional models,
   3) Following instructions of the manufacturer on selecting the programmes since they may have different water and energy demands.

e) *Dish washers*: Water saving concern:

   1) Selection of the machines that require less volume of water to perform the same operations; some modern machines use about one third of the water compared to older models.
   2) Load the machines to full capacity; modern dishwashers have somewhat reduced demand when half charged,
   3) Follow instructions of the manufacturer on selecting the programmes since they may have different water and energy demands.

f) *Control of indoor leaks*: Leaks in pipes, plumbing fixtures, showers, taps, and toilets are very frequent and may be difficult to detect. One way to solve the problem is to use leak-proof fittings and other materials whose specifications correspond to pressures to be used as well as to the temperature of the water. Hot water pipes and plumbing often leak. After installation of the pipes, a test must be performed to check joints and valves before walls are closed. Safety valves must be installed for easy repair operations.

10.2.2 *Water Saving in Outdoor Applications*

Outdoor watering accounts, on average, for 32% of the total water used by households in USA. In summer the average household outdoor water use can go as high as 70%. Aiming at water saving, a group of irrigation industries and water supply agencies created *Smart Water Application Technology* (SWAT), which provides information and technology to designers and household users for outdoor water saving. A variety of websites produce related information, e.g., the *Irrigation Association* (http://www.irrigation.org/smartwater/). Also a large number of websites from many countries give information on turf and landscape irrigation management, equipment and controllers.

The main water saving opportunities are:

a) *Lawns and gardens*: Water saving practices summarized under Section 10.3 (a) can be easily adopted for house gardens and lawns. These concern irrigation methods and scheduling, and irrigation practices when restrictions in water supply are enforced.

    The conversion from turfgrass to a xeriscape (low water-use landscape) in house gardens has been studied in Nevada, USA, where typically 60–90% of potable water drawn by households is used for outdoor irrigation. A xeriscape is characterised by sound landscape planning and design, limited use of turf, use of water-efficient plants (usually those native to the region), efficient irrigation, soil amendments and mulches, and proper landscape maintenance procedures. The study, concerning 499 properties, shows that the adoption of xeriscape reduced the water use for outdoor irrigation by 33% relative to a turf landscape, that proportion being 39% during the summer months (e.g., Sovocool and Rosales 2001). This may represent an enormous saving of good quality municipal water for areas of hot, dry extreme climates. Further information on xeric landscapes and water-wise landscapes is available through a large number of websites.

b) *Sidewalks*: should be cleaned dry. To wash stone or concrete sidewalks with a flexible hose is wasteful of water.

c) *Swimming pools*:

  1) Water used to fill a swimming pool can be continuously recycled, clarified and purified by using easy to handle portable equipment and appropriate chemicals. It may then be renovated after one year or so,

  2) Leakage may be prevented by using appropriate construction materials and techniques. When empty it is advisable to check the state of the sides and bottom and perform any appropriate repairs,

  3) Evaporation losses can be minimised by covering the pool when not in use.

d) *Car washing*:

  1) Washing cars with a hose is water wasteful. It is recommended that just a bucket and cloth be used. For washing, cars should be stood on lawns or other surfaces that could benefit from the surplus water.

  2) Car wash services should recycle or reuse water.

Most of the fixtures discussed above, both indoors and outdoors increase their rate of discharge in direct proportion to the water pressure. In areas where pressure is high, the use of pressure-reduction valves is recommended.

10.3 Water Conservation and Saving in Landscape and Recreational Uses

Landscape and recreational water uses include a variety of outdoor and indoor uses and can be managed by municipal, governmental or private institutions, but all have in common the fact that they are for public use. Landscape irrigation, including golf courses represents more than 20% of irrigation water use in the United States. Golf course irrigation is increasing enormously worldwide, including in arid regions. New manuals and equipment were recently made available to encourage water saving landscape design and irrigation management. The above mentioned SWAT provides relevant information for users and designers. Advances have been made on best management practices for turf and landscape irrigation, as well as for estimating plant water requirements (Fig. 10.2) and irrigation scheduling (Irrigation Association 2005a,b, Allen et al. 2007a). Numerous websites provide related information worldwide.

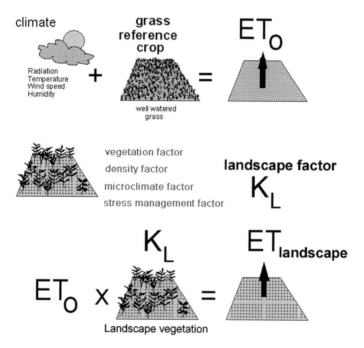

Fig. 10.2 New approach to estimate landscape evapotranspiration ($ET_{landscape}$) from reference evapotranspiration ET_o and landscape factors bulked into a landscape factor K_L

Table 10.3 Issues in water conservation and saving in recreational water uses

Water use	Issues	Applicability
Gardens and lawns	• Selection of ornamental plants less sensitive to water stress • Adopt water reuse • Adopt well designed and automated drip and micro-spray irrigation systems • Establish irrigation management strategies to cope with increased levels of water scarcity • Irrigate by night • Use soil mulch • Control of weeds	Requires technological support and training of personnel to be effective
Pools and ponds	• Use water purification tools when not used to support life, such as fish • When having fish, use a technique similar to fish ponds, or flowing water for other uses, e.g. as irrigation tanks	Requires planning and technical support to personnel
Golf courses	• Adopt an irrigation design oriented for easy management under scarcity • Reuse treated wastewater for irrigation • Establish irrigation management strategies to cope with increased levels of water scarcity	Requires new approaches in design and management, and training of personnel
Parks and Lakes	• Enforce water quality policies • Adopt integrated resource management	Needs involvement of public authorities
Sport areas	• Sprinkler irrigation of green areas with high uniformity • Adopt precise irrigation	Requires technical support
Public swimming-pools	• Use purification tools and chemicals • Enforce health prevention measures	Easy
Indoor facilities	• Adopt controlled flush toilets • Adopt limited flow showers • Use time controlled hand-wash taps • Use controlled discharge kitchen taps • Care for leak detection and repair	For most of the issues, they follow common sense practices
General issues	• Publicise water saving measures • Advertise to make water users aware • Educate children and youth	Favours public behaviour

Water conservation and water saving issues are summarised in Table 10.3. Many of the irrigation aspects are dealt in Sections 10.6.4 and 10.6.5 for sprinkler and microirrigation, which apply to landscape irrigation, and in Section 10.6.6 relative to irrigation scheduling.

a) *Gardens and lawns*: Water saving in garden and lawn irrigation can be achieved through several practices:

1) Adopting appropriate garden watering practices, particularly avoiding excess water application. The amount of irrigation water that should be applied varies with the climate. Because most garden plants are shallow rooted, the recommended depth of soil moistening during the watering period is only 15 cm. Irrigation should stop when ponding or runoff occurs, which points out that water applied is in excess of infiltration.

2) Using irrigation equipment that helps reduce irrigation water use, preferably micro-irrigation: localised drip irrigation for trees, shrubs and flowers in a row; microsprayers for lawns and flowers in a wider space; bubblers or drippers for small basins. Drip lines may have drippers at different distances according to plant spacing, and may be laid on the surface or buried to follow the layout of the plants. Microsprayers may have a variety of spray heads to irrigate in circles, half circles and corners with different angles; they also have a variety of discharge, wetted diameters, wetted patterns and drop sizes, including mist, which can be adapted to every kind of garden plants. Innovative sprinklers designed for landscape irrigation are available with a variety of sprinkler heads that allow adapting to part circle wetting and for various throw angles.

3) Using simple information for scheduling irrigation: plant appearance, soil touch and feel, broadcast information on evapotranspiration, or instruments for indicating the soil water depletion in the upper 15 cm soil layer. These instruments, although not very precise, are very helpful to city water consumers.

4) Using modern automated irrigation controllers, which may incorporate various sensors and may even cut or delay irrigation when rainfall is detected (the so called SMART sensors developed with the already mentioned program SWAT). Adopting automation makes it easier to adopt the water saving irrigation practices referred above, and allows for irrigating by night, when wind drift and evaporation losses are minimal, and water demand to the supply system is also reduced, so minimising impacts on the functioning of the supply system. The best time to irrigate is between 4 and 8 a.m. or, alternatively between 8 p.m. and midnight.

5) When strong restrictions on water supply are enforced for outdoor irrigation, the following strategies may be adopted, in order of increasingly stringent restrictions:

   • To decrease the frequency of irrigation but apply the same amount of water at each irrigation; to decrease both the frequency and the irrigation volumes;
   • To abandon the irrigation of annual plants, so maintaining the irrigation of perennials only;
   • To abandon the irrigation of lawns but watering trees and shrubs to keep them alive when they are difficult to replace.

6) Using mulch for soil water conservation, particularly to control evaporation. Various types of mulch can be used in gardens: organic mulch made from

Fig. 10.3 Mulch of tree residuals for the area with flower bushes in a city park

wood residuals (Fig. 10.3), stone mulch, namely of gravel type or riverbed stones, coarse sand, and straw mulch. Other products for mulch may be available in gardening shops, or be prepared in public gardens with available materials.

7) Adopting weed control is also important because weeds compete with garden plants for water and nutrients.

8) Selecting garden plants native to the region or well adapted to the prevailing environmental conditions, including low water demand, is a potential condition for successfully coping with water scarcity. On the contrary, choosing ornamental shrubs and trees from temperate and humid climates creates additional water requirements and greater management difficulties since these plants may be very sensitive to water stress. An alternative to native plants, or in combination with them, are the ornamental xerophytes, such as cactus, which are extremely efficient in water use.

9) In water scarce areas, it is advisable to consider the use of water of inferior quality for irrigation, including treated wastewater. Microirrigation systems (drippers and bubblers) are the most adequate when reused wastewater is applied in gardens. Sprinklers are used in golf courses but water is treated to a third or, desirably to a fourth level. Filtering should then be carefully designed. The direct contact with public shall be avoided.

10) Careful layout and pipe network design as well as appropriate selection of emitters and filters are required. A good design and good quality of materials avoid uneven pressure distribution, non-uniform discharges, and frequent system failures. It makes automation easy and favours the implementation

of water saving strategies. Poor design and low quality equipment cause irrigation management to be difficult and expensive, leads to waste of water and for plant stress to occur frequently.

b) *Pools and ponds*: two situations can be considered:

1) When pools do not support life, i.e. no fish are present, the best water saving practice is to avoid the need for frequent replacement of the water. This can be achieved by applying a treatment technique for water clarification to combat water eutrophication.

2) When pools and ponds support life, such clarification treatment is not viable because chemicals may affect the fish. If a prime objective is to have fish, then technologies similar to those for fish-ponds may be adapted. Life and clear water may be achieved if pools and ponds are used as irrigation tanks with permanent flowing water. However, it may be necessary to reduce the stored volumes in proportion to the volumes used in irrigation to avoid eutrophication, i.e. to keep the water moving. To treat the water against eutrophication may be expensive and not justified when heavy restrictions are applied. Ponds with standing water have to be avoided for environmental and heath protection since they constitute an attractive base location for vectors of water borne diseases.

c) *Golf courses*: large volumes of water are required to fully irrigated golf courses. In general good design and equipment are used since this is a high-income generation activity. Water saving can be attained when:

1) Sprinkler or, less often, micro-sprinkler irrigation systems are utilised. Irrigation is better performed at night, so evaporation and wind drift losses are minimised. Different requirements in the various parts of the area may be satisfied when the system is designed to allow for different irrigation management in each of the distinct grassed areas that constitute a golf course since they are covered with different grasses and respective aspect and use are different.

2) Grass may be selected, at least for certain parts of the golf course, from species that have less water requirements and are less sensitive to water stress.

3) Under-irrigation problems may be avoided when treated wastewater is used but high standards of treatment need to be met. Favourable conditions for its use exist because these irrigation systems are automated, night irrigation is the rule, and the direct contact of humans with the irrigation water is minimal. However, monitoring is advisable to prevent any negative health impacts. Filters are required for safe operation of the irrigation system depending on the characteristics of the water.

4) When limited water is available, the following deficit irrigation strategies may be adopted as supply restrictions increase:

- Abandon irrigation outside of the fairway areas,
- Reduce irrigation in the fairway area but keeping the greens fully irrigated,
- Under-irrigate all areas except the greens near the holes,
- Under-irrigate to just keep the grass alive (to play under these extreme conditions may be not possible)

d) *Parks and Lakes.* Water uses concern the maintenance of ponds, the irrigation of gardens and lawns, and the irrigation of plant nurseries and greenhouses. Issues relative to the first two have been described above. For the latter, approaches outlined below for irrigated agriculture may be adopted. Lakes are usually at low levels or dry when high water scarcity occurs and, often, recreational lakes are exploited for adding to the limited available resource. The main issues concern water quality preservation since the concentration of pollutants increase when flows supplying those lakes decrease. Therefore, the focus of attention should be on the enforcement of water quality policies in the upstream areas, aiming at limiting point-source pollution and at expanding the treatment of effluents to be returned to nature. Control of water borne disease vectors is also essential.

e) *Sport areas and swimming pools.* They consist of indoor facilities, which are discussed below, under item (f), and of outdoor support areas. In the latter, water consumption mainly concerns the irrigation of green areas, washing of stadium seats and floors and parking lots:

1) Irrigation of green areas: Sprinkler irrigation with portable pipe laterals or hand pull hose systems are often practised. To keep fields in good conditions for sports practice, grass water stress cannot be allowed. To avoid health hazards, using treated wastewater is not viable. Then water saving practices concern:

- Adoption of an optimal irrigation schedule, which can easily be performed if information on evapotranspiration is made available,
- Use of a well designed system having high uniformity of application, and adopting proper equipment and maintenance,
- Irrigating at the early hours of the day, when wind and evaporation are small, and pressure available from the supply system has reduced variability.

2) Washing: Saving water in cleaning activities is mainly possible by replacing washing by cleaning dry, using cleaning machines in large areas, and avoiding hose washing, which is always wasteful of water.

3) Swimming-pools: water saving consist of:

- Using purification equipment and chemicals that avoid frequent refills of the pools,
- Adopting health preventive measures for users,
- Caring about leak detection and repair.

f) *Indoor facilities*: water uses in indoor public facilities mainly concern the toilet area in public buildings and in support of outdoor activities, the shower area in sports facilities, food areas in support of recreational activities, and washing of floors and special rooms. Corresponding issues for water saving are common to those analysed for domestic uses (Section 10.2.1). In addition to general recommendations there, it is opportune to mention:

1) Toilets: adopt toilets with double flushing or toilets with flushing controlled by sensors.
2) Showers: use controlled discharges and pressure.
3) Basin taps in bath rooms: adopt valves with sensors that activate the water flow only when the hands are placed beneath the taps, or valves for limited flow time.
4) Basin taps in kitchens: use flow reduction devices, easy manoeuvring taps, and care to keep taps closed when not being required.
5) Dish washers and clothes washers: recommendations under Section 10.2.1 generally apply.
6) Control of indoor leaks: This includes:

   - Care to detect and repair of leaks in pipes, plumbing fixtures and toilets,
   - Using leak-proof fittings and other materials with specifications which correspond to pressures to be used as well as to the temperature of the water,
   - Adopting good construction techniques and materials, as well as testing at the installation phase.

10.4 Water Conservation and Saving in Industrial and Energy Uses

Water conservation in industry and energy production mainly refers to the managerial measures and user practices that aim at preserving the water resources, essentially by combating the degradation of its quality. Water saving should be adopted in every managerial measure or user practice that aims at limiting or controlling the water demand and use for any specific process, including the avoidance of water wastes and the misuse of water.

In general, industrial processes and related support services have a large demand for water but, very often, the non-consumed fraction is very high and is returned to nature or to sewage systems. However, industrial demand is much smaller than the agricultural demand, particularly in arid and semi-arid regions. This industrial demand tends to be of the same order of magnitude as the domestic demand except in tourist areas. On the other hand, in developed and industrialised regions, it largely exceeds the domestic demand. The great importance of the industrial water uses also result from the fact that they occur in populated areas, are in direct

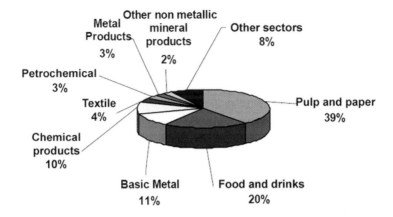

Fig. 10.4 Relative importance of water use in various industries (courtesy by Elisabeth Duarte)

competition with domestic uses, and very often are supplied by the same urban networks.

The volumes and quality of water required varies with the type of industry and process (Fig. 10.4). Moreover, the same industry has different requirements for the quality of water supplied, and produces effluents with different qualities, which often require different treatment processes.

Water recycling and reuse are the main water saving processes in industry (Fig. 10.5). Recycling consists in utilising water in the same process as it was originally used. The physical and chemical properties of the water may change

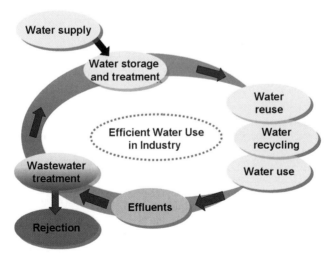

Fig. 10.5 Efficient water use in industry implies water reuse and recycling, and a smaller amount of rejected but treated effluents

after being used, requiring some simple treatment before being used again. Water reuse consists of using treated water in a different process if it cannot be utilised again for the same process. Wastewater treatment is the main conservation issue. It must be designed according to the quality required for reuse and recycling and effluents rejected have also to respect standard quality requirements. Reuse and recycling require that specific water pathways be adopted in relation to the quality requirements of industrial processes, and the quality of effluents.

Water for energy production is generally not consumed, as for hydropower, though for evaporative cooling larger consumption is involved. However, in the latter case the disposed water may have higher temperature than required for some uses, including for nature in river ecosystems.

Opportunities for water uses can be grouped in various water use activities as summarised in Table 10.4. The following aspects are further considered:

a) *Temperature control*: Water is used for both heating and cooling. The former usually involves the generation of steam in boilers that burn coal, oil, gas or waste products. Cooling processes employ the circulation of cold water through

Table 10.4 Issues in water conservation and water saving in industry

Water uses	Issues
Industry	
Temperature control	• More efficient recovery • Recycling • Reuse in other activities
Manufacturing processes	• Reduced demand for water • Adjust water quality to process requirements • Replace water use processes by mechanical ones • Recycling • Reuse after treatment
Washing	• Use cleaning machines • Avoid hose washing • Water reuse
Indoor water uses	• Devices to control discharges • Care in use
Outdoor water uses	• Water reuse • Gardens with plants not sensitive to water stress • Adopt micro-irrigation and water saving irrigation methods • Use cleaning machines and avoid hose washing
General issues	• Water metering • Water quality monitoring • Leak detection and repair • Water auditing to find and support efficient water use and innovation
Energy generation	
Hydropower	• Integrated basin planning
Thermoelectric and nuclear	• As for temperature control

cooling towers, pipes or pools. Water conservation and saving concerns, among other things, the following issues:

1) Heating processes are generally in closed circuits and do not require high quality water, except it needs low calcium, but mainly that the water does not produce detrimental residuals. The amount of residuals admissible depends upon the frequency and processes applied to clean the boiler. Treated water, inclusive of previous industrial uses, may be used. For steam production the recovery of water by condensation may be expensive but may be justified when water is scarce and could be reused.

2) Cooling requires water that has low temperature and is non corrosive or is not likely to attack the equipment or materials with which it is in contact. After cooling, the temperature of the water is elevated but can be recovered by open air-cooling and aeration, or be directly reused such as for greenhouse heating. Evaporative cooling towers consume quite large amounts of water. Evaporation losses are generally higher when cooling is practised in pools.

b) *Manufacturing processes*: industrial processes requiring water are very diverse. Among the more important, water conservation and saving opportunities are:

1) Incorporation in the product, as is the case for drinking and food industries: extremely good quality water is required. The non-consumed fraction is small and may be added to reusable effluents.

2) Water used in physical extraction processes, such as in the vegetable oil industry, must be of high quality. However, effluents usually have too much organic inclusions but can often be reused for irrigation after treatment.

3) Effluents from sugar cane mills should not be rejected to water courses but reused for irrigation.

4) When chemical extraction processes are used, the water quality requirements are related to the chemicals utilised, the type of raw material, and the type of manufactured product. Generally, water savings relate to changes in the processes and water conservation to the treatment of effluents, as is the case for the paper industry.

5) Textile dyeing requires huge amounts of water and reuse is limited due to difficulties in removing the dye. Water conservation in this industry depends upon further developments in research relative to that specific water treatment problem, such as is being pursued in China.

6) Washing raw food materials requires high quality water, which may be reused for cooling, heating or indoor washing if the size of the industry is large enough to reuse the water. Otherwise, effluents may be treated for external reuse.

7) Washing raw non-food materials generally do not require high quality water such as in the mining industry. Recycling may be adopted when effluents mostly carry materials in suspension, which could be easily filtered. In this case, sedimentation ponds may be used for final treatment.

8) Water used for transportation of materials, in pipes or channels, requires variable water quality according to the nature of material to be transported, but almost always demands large volumes of water. Water demand is reduced when mechanical transport is adopted, when recycled water could be used and when both the material transported and the water are needed at the downstream end.

9) Water is also used for cooling equipment such as in wood and stone cutting and polishing. Water quality requirements relate then to not having materials in suspension nor corrosive chemical components. Generally water saving depends on advances in the equipment itself. Effluents often require extraction of sediments produced during the processes of cutting and polishing the materials.

10) In the construction industry water uses include incorporation in the product such as in concrete and other masses, cooling of stone and mosaic cutting equipment, and earth compaction. Non-saline water should be used. Water saving mainly concerns the control of leaks from tanks and pipes utilised, and care in handling them to avoid unnecessary wastes.

c) *Power generation*:

1) Water energy can be directly used to generate electricity by running turbines, so the water used is not consumed. However, water storage is often required, which may create a competition with other uses when the stored water could be used by other sectors. This problem can only be solved under a water conservation perspective when water resources are planned and integrated management is adopted at the basin scale, which needs to include allocation of the resource under competitive demands and multiple uses of stored water from reservoirs.

2) When reservoirs are required for energy production, losses by evaporation occur, as well as deep percolation. The latter is mainly controlled at construction. Evaporation losses are extremely difficult to control since the application of chemicals on the water surface may affect aeration of the water and is often detrimental for other uses, such as fish production, human consumption and recreational and environmental uses, which are common in these reservoirs.

3) Water is used in thermoelectric and nuclear plants for cooling. A large fraction of the water used is consumed because it vaporises, while another fraction is recovered by condensation and returned to nature, preferably after lowering its temperature. It could also be reused for heating in a different activity, such as greenhouse heating.

d) *Washing*. A large variety of industrial residuals and other materials accumulate every day on the floors of industrial installations and buildings, as well as dust on the industrial equipment. Cleaning by washing is common. Water saving may be achieved in several ways:

1) Using dry cleaning machines when residuals and the floor are mostly non wet.
2) Using water cleaning machines, which require less water and detergent than water cleaning with a hose.
3) Separating heavily polluted washing effluents, such as with oils, from less polluted effluents. The first may not be reused while the latter may be reused in washing the floor.
4) Dry clean outside areas such as sidewalks and parking lots.

e) *Water use in toilets, hand washing basins, showers, kitchens, and other services*: as for other indoor water uses in Sections 10.2.1 and 10.3.
f) *Outdoor water uses*: as for other outdoor water uses in Sections 10.2.2 and 10.3.

Recycling, reuse, and demand reduction in industry needs appropriate planning (Duarte et al. 2005), often based in water use audits (Fig. 10.6). Inclusion of demand minimisation at original construction of facilities always costs less than retrofitting. To develop and implement such a plan requires:

• Appropriate knowledge of the industrial processes and respective operational fluxes,
• Understanding and characterising the fluxes of residuals, mainly those of effluents, including those from supporting activities such as bathing, kitchens and others,
• Measuring the volumes of water used both in the industrial processes and in other indoor and outdoor activities, and assessment of the quantities and quality of effluents produced,
• Identification of procedures that lead to minimising the effluents produced,
• Identification of processes where recycled water may be used,
• Identification of processes and activities where treated water may be used,
• Revision of the technologies adopted and modernisation of processes used,
• Selecting technologies for recycling and for water treatment,
• Performing an economic evaluation of alternatives to select the one to be adopted.

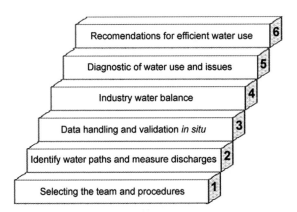

Fig. 10.6 Phases of water auditing in industry aimed at adopting efficient water use practices (from Elisabeth Duarte)

Water metering is essential for implementing and managing a water saving program in industry. Metering is required to determine the water used and consumed in each of the main processes and globally, adopting a time step for evaluation from the hour to the month according to the type and size of the industry. Metering helps to evaluate how efficient are the processes adopted and for the water saving program itself, how and where to introduce improvements, and how to more easily involve and motivate the employees in saving water.

Monitoring the water quality is required to evaluate the efficiency of recycling and reuse treatments, and to evaluate the introduction of any modification in the respective processes that could be necessary. Monitoring is also necessary for the treatment of effluents that are to be rejected, which should respect target standards of quality, as is compulsory when the "polluter pays" principle is applied for rejection of inferior quality water.

As for other urban water uses, leak detection and repair assumes an important role in industry, particularly when large water uses are required, and recycling, reuse and water treatment are practised.

10.5 Water Conservation in Dryland Agriculture

10.5.1 Introductory Concepts

Dryland agriculture, also called dry farming or rainfed agriculture, is crop production without irrigation. Crop water requirements are satisfied by the rainfall occurring during the crop season. However, distinction should be made between rainfed agriculture in areas where rainfall is abundant, in humid and temperate climates, and crop production in water scarce regions where rainfall is low and erratic in quantity and temporal distribution and evapotranspiration largely exceeds precipitation during much of the year. The terms dryland agriculture and dry farming apply to the latter.

Dryland agriculture is traditionally practised in water scarce areas when rainfall during the rainy season is sufficient for the crop to develop and produce. Yields are commonly lower than for irrigated agriculture and vary in a large range from one year to the other, following the trends in the temporal variation of precipitation. The use of water conservation practices is usually required for successful crop production under water scarcity conditions.

Water conservation in dryland agriculture may assume quite different facets and include the selection of crops that are less affected by rain water deficits, the adoption of cropping practices that favour the ability of crops to escape water stress, and soil management practices that help to conserve water in the soil for crop use. A review has been produced by Unger and Howell (1999) and various case studies are described by Liniger and Critchley (2007). In semi-arid and arid conditions and when drought occurs the water deficit periods during the crop season may be quite long, and supplemental irrigation and water harvesting become part of dry farming practices.

Typically, the term water conservation in dryland agriculture is used for the technologies which aim to improve the amount of water infiltrated and stored in the soil. However it is used here to include other technologies that improve the water productivity of the crops, i.e. the yield quantity produced per unit quantity of water as discussed in Section 9.3.2.

10.5.2 Crop Resistance to Water Stress and Water Use Efficiency

In dryland agriculture, crops and crop varieties are selected taking into consideration their tolerance to the water stress conditions that characterise the environments where they are cultivated. In general, these crops correspond to centuries of domestication of plants native to these environments, but new varieties have been introduced in the last decades following scientific plant breeding and improvement programmes. The most common food crops are wheat, barley and millet among cereals, and beans, cowpea and chickpea as legumes (pulses), as well as mustard and sunflower.

The mechanisms of crop resistance to water stress are numerous and vary according to the nature of physiological and morphological processes and characteristics involved. These resistance mechanisms refer to three main groups: drought escape, drought avoidance and drought tolerance.

"Drought escape" refers to plants that develop very fast and have a quite short biological cycle, or that have morphological characteristics that favour reduced impacts of insufficient water availability. These are characteristics of herbaceous plants and shrubs native to arid areas.

"Drought avoidance" consists of plant characteristics that provide for reduced transpiration and improved conditions to extract water from the soil. However, reduced transpiration affects photosynthesis and yields are reduced. Better capabilities to extract water from the soil are associated with more developed and deeper root systems, that may extend to several metres in steppe shrubs. Some plants also have higher water potential adjustment favouring the water fluxes from the soil to the roots and then to the leaves. Several of these traits are currently used in plant improvement programs, including for pastoral areas.

"Drought tolerance" consists of plant characteristics that make them able to cope with water deficits. These include physiological adjustments for maintaining photosynthesis despite reduced transpiration, or controlling the senescence and abscission of leaves, as is the case for many shrubs in arid areas. These traits are useful in plant breeding.

Despite progress in plant breeding research, results in creating crop varieties resistant to water stress are still limited. On the one hand, the number of genes and characteristics with potential for altering the plant responses to water stress is too large, thus making genetic manipulation difficult and only slowly responsive. This is contrary to the success of plant improvement programs focusing on resistances to given diseases or salinity, at least for some species. On the other hand, benefits of plant responses to water stress are commonly contradictory to crop yield responses, i.e. a water stress resistant variety may be able to yield regularly with less water use than other common varieties of the same crop but often yields less than a more sensitive variety. This behaviour creates additional difficulties for plant breeding and improvement programs since the yield potential is very important. When varieties resistant to water deficit are less productive than more common varieties, it becomes quite difficult to get them adopted by farmers.

Crop responses to water stress are influenced by other environmental factors such as wind and temperature, as well as by the cropping conditions (Pessarakli 1999, Wilkinson 2000). Therefore, field tests to examine the resistance of crop varieties to water deficits are very demanding in both the setting up and evaluation of the experiments, and the tests are generally difficult to conduct.

Research is also oriented to evaluate and select plants and crop varieties having high water use efficiency (WUE), i.e. a favourable ratio between the harvestable yield and the amount of water consumed. Experiments often combine the assessment of WUE performance with the evaluation of improved crop management practices such as fertilizing, soil management, planting dates and water conservation. Yield responses to supplemental irrigation are also considered when irrigation water may be available during the critical periods for crop water stress.

For small grains, data in Fig. 10.7 show that barley reacts better than wheat to water conservation programmes but differences are not extreme. However, the advantages for barley increase when salinity stress is also considered. Differently, when comparing the water use performance of maize, sorghum and millet, the advantages for millet and, to a lesser extent, for sorghum are evident (Fig. 10.7). Moreover, it can be observed that the most common crops traditionally used by farmers in the semi-arid and arid zones of the Middle East and North Africa are among the best performing crops in Fig. 10.7.

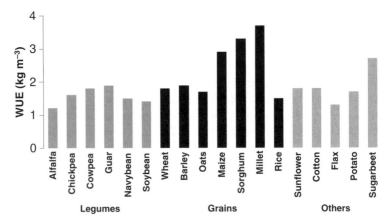

Fig. 10.7 Indicative values of water use efficiency (WUE) for selected crops (adapted from Tarjuelo and de Juan 1999)

Reducing water consumption by crops cannot be an objective of water saving or water conservation because this would often mean a reduction in plant transpiration and, therefore, a reduction in photosynthesis and biomass production, which is approximately proportional to transpiration. Reducing crop consumption is therefore only acceptable under drought conditions as a way to minimise crop production losses.

Instead of attempting to reduce the water consumption of a crop, water conservation and water saving programmes consider the selection of crops, crop varieties and crop patterns that consume less water than others. In particular, this may be achieved by looking for those crops and crop patterns that lead to higher water productivity, and to the selection of crop and irrigation practices that provide for reducing the water use, i.e., that provide for better control of the non-beneficial water uses and the non-consumed fraction of water use as reported in Sections 9.2 and 9.3. These aspects are discussed below.

10.5.3 Crop Management for Coping with Water Scarcity

Water conservation in dryland agriculture mainly refers to crop management techniques and to soil management practices (Edwards et al. 1990, FAO 2000, Liniger and Critchley 2007). Techniques for crop management to cope with water scarcity are summarised in Table 10.5. They relate to three main approaches: crop risk management, controlling the effects of water stress, and water conservation cropping techniques.

Techniques to manage crop risk concern crop management techniques designed to minimise the risks of crop failure and to increase the chances for beneficial crop yield using the available rainfall. The selection of crop patterns, taking into consideration the seasonal rainfall availability and variability and the water productivity of the crops and crop varieties, is a major issue. It is important to select crops which are least subject to water stress since stress affects crop development and yields, and is an important determinant of how drought conditions affect yields. That selection may involve the adoption of water stress resistant crop varieties instead of highly productive but more sensitive ones, and the use of short cycle crops or crop varieties, thus having smaller crop water requirements, particularly when there is the possibility of a drought. It may also be important to adapt planting dates such as to plant after the onset of the rainy season to ensure more effective conditions for crop establishment, or to use early seeding to avoid terminal stress of the crop. When the degradation of the stressed crop is expected, then early cutting of forage crops or grazing of drought damaged field crops can be practiced to avoid total crop loss and minimize cattle feeding problems under drought. If water is available, it is wise to adopt supplemental irrigation of dryland crops at critical crop growth stages to avoid loosing the crop yield when a drought occurs. This can be a highly effective technique when water may not be sufficient to adequately irrigate a dry season crop.

Table 10.5 Techniques for crop management to cope with water scarcity

Crop management techniques	Benefits	Effectiveness
Crop risk management		
• Selection of crop patterns taking into consideration the available season rainfall	• Lesser water stress effects, including under drought	• High
• Choice of water stress resistant crop varieties when drought and high water deficiency are likely	• Lesser impacts on yields	• High (under drought)
• Use of short cycle crops/varieties	• Lesser crop water requirements	• High
• Planting after the onset of the rainy season	• Ensure crop establishment	• High
• Early seeding	• Avoidance of terminal stress	• Variable
• Early cutting of forage crops	• Avoiding larger stress impacts	• Limited
• Grazing drought damaged fields	• Alternative use for livestock when grain yield is lost	• Limited to drought only
• Supplemental irrigation of dryland crops at critical crop growth stages	• Avoid losing crop yield when a drought occurs	• High
Controlling effects of water stress		
• Use of appropriate soil management techniques	• Increase available soil water	• High
• Adaptation of crop rotations to the environmental constraints, namely to include covered fallow for grazing	• Coping with water stressed environment	• High
• Include a tilled bare soil fallow in the crop rotation	• Increase in soil moisture	• Variable
• Use of mixed cropping and inter-cropping, namely for legumes	• Better use of resources	• Variable
• Adopt large plant spacing of perennials (tree crops), large row spacing for annual crops, and low density seeding for cereals	• Higher exploitable soil volume by tree or plant, increased available soil water	• High
• Reduced fertiliser rates	• Adaptation to reduced yield potential	• Variable
Water conservation cropping techniques		
• Conservation tillage	• Control of soil evaporation losses	• High
• Adequate seed placement	• Prevention of rapid soil drying around the seed	• High
• Seed placement in the furrow bottom and planting in soil depressions	• Place root systems where soil moisture is better conserved	• High
• Pre-emergence weed control	• Alleviating competition for water, avoiding herbicide effects on crop	• Variable
• Early defoliation	• Decrease crop transpiration surfaces and use leaves for animal feeding	• Limited
• Windbreaks	• Decrease wind impacts on evaporation,	• Variable
• Anti-transpirants and reflectants	• Reduction of plant transpiration	• Costly
• Growth regulators	• Control physiological processes in relation to water availability	• Costly and difficult
• CO_2 enrichment (controlled environments)	• Increased yields per unit of water consumed	• Limited and costly

Controlling the effects of water stress may be achieved by adopting techniques and practices that reduce the impacts of water deficits on crop development and yield. In addition to soil management techniques that provide for increasing the available soil water, which are discussed in Section 10.5.4, these include a variety of practices, mainly traditional ones and commonly practiced in the Mediterranean region and Australia. It is the case of including in crop rotations several years with covered fallow for grazing, or the inclusion of a bare tilled soil fallow period in the crop rotation in order to control soil water evaporation losses and provide for enough soil moisture at the next crop planting. Also adopted worldwide, is the use of mixed cropping and inter-cropping with legumes (pulses) to better use the available soil and water resources.

The adoption of large plant spacing for perennial crops, mainly tree crops, as is widely practiced for olives in North Africa and the Middle East, provides for a large soil volume to be explored by the crop roots in low rainfall areas. Similarly, using large row spacing for annual crops and low density seeding for cereals, also aims at increasing the soil volume exploitable by each plant.

Water conservation cropping techniques are designed to increase the available soil water, to control soil water losses by evaporation, to minimise the transpiration by weeds and their competition for water, as well as to reduce crop transpiration when water stress is extreme. Conservation tillage is highly effective for increasing soil infiltration and controlling soil evaporation losses as discussed in the next Section. Other techniques refer to:

1) *Adequate seed placement* to prevent rapid drying of the soil around the seed. In arid climates, seed placement in the furrow bottom and planting in soil pits provides for the development of the root systems where infiltration is higher and soil moisture may be better conserved.
2) Appropriate *weed control* to alleviate competition for water and transpiration losses by weeds. Adopting pre-emergence treatments avoids herbicide effects on stressed crop plants as may happen with later treatments.
3) *Delayed fertilisation* to favour root development *and splitting applications* to improve fertilisers use, helps to control their residual accumulation in the soil and reduces their leaching and transport to the groundwater when excess rainfall infiltrates.
4) *Early defoliation* to decrease the leaf area contributing to transpiration, which is a traditional practice in water scarce areas, especially where leaves can then be used as fodder for livestock.
5) Using *windbreaks* that, in addition to controlling wind stress on the crops, decrease evaporation, particularly when advective conditions prevail.
6) Using *anti-transpirants* to reduce plant transpiration, *reflectants* to decrease the energy available for transpiration by increasing the fraction of the incoming solar energy that is reflected by the crop canopy, *and growth regulators* to adjust the plant physiological processes to the water availability. However, these techniques only apply to farms having appropriate technological capabilities, and in general crop yield is reduced proportionally to transpiration reduction.

10.5.4 Soil Management for Water Conservation

Soil management practices for water conservation refer to tillage and land-forming practices that favour rainfall infiltration into the soil, water storage in the soil zone explored by roots, capture of runoff to infiltrate the soil, control of evaporation losses from the soil and weeds, extraction of water by plant roots, and crop emergence and development (Unger 1994).

These practices have long been known to have positive impacts on water conservation in dryland farming. However, results of any soil management technology depend upon the soil physical and chemical characteristics, the land-forms and geomorphology, the climate and the kind of implements used. All these factors interact, creating variable responses in terms of crop yields. In addition, economic results have to be considered. When a technique is to be introduced in a given environment and it is substantially different from the traditional and well-proved practices adopted by local farmers, it is advisable to perform appropriate testing before it is widely adopted. However, the principles of soil management for water conservation are of general application, regardless of the size of the farm, the traction used, or the farming conditions.

Soil management practices for water conservation, are often common to the practices for soil conservation, i.e. they not only provide for augmenting the soil moisture availability for plant growth but they also contribute to the control of erosion and soil chemical degradation. In many cases they contribute to improving the soil quality. Because these practices produce changes in soil infiltration rates and amounts, soil water storage and runoff volumes, they may produce relatively important changes in the hydrologic balance at the local field scale and, when widely adopted, they may affect the hydrologic balance at the basin scale. The soil management practices for water conservation are summarised in Table 10.6.

Runoff control and improved water retention on the soil surface refers to several techniques aimed to provide for a higher amount of rainwater to infiltrate into the soil and a larger time opportunity for the infiltration to occur. These effects are produced by creating a higher roughness of the cropped land where slopes are flat, but do not apply to sloping landscapes, which are discussed later. Practices, many of them traditionally used in dry farming, include:

1) *Soil surface tillage*, which concern shallow cultivation tillage practices to produce an increased roughness on the soil surface permitting short time storage in small depressions of the rainfall in excess to the infiltration, i.e. limiting overland-flow to give a larger time opportunity for infiltration. This practice is effective for slightly sloping land cropped with small grains in soils having stable aggregates. In addition, clods at the soil surface decrease evaporation of soil water as discussed below.

2) *Contour tillage*, where soil cultivation is made along the land contour and the soil is left with small furrows and ridges, prevents runoff formation and creates conditions for the water to be stored at the surface until infiltration can be completed. This technique is also effective in erosion control and may be applied to

Table 10.6 Soil management techniques for water conservation in agriculture

Soil management techniques	Benefits	Effectiveness
Water retention on the soil surface and runoff control		
• Soil surface tillage for increased surface roughness	• Ponding of rainfall excess in depressions, • Larger time opportunity for infiltration	High (flat lands)
• Tillage for contour (and graded) furrows and ridges	• Runoff and erosion control, storage in furrows, increased time for infiltration	High (low slopes)
• Residues and crop mulching	• Runoff retardation and higher infiltration	High (low slopes)
• Furrow dikes	• Rain water storage in furrow basins/pits and increased infiltration amounts	High
• Bed surface cultivation	• Runoff control and increased infiltration	Variable
Increasing soil infiltration rates		
• Organic matter	• Improved soil aggregation and infiltration rates	High/very high
• Conservation tillage	• Preserve soil aggregates and infiltration	High/very high
• Mulches, crop residues	• Soil surface protection, better aggregates and higher infiltration rates	High/very high
• Traffic control	• Less soil compaction and improved water penetration in the cropped area	Variable
• Chemicals for aggregates	• Favours soil aggregates and infiltration	Medium/high
Increasing the soil water storage capacity		
• Loosening tillage	• Increased soil porosity, soil water transmission and retention	High
• Subsoiling to open natural or plough made hardpans	• Improved soil water transmission and storage, and increased soil root zone depth	High/very high
• Deep tillage/profile modification in clay horizons	• Increased water penetration and soil depth exploitable by roots	High but costly
• Chemical/physical treatments of salt-affected soils	• Increased infiltration and available soil water	High/very high
• Hydrophilic chemicals to sandy/coarse soils	• Increase water retention in the soil profile	Economic limits
• Mixing fine and coarse horizons	• Increase water transmission and retention	Economic limits
• Asphalt barriers in sandy soils	• Decrease deep percolation	Limited, costly
• Compacting sandy soils	• Lower deep percolation, higher retention	Variable
• Control of acidity by liming, and of salinity by gypsum	• More intensive and deep rooting, and improvement of aggregation	High
Control of soil evaporation		
• Crop residues and mulching	• Decrease energy available on soil surface for evaporation	Very high
• Shallow tillage	• Control soil water fluxes to soil surface	High
• Chemical surfactants	• Decrease capillary rise	Economic limits

Table 10.6 (continued)

Soil management techniques	Benefits	Effectiveness
Runoff control in sloping areas		
• Terracing, contour ridges and strip cropping	• Reduced runoff, increased infiltration, and improved soil water storage	High/very high
Water harvesting (arid lands)		
• Micro water-harvesting	• Maximise rainfall infiltration, plant scale	High/very high
• Micro-watersheds	• Maximise runoff collection at tree scale	High/very high
• Runoff farming	• Maximise runoff collection at field scale	High/very high
• Water spreading	• Maximise the use of flood runoff through diversion for infiltration in cropped fields	High/very high

row crops and small grains provided that field slopes are small. When rain could create waterlogging graded furrow tillage across the slope may be more helpful than contour tillage.

3) Using *mulches* from crop residues, or straw applied on the soil surface, or stones from deep soil loosening, increases surface roughness that slows the overland-flow and improves soil infiltration conditions; in addition, mulches control soil evaporation.

4) *Furrow diking* to permit the storage of rain water in small basins or pits created along the furrows until infiltration is completed. It is useful in sprinkler irrigated land to control runoff and store the excess applied water until infiltration can occur (Fig. 10.8). This practice is highly effective for sprinklers on moving laterals that often apply water at rates higher than the soil infiltration rate. However the effectiveness of this method depends upon the slope of the land.

5) *Bed surface profile*, which concerns cultivation on wide beds and is typically used for horticultural row crops. Often beds are permanent and traffic is only practiced in the furrows between beds, which should follow the land contours. The soil aggregation and infiltration are kept undisturbed on the bed and water is captured, stored and infiltrated into the furrows

The improvement of soil infiltration rates may be achieved through a variety of practices that aim at increasing water penetration into the soil and maintaining high rates of infiltration. This limits the rain water that may run off or that evaporates when stored at the soil surface. These practices refer to:

1) Increasing or maintaining the amount of *organic matter* in the upper soil layers because this provides for better soil aggregation, which is responsible for the macropores in the soil. Organic matter preserves soil aggregates, thus avoiding crusting or sealing at the soil surface, which closes soil pores and significantly

Fig. 10.8 Furrow diking for controlling runoff and favouring infiltration in sprinkler irrigation

decreases the infiltration rates. Maintaining the continuity of macropores through which water penetrates and redistributes within the soil profile is essential.

2) *Conservation tillage*, including no-tillage (also called direct seeding) and reduced tillage, where residuals of the previous crop are kept on the soil at planting (Fig. 10.9). As referred above, mulches protect the soil from direct impact of

Fig. 10.9 No tillage mulching in soybeans cultivation in Brazil, where direct seeding is adopted on several million hectares, mainly rainfed. Note the good soil cover by mulch (courtesy by Reimar Carlesso)

rain drops, thus controlling crusting and sealing processes resulting from dispersion of aggregates by the rain drop impacts. Conservation tillage also helps to maintain high levels of organic matter in the soil. The soil is also less disturbed by tillage operations which for silty soils could affect soil aggregates. Therefore, conservation tillage, which is now practised world-wide, is highly effective in improving soil infiltration and controlling erosion in dryland farming.

3) Application of *mulches* in tree and shrub fruit crops, or use of weed residuals when orchards are not tilled, also improves soil infiltration similarly to the processes described above. The effectiveness of this technique is particularly relevant for tropical soils where organic matter is rapidly mineralised and rains may be very intense.

4) *Traffic control*, i.e. adopting permanent paths for tractors and other equipment, limits soil compaction to these zones, resulting in improved water penetration in the remainder of the cropped area.

5) Application of *chemical additives* to the soils for strengthening the soil aggregates contributes to avoid soil sealing, and to preserve soil porosity and water pathways in the soil. These additives may be highly effective for improving infiltration conditions in soils with less stable aggregates.

Increasing the soil water storage capacity by improving the soil water-holding characteristics, increasing the depth of the soil root zone, or favouring soil water conditions for water extraction by plant roots. Several practices may be considered:

1) *Loosening tillage*, which is applied to naturally compacted soils such as heavy silt soils, or to soils compacted by frequent tillage operations and traffic of tractors and equipment. It provides for increased soil porosity, thus enhancing conditions for water transmission and retention in the soil. Conditions for root development are also improved. This technique has variable effectiveness and often has to be repeated quite frequently.

2) *Subsoiling* for destroying natural or plough made hardpans, which may significantly improve conditions for water movement downwards and, consequently increase soil water storage. It also provides improved pathways for the roots to develop into deeper layers, thus enlarging the soil depth exploitable by crop roots. Overall, this practice may appreciably increase the amount of water stored in the soil which then becomes available for crop use.

3) *Deep tillage* or soil profile modification when clay horizons overlay more coarse soil horizons. This mixing or even inverting of relative positions of layers contributes to increasing infiltration, deepening of crop roots, improving the availability of soil water for crop use, and provides for larger water storage volumes.

4) In cold regions, loosening the soil by refilling the soil water storage prior to the soil freezing season. Water changes in volume when freezing and melting occurs, which increases soil porosity and as a result, infiltration and water retention in the soil profile. This is commonly practiced in North China silty soils.

5) Chemical and physical *treatments of salt-affected soils*, i.e. treating saline and sodic soils to reduce salt concentration and toxicity to plants. This treatment also provides for reducing the osmotic potential with consequent improvement in the availability of soil water for crops use. It also modifies the physical conditions for water penetration and movement in the soil, thus increasing infiltration, and favours the development of crop roots into deeper layers. Overall, the amount of water stored in the soil usable by crops can be greatly enhanced.

6) Adding fine materials or hydrophilic chemicals to sandy/coarse soils slows the water transmission downwards, increases water retention, controls deep percolation and therefore increases the availability of water in soils with natural low water holding capacity. Similarly, asphalt barriers in sandy soils may be used to decrease deep percolation, but their use is more costly and application is difficult.

7) Compacting coarse textured soils may help reduce infiltration and deep percolation but results are variable and uncertain.

8) Control of acidity by liming, similarly to gypsum application to soils with high pH. This treatment favours more intensive and deep rooting, better crop development and contributes to improved soil aggregation, thus producing some increase in soil water availability.

An example comparing various soil management techniques for soil management in rainfed maize cultivated in a silty soil in Daxing, North China, is given in Fig. 10.10. Results indicate the superiority of mulching in terms of soil water conservation impacts on yields. It can be seen that sub-soiling is the second most effective treatment. Changing from flat soil as traditionally practiced in the area to tied ridges (a type of furrow-bed system) did not produce improvements (Pereira et al. 2003a).

The control of soil evaporation may be achieved in different ways:

1) *Mulching* (Fig. 10.9) with crop residues, straw, and several other materials including plastic and stones, is aimed at decreasing the amount of energy available at the soil surface for soil water evaporation. Mulching, which shades

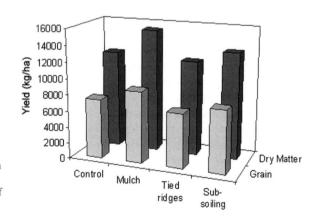

Fig. 10.10 Summer maize dry matter and grain yields (kg ha-1) relative to various soil management practices in a silty soil at Daxing, North China (averages of 3 years of experiments)

the soil, also contributes to control of weeds and therefore of non-beneficial water use.

2) *Plastic mulching*, as used for horticultural field crops to speed up crop emergence and the first stages of crop development, also provides for control of evaporation from the soil. This is particularly true for non transparent plastics that decrease net radiation available at the soil surface.

3) *Shallow tillage*, also called dust mulching or soil mulching, is a practice consisting of tilling the soil surface, only to a shallow depth, to create a discontinuity between the surface, where energy is available, and the deeper soil layers, where water is retained and from where it would move upwards by capillarity if continuous pathways existed. Shallow tillage is a very common traditional practice in orchards and for the bare fallow period antecedent to crop planting. Besides its effects in directly controlling soil evaporation, it also provides mechanical control of weeds, and therefore of their transpiration losses. Shallow tillage is highly effective when the high evaporation season is the dry season, i.e. when tillage can be performed before the high evaporation season. In addition, it is effective in controlling wind erosion due to creating soil roughness.

4) *Chemical surfactants* that limit upwards capillary fluxes are also stated to have potential to limit soil water evaporation. However, there is not enough evidence on the effectiveness of this technique.

Runoff control in sloping areas, which correspond to well known erosion control measures, also provide for water conservation in sloping landscapes. Most of them are associated with modifications of the land forms, reducing both the slope lengths and the slope angles. Generally they consist of:

1) *Terracing*, which may assume different forms as terraces are designed for both low slope lands cultivated by non-intensive field crops, or for medium to steep slopes, usually adopted for small farms. The first may be designed for infiltration of excess water in the downstream part of the cultivated area, as is common in low rainfall regions, or to drain off that excess water that may be stored in small reservoirs for use by cattle. The second may be designed to naturally drain excess water in the direction of the land slope, or may be designed to infiltrate most of excess water and then have horizontal or near horizontal surfaces. In every case, terraces slow down the overland flow and increase the amount of rainwater that can infiltrate the soil. In low rainfall areas only a small part of the storm rain flows out as runoff. Therefore, the amount of water that infiltrates the soil is significantly increased. Terraces also build landscapes as in the example of Fig. 10.11, showing a hillside in the Port wine area, nowadays part of the Unesco World Heritage, an area of low rainfall and high evaporative demand, where vines are cropped rainfed due to terracing water conservation.

2) *Contour ridges*, which correspond to a simplification of large terraces on low slope landscapes. Here the natural slope is divided by ridges along the contour lines at spacings which vary inversely with the slope angle. Ridges constitute

Fig. 10.11 Terracing
landscape in the Port wine
producing area, Portugal

obstacles to the runoff, that may be drained out by surface drains located immediately upstream of the contour ridges, or that may infiltrate from shallow ditches
also located near the ridges. Alternatively, the land slope may be divided by stone
walls that decrease the overland flow velocity and favour infiltration. As a result,
whatever the solution that is adopted, runoff is controlled and the fraction of
rainfall that infiltrates is increased.

3) *Strip cropping*, which is a technique for retardation of runoff and enhancing
   infiltration in low sloping areas. The land slope is divided into strips where different crops are cultivated using the contour tillage practice referred to above.
   Because different crops are used in successive strips they create different conditions for runoff and infiltration (particularly resistance to overland flow),
   that overall decreases the fraction of rainfall lost to the crops as
   runoff.

Water harvesting (see Section 6.4), which are techniques used in arid lands to
maximise the fraction of rainfall that is used for crops (Critchley and Siegert 1991,
FAO 1994, Prinz 1996, Oweis et al. 2004). Water harvesting has many different
forms that may be grouped as:

1) *Micro water-harvesting*, referring to a technique where planting is performed
   on the bottom of wide spaced furrows, so that the large surfaces between
   the ridges act as rainfall collection areas to maximise infiltration near the
   plants.
2) *Micro-watersheds*, where a fraction of the available area is used to collect the
   rainfall, which flows downslope to infiltrate into the cropped area. The ratio
   between the collection area and the cropped area generally decreases as the
   rainfall increases.
3) *Runoff farming*, corresponding to techniques that maximise runoff from small
   areas, which is then stored in small reservoirs for supplemental irrigation or
   infiltrated in the cropped area.

4) *Water spreading*, also known as spate irrigation, which consists of diverting flood runoff to the cropped fields where it infiltrates.

The analysis above shows that there are a large variety of traditional and modern soil management practices for water conservation in dryland agriculture. Some of these practices are specific to given environments but others are of more general use, particularly mulching, which may be associated with conservation tillage, and organic matter incorporation. Many of these practices are for dryland farming only but some may also apply to irrigated agriculture. This is the case for the techniques that enhance soil water storage, that improve infiltration conditions, or that help in controlling soil evaporation.

The analysis that follows, which refers to irrigated agriculture does not specifically mention most of the soil management practices. However, their consideration must be included when fully assessing strategies to improve water conservation and saving in water scarce agricultural ecosystems.

10.6 Water Saving and Conservation in Irrigated Agriculture

10.6.1 The Overwhelming Importance of Saving Water in Irrigation

Irrigation is the main water user in water scarce regions as discussed in Section 2.2. Demand for irrigation largely exceeds that of all other user sectors which makes water conservation and saving in irrigation to be of vital importance. This justifies the dominant place given in this chapter to demand management and control in irrigation. Demand management in irrigation, or water conservation and saving, refers essentially to two domains: (1) appropriate selection and use of the irrigation systems and (2) the adoption of irrigation scheduling strategies that both allow the satisfaction of the crop water requirements when minimising water use and at the same time maximising crop profit under the constraints of water scarcity.

There is a common idea that has unfortunately spread everywhere – that surface irrigation is highly inefficient and water consuming and thus should be replaced by modern methods such as sprinkler and drip irrigation. Proponents have ignored some facts: surface irrigation has supported sustainable farming in water scarce regions for centuries or millennia and users developed local management and engineering skills that adapted the irrigation technologies to solve the dominant problems. There is evidence of this in many areas around the world, from the plains of China, the Indian sub-continent or Mesopotamia to the mountains of the Andes. In many regions, mainly in developed countries, surface irrigation has been replaced by sprinkling or drip due to economic and managerial reasons, mainly to meet the high costs of manpower. In other regions, where fields are large enough and farmer advisory services are effective, surface irrigation was modernized, mechanized and

automated. An example is the western USA. However, these technological developments have not spread everywhere because, with a few exceptions, the market does
not offer this modern equipment and technology and the related advisory and maintenance services are lacking. However, when they are occasionally made available,
farmers easily adopt them as has been observed in several irrigated areas around the
world.

Another common idea is that drip irrigation is the only irrigation system that
provides for effective water saving. Unfortunately, this is not true. For drip irrigation to produce water saving and conservation it is necessary that the irrigation
systems are well designed and operated, equipment is selected that is appropriate for
the crops, the soil, water quality and environmental conditions, and that equipment
maintenance is adequate. Drip systems are highly expensive and can only be adopted
when there are capital resources and crop yields are good and of high value. These
conditions occur in developed countries were farmers have access to information, to
credit facilities and to advisory services, often paid for by the farmers themselves.
Unfortunately drip systems are being sold by shops and people without qualifica-
tions, farmers never receive adequate instruction, such as that a filter is required or it
must be cleaned. In many of these circumstances farmers use more water with drippers than they would with surface irrigation. Drip irrigation is often inappropriately
used in sandy soils, where drip bulbs are narrow and water easy infiltrates below the
roots, producing rapid contamination of the groundwater while keeping the crops
stressed and low yielding.

Irrigation scheduling is another area of concern. It is well known that crop
responses to irrigation water depend upon the timeliness of water applications.
Farmers have always known this and developed appropriate skills to irrigate their
common crops within the constraints of their soils and climate as well as of the irrigation systems they use. Scientific knowledge develops new tools to estimate crop
water requirements and to develop upgraded calendars for irrigation to avoid crop
water stress at various phenological stages and to maximize crop response to water
application. However, the use of these new tools requires updated knowledge. Advisory services play an important role. Their adoption is easy when farming systems
and crop yield values are able to cover the costs of equipment and advising. However
they can only be effective when farmers are in full control of timings and amounts of
irrigation. This is therefore possible for large farms having their own water sources
or for farms supplied by pressurized on-demand networks. Small farms with supply
from surface canal systems that get water according to the managerial rules of the
canal systems have no access to these developments.

Pressures to reduce irrigation demand include those for adopting deficit irrigation, i.e., adopting under-irrigation in such a way that yield decreases due to water
deficits would not be economically detrimental. However, deficit irrigation requires
upgraded irrigation scheduling, thus knowledge, tools, and absolute freedom to
decide when and how much to irrigate. This applies well to large farms where water
use may be optimised, and where the farmers are in control of the water supplies
and manage information well. It is virtually impossible to apply deficit irrigation to
small farms supplied by canal systems. Under these conditions, demand for water

may be reduced but negative impacts on productivity and the farm gross margins may be excessive.

This brief analysis shows that water saving and conservation in irrigated agriculture faces two different worlds: one is that of large and commercial farms, oriented for the market and having access to the market of irrigation equipment (for both the irrigation system and irrigation scheduling), to credit facilities, and to irrigation advisory services; the other is that of small farms, not in control of water supplies, under-funded, and not having easy access to information and advising. These latter farms are in the majority among world-wide irrigation activities.

In the following, demand management for coping with water scarcity is developed to show that efficient water use may be adopted for surface, sprinkler or drip irrigation or through appropriate irrigation scheduling. A variety of technological solutions have become available and they are discussed in the following sections. Particular attention is paid to their limitations regarding the need for investment, operational and maintenance costs and adaptability to given types of farms. Our intention is to show that there are tools available for demand management that may apply to every condition. However, there is a need for upgrading of the knowledge and skills of engineers, agronomists, managers and policy and decision makers to provide conditions for their effective application. There is also the need for effective, available advisory services, improvement of markets, training and capacity building, incentives and credit facilities. Otherwise new developments will not assist or be of advantage to the many millions of needy small farmers. Poverty will continue to be difficult to eradicate if these small farmers continue as peasants who have little or no control over their farming activities. Then the general population will continue to claim that irrigation is synonymous with water wastage.

10.6.2 Demand Management: General Aspects

There are some current irrigation regions where water wastage is so large, or the productivity of water use is so small, that irrigated agriculture should be strongly discouraged in those regions. This strong statement is made here because it applies to a number of regions, but policy makers or governments in these regions are unfortunately unwilling to take the decisive action needed to bring about the large water savings and increase overall productivity that could improve the lot of the local population. For these regions most of the material discussed below is not relevant.

Demand management for irrigation to cope with water scarcity consists of reducing crop irrigation water requirements, adopting irrigation practices that lead to higher irrigation performance and water saving, controlling system water losses (i.e., N-BWU as defined in Section 9.2.2), and increasing yields and income per unit of water used (see Section 9.3). It includes practices and management decisions of an agronomic, economic, and engineering nature (Hoffman et al. 2007).

The objectives of irrigation demand management can be summarised as follows:

- *Reduced water demand* through selection of low demand crop varieties or crop patterns, and adopting deficit irrigation, i.e. deliberately allowing crop stress due to under-irrigation, which is essentially an agronomic and economic decision.
- *Water saving/conservation*, mainly by improving the irrigation systems, particularly the uniformity of water distribution and the application efficiency, reuse of water spills and runoff return flows, controlling evaporation from soil, and adopting soil management practices appropriate for augmenting the soil water reserve, which are technical considerations.
- *Higher yields per unit of water*, which requires adopting best farming practices, i.e. practices well adapted to the prevailing environmental conditions, and avoiding crop stress at critical periods. Improvements in water productivity result from a combination of agronomic and irrigation practices.
- *Higher farmer income*, which implies to farm high quality products, and to select cash crops. This improvement is related to economic decisions and market opportunities.

The agronomic aspects of irrigation demand management refer essentially to those described in the previous section. They concern crop improvement relative to resistance to water stress and improving water productivity, cropping techniques that favour coping with lesser water availability, and soil management for water conservation. Economic decisions, not dealt with here, concern the decision making processes relative to the selection of crop patterns and farming practices that reduce the crop irrigation demand. These include the evaluation of the economic returns and feasibility of water saving and conservation practices. The technical aspects of demand management which concern the various practices within irrigation are dealt with in this section.

Issues for irrigation demand management often refer only to irrigation scheduling, giving to irrigation methods a minor role. However, an integrated approach is required (Pereira 1999, Pereira et al. 2002b, Plusquellec and Ochs 2003). Irrigation scheduling is the farmers decision process relative to "when" to irrigate and "how much" water to apply at each irrigation. The irrigation method concerns "how" that desired water depth is applied to the field. The crop growth phase, its sensitivity to water stress, the climatic demand by the atmosphere, and the water availability in the soil determine when to apply an irrigation or, in other words, the frequency of irrigation. However, this frequency depends upon the irrigation method, i.e. on the water depths that are typically associated with the on-farm irrigation system. Therefore, both the irrigation method and the irrigation scheduling are inter-related.

Irrigation scheduling requires knowledge of (1) crop water requirements and yield responses to water, (2) the constraints specific to the irrigation method and respective on-farm delivery systems, (3) the limitations of the water supply system relative to the delivery schedules and (4) the financial and economic implications of the irrigation practice. To improve the irrigation method requires the consideration of the factors influencing the hydraulic processes, the water infiltration into the soil, and the uniformity of water application to the entire field. Therefore, irrigation demand management to cope with water scarcity is discussed here with respect to both the irrigation systems and scheduling.

It should be noted that vast irrigation regions, such as through much of the Indian Sub-continent, have fixed irrigation schedules, both in terms of time between water deliveries and volume delivered (the "warabandi" system) and this system leads to much inefficiency of water use (Zardari and Cordery 2007). In that water-scarce region there is a pressing need for reform to allow farmers the flexibilities of scheduling, discussed here, which has the potential to much more productivity per unit of scarce water resource in those regions.

Several performance indicators are currently used in on-farm irrigation. The uniformity of water application to the entire field is commonly evaluated through the distribution uniformity (DU), which is the ratio between the average infiltrated water depth (mm) in the low quarter of the field and the average infiltrated water depth (mm) in the entire field (Burt et al. 1997, Pereira 1999). The distribution uniformity essentially depends upon the characteristics of the irrigation system and only to a small degree on farmer management. In other words, high DU can only be achieved when the farmers manage the irrigation system well (Fig. 10.12) and it is well designed and maintained, whilst poorly designed and/or maintained irrigation systems will almost always lead to low DU (Pereira et al. 2002b).

The main farm efficiency indicator is the application efficiency (AE), the ratio between the average water depth (mm) added to root zone storage and the average depth (mm) of water applied to the field. AE is a measure of the quality of irrigation management by the farmer and is strongly related to the appropriateness of decisions on when and how much water is applied. Due to the limitations imposed by the system characteristics, the application efficiency depends upon the distribution uniformity. In general, when the distribution uniformity is high, the application efficiencies will also be good provided irrigation scheduling is appropriate. Figure 10.12 depicts various situations referring to DU and AE in relation to appropriate and under- and over-irrigation.

Useful relations between irrigation uniformity and crop yields have been made available and may be helpful to practitioners. These relationships show that attaining

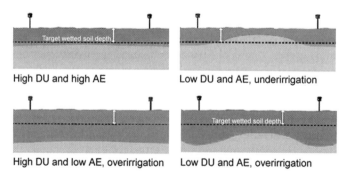

High DU and high AE   Low DU and AE, underirrigation

High DU and low AE, overirrigation   Low DU and AE, overirrigation

Fig. 10.12 Comparing various examples of high and low distribution uniformity (DU) and application efficiency (AE), with under- and over-irrigation

high DU is a pre-condition to achieve high application efficiencies, and therefore to obtain a good match between the amounts of water applied and the crop use requirements. Therefore, DU is an indicator that relates well to the system characteristics that favour water conservation and saving, as well as to higher water productivity. Therefore improvements in performance of farm irrigation methods and systems will be mainly discussed relative to their effects on distribution uniformity.

10.6.3 Demand Management: Improving Surface Irrigation Systems

Several surface irrigation methods are used in practice (Pereira and Trout 1999). The main ones are basin, furrow and border irrigation.

Basin irrigation, which is the most commonly used irrigation system world-wide. Basin irrigation consists of applying water to levelled fields bounded by dikes, called basins. Two different types are considered, one for paddy rice irrigation, where ponded water is maintained during the crop season, and the other for other field crops, where the ponding time is short, just until the applied volume infiltrates. For non-rice crops, basin irrigation can be divided into two categories: traditional basins, with small size and traditional levelling; and modern precision-levelled basins, which are laser levelled and have large sizes and regular shapes. Especially with traditional basins, shape depends on the land slope and may be rectangular in flat areas and follow natural land contours in steep areas. For row crops, and especially horticultural crops, the basins are often furrowed with the crops being planted on raised beds or ridges (Fig. 10.13). For cereals and pastures, the land is commonly flat inside the basin. Tree crops sometimes have raised beds around the tree trunks for disease control. Basin irrigation is most practical when soil infiltration rates are moderate to low and soil water holding capacity is high so large irrigations can be given. Basin irrigation depths usually exceed 50 mm. Inflow rates for basin irrigation have to be relatively high ($>2 l s^{-1}$ per metre width) to achieve quick flooding of the basin and therefore provide for uniform opportunity for infiltration along the basin length. Basins must be precisely levelled for uniform water distribution, because basin topography determines the recession, or removal from the surface, of the ponded water.

Furrow irrigation: water is applied to small and regular channels, called furrows, which serve firstly to direct the water across the field and secondly act as the surface through which infiltration occurs. There is a small discharge in each furrow to favour water infiltration while the water advances down the field. Furrow irrigation is primarily used for row crops. Fields must have a mild slope, and inflow discharges must be such that advance of water along the furrows is not too fast, nor too slow, i.e. the time elapsed since inflow starts at the upstream end until the water arrives to the other end must be in equilibrium with the infiltration to avoid either excess runoff at the downstream end, or excess infiltration in the upstream zone. Efficient furrow irrigation nearly always requires irrigation times longer than advance times. Runoff at the downstream end typically varies from 10 to 40% of the applied water, which

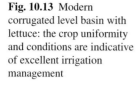

Fig. 10.13 Modern
corrugated level basin with
lettuce: the crop uniformity
and conditions are indicative
of excellent irrigation
management

should be collected, stored, and reused. Irrigation furrows are usually directed along
the predominant slope of the field. Furrows are used on slopes varying from 0.001
to 0.05 m m^{-1}. Low slopes require soils with low infiltration rates. Slopes greater
than 0.01 m m^{-1} usually result in soil erosion.

Border irrigation: water is applied to short or long strips of land, diked on both
sides and open at the downstream end. Water is applied at the upstream end and
moves as a sheet down the border. Border irrigation is used primarily for close
growing crops such as small grains, pastures, and fodder crops, and for orchards and
vineyards. The method is best adapted to areas with low slopes, moderate soil infiltration rates, and large water supply rates. Borders are most common and practical
on slopes less than 0.005 m m^{-1} but they can be used on steeper slopes if infiltration
is moderately high and the crops are close growing. Irrigation to establish new crops
on steep borders is difficult because water flows quickly, is difficult to spread evenly,
and may cause erosion. Design and management of very flat borders approximates
conditions for level basins. Precise land levelling is required, and inflow rates should
be neither erosive, nor producing too slow or too fast advance.

Farm irrigation systems may be distinguished according to their degree of modernisation. In traditional systems, the water control is carried out manually according to the ability of the irrigator. In small basins or borders and in short furrows, the
irrigator cuts off the supply when the advance is completed. This practice induces
large variations in the volumes of water applied at each irrigation, and it is difficult
to control "how much" water is applied. Over-irrigation is then often practiced. In
modernised systems, some form of control of discharge and of automation is used.
The fields are often precision levelled, and the supply time and the inflow rate can
be known and controlled. In modernised systems it is much easier than in traditional
systems to control "how much" water should be applied, which favours the adoption
of controlled and reduced irrigation demand.

In surface irrigation, the distribution uniformity, DU, mainly depends upon the
system variables controlling the water advance along the field and the way in which
the water is distributed and infiltrates throughout the field. These variables include

the inflow discharge, field length and slope, uniformity of the land surface topography in relation to land levelling conditions, and the duration of water supply to the field. The application efficiency, in addition to these variables, is influenced by the soil water deficit at the time of irrigation. The farmer's skill plays a major role in controlling the duration of water application, in applying the water at the appropriate soil water deficit and in maintaining the system in good operational conditions. However, his capability to achieve higher performance is limited by the system characteristics and, often, by the delivery rules relative to supply timing, duration and discharges, which are often dictated by the canal network managers. Therefore, under these circumstances, adopting improved scheduling and management rules is often beyond farmers control because the off– and on-farm system limitations are the prime constraint.

Improvements in surface irrigation systems that help to cope with water scarcity are numerous and depend upon actual field conditions (Pereira and Trout 1999, Pereira et al. 2002b). They relate to improvements in water advance conditions, increased uniformity of water distribution over the entire field, easy control of the water depths applied and resulting reduction of deep percolation and runoff return flows. These improvements are briefly referred to in Table 10.7, where benefits expected and common limitations for implementation are also included. Benefits generally relate to irrigation performance and to the water productivity (WP) as discussed in Section 9.3.

The ease of control of the required leaching fraction (LF) when using saline waters or irrigating saline soils, discussed in Section 8.3.4, is also considered because the application of a leaching fraction is commonly necessary in arid and semi-arid climates where the seasonal rainfall is insufficient to produce natural leaching of salts.

Improvements in surface irrigation systems aimed at reducing the water volumes applied and increasing the water productivity to cope with water scarcity can be grouped as follows:

a) *Land levelling* is a very important practice to improve surface irrigation performance. It provides conditions for reducing the advance time and water volumes required to complete the advance, higher irrigation performance (DU and AE), as well as better conditions for adopting deficit irrigation and easier control of the required leaching fraction (LF). This technique is easy to apply in large farms, including precision levelling as a maintenance procedure, but is costly and requires appropriate support when it is to be applied on small farms.

b) *Irrigation with anticipated cut-off*, i.e. cutting the inflow to basins or borders before the advance is completed, or to furrows before the downstream area is irrigated. This technique reduces water application volumes but the last quarter or half of the field is often under-irrigated or, as for furrows, may be just rain-fed. This practice avoids runoff and minimises percolation. It is easily applicable under conditions of limited water availability but may have important impacts on yields and economic returns.

Table 10.7 Improvements in surface irrigation systems aimed at (1) reducing water volumes applied and (2) increasing water productivity to cope with water scarcity

Techniques	Benefits	Applicability
Land levelling (precision levelling)	• Less water to complete advance, better conditions for adopting deficit irrigation and control of the leaching fraction	• Requires support to small farmers
Irrigation with anticipated cut-off	• Reduced water application, runoff is avoided and percolation is minimal	• When available water is limited
Basin irrigation		
• Higher discharges, reduced widths and/or shorter lengths	• Fast advance time, reduced volumes applied, easier application of deficit irrigation and control of the leaching fraction	• Easy; limitations are due to field size and geometry
• Higher basin dikes to catch storm rainfall	• Providing for full conjunctive use of irrigation and rainfall	• Easy to implement, mainly for paddy fields
• Corrugated basin irrigation for row crops	• Faster advance, improved emergence and rooting of the crops planted on the bed, easy introduction of row crops in rotation with rice	• Easy to implement
• Maintaining low water depths in rice basins	• Lower seepage and percolation and better conditions to store any storm rainfall	• Limitations when land levelling is poor and delivery is infrequent
• Non-flooded paddies, i.e. maintaining the soil water near saturation	• Lower seepage, deep percolation and evaporation losses, and better conditions to store any storm rainfall	• Only for paddies in warm climates and when deliveries are frequent
Furrows and borders		
• Irrigation with alternate furrows	• Reduced water application to the entire field; deep rooting of the crops is favoured	• Easy to apply
• Reuse of tail water runoff	• Avoidance of runoff, increased systems efficiency and control of quality of return flows	• Need for collective facilities with small farms
• Closed furrows and borders	• Avoiding runoff at the downstream end	• Easy (in flat lands)
• Contour furrows	• Runoff and erosion control in sloping land	• When fields are not oriented down slope
• Surge flow	• Faster advance, reduced percolation and runoff, higher performance and provides for system automation	• Easy for large farms but difficult for small ones. Needs slight slopes
• Continuously decreased inflow rates, (cablegation)	• Control of percolation and runoff by continuously adjusting flow rates to infiltration, and provides for system automation	• Requires technological support and is difficult for small farmers
• Improved furrow bed forms	• Improved furrow hydraulics and infiltration that favour other improvements for water saving	• Requires appropriate implements; large farms

Table 10.7 (continued)

Techniques	Benefits	Applicability
On-farm water distribution		
• Gated pipes and layflat pipes	• Easier control of discharges, control of seepage and provides for automation	• Easy to adopt but require farmer investment
• Buried pipes for basins and borders	• Easier control of discharges, control of seepage and easy to be automated	• Less appropriate for small farms
• Lined on-farm distribution canals	• Easier control of discharges when siphons and gates are used, and control of seepage	• Only for large farms
• Good construction of on-farm earth canals	• Easier control of discharges when using siphons and some control of seepage	• Limitations due to farm implements available
• Automation and remote control of farm systems	• Improved conditions for operation, easier application of improved irrigation scheduling, including in real time, and precise irrigation management techniques	• Application only to large farms with high technological and financial capabilities

c) *Field evaluation*, which consists of monitoring irrigation events in farmers' fields. It does not provide water saving by itself but produces detailed information to the farmers and extension staff that permit the selection and further adoption of corrective measures. It can provide basic information for controlling percolation and runoff, improving DU and AE, and adopting water saving practices, as well as improving the timeliness and duration of irrigation events leading to better system management and salinity control. Appropriate field monitoring programmes need to be implemented. Field evaluations play a fundamental role in improving surface irrigation systems, as they provide information for systems design and, mainly, for advising irrigators on how to improve their systems and practices.

d) *Improved design and modelling*. As for field evaluation, it does not produce water saving but provides for it through the selection of the best combination of field sizes, slope, inflow discharges and time of application that optimise conditions for controlling deep percolation, runoff, leaching fraction applications, and deficit irrigation. It requires support by extension services and should be based on data from field evaluation and monitoring programmes.

e) *Basin irrigation improvements* include:

1) Adopting higher discharges, reduced widths or shorter lengths to permit a fast advance time, improved uniformity of infiltrated depths along the field, which makes it possible to reduce the volumes applied and an easier control of the leaching fraction.

2) Corrugated basin irrigation for row crops, i.e. a system of furrows and beds installed inside the basin (Fig. 10.13, above). These furrows favour a faster

advance and water saving while cultivation of the beds improves the emergence and rooting of the crops. This practice is easy to implement in both modern and traditional systems.

f) Paddy rice irrigation

1) Replacing permanent basin flooding by temporary, intermittent flooding where the soil water is maintained near saturation for most of the time. These conditions decrease seepage, deep percolation and evaporation losses, and create conditions that favour storing any storm rainfall. However, this practice is limited to regions having warm to hot climates where the water is not playing a role for temperature regulation. It cannot be practiced in areas where deliveries are not frequent enough to keep the soil water above the critical threshold.

2) Where permanent flooding is practiced, water saving may be achieved by maintaining low water depths in the rice paddies because seepage and percolation are then controlled (Mao et al. 2004). Low flooding levels require appropriate land levelling. Note also that much of the seepage loss often occurs through the uncompacted soils below the bunds.

3) Adopting basin dikes high enough to catch storm rainfall provides for full conjunctive use of irrigation and rainfall, which can be important for water conservation if the irrigation season coincides with a period when rain storms occur.

g) Improvements in furrow and border irrigation, including:

1) Irrigation with alternate furrows (mainly during vegetative crop stages) to reduce water application to the entire field and favouring deep rooting of the crops. In cases of limited water availability and/or water stress resistant crops, this technique may be applied for the entire crop season.

2) Closed furrows and borders to avoid tail end runoff. However, waterlogging at the downstream end may result if excess water is applied.

3) Reuse of tail water runoff, which is the only way to make efficient use of water where tailwater runoff occurs, i.e. open furrows and borders. Reuse provides for water saving that may be as high as 40% of the applied volumes and preserves the quality of the downstream water bodies by avoiding runoff from the field system. It is easy to adopt in large farms but collective reuse facilities are generally required for small farms.

4) Contour furrows, to be adopted when land has a gentle slope and where down slope furrows would produce high runoff and erosion. However, it is not applicable when fields are narrow and down slope oriented.

5) Surge flow, i.e. intermittent, cycling of water application to furrows and, less frequently, to borders. Surging produces changes in the soil surface conditions that provide for a faster advance in moderate to high infiltration soils and for reduced percolation and runoff. For control on the cycle times, surging usually requires system automation. It is easy to apply in large and commercial farms but is difficult for small farms.

6) Application of progressively smaller inflow rates to furrows to adjust them to the diminishing soil infiltration rates, e.g. adopting the automated cablegation system. This technique controls percolation and runoff and may lead to reasonable water saving. It requires an automation system and technological support; it is difficult for small farms.

h) *Water efficient systems to deliver water to basins, furrows and borders*:

1) Gated pipes and layflat tubes to convey and deliver the water to basins, furrows and borders. This equipment, relative to traditional earth canals or ditches, provides easier control of discharges, reduces seepage, and is easy to automate, especially for surge flow and cablegation referred to above. Gated pipes and layflat tubes are easy to adopt, even by small farmers.
2) Buried pipes for delivering water to basins and borders also provide easy control of applied discharges, reduced seepage and avoid runoff. They are easy to automate but require higher investments than gated pipes.
3) Lined farm distribution canals permit good control of discharges applied to furrows when siphons are used, and to borders and basins through gates or valves. Lining controls seepage. Limitations in use are caused by investment costs and disturbance of other farming operations because these structures are permanent.
4) Improved farm earth ditches and canals are an alternative to lining, but resulting reductions in seepage are small and they require special implements for their reconstruction every year, which may be unrealistic for small farmers.
5) Automation and remote control of farm systems can improve operations, mainly by applying improved irrigation scheduling, including in real time, and precise irrigation management techniques. However, these technologies can only be adopted by large farms which have high technological and financial capabilities.

Some of the listed improvements are of general application. However, they are important for water conservation and saving in water scarce areas. Others are specific for situations when there are restrictions in water supply volumes. Improvements not directly related to water scarcity are based upon attaining higher distribution uniformity. The importance of high DU for achieving higher yields with less water in surface irrigation is well evidenced in the literature.

Improving surface irrigation systems requires not only knowledge and technology but economic feasibility. In a study performed at Fergana, in the Aral basin, several improvements relative to the current furrow irrigation practices for cotton were analysed (Horst et al. 2007). Results in Fig. 10.14 refer to the performance achieved using several alternative methods: irrigation in every furrow with continuous flow (EC) and surge-flow (ES), and irrigation of alternate furrows with continuous flow (AC) and surge flow (AS). Performance indicators are water productivity of irrigation water (WP_{irrig}), the consumed fraction (CF), the beneficial water use fraction (BWUF) and the ratio of actual to maximal yield (Y_a/Y_{max}).

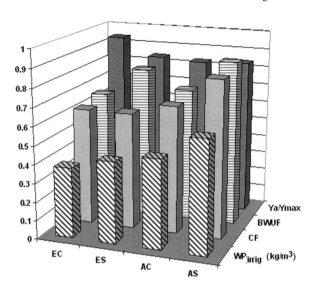

Fig. 10.14 Furrow irrigation of cotton in the Aral Sea basin: comparing irrigation in every furrow with continuous flow (EC) and surge-flow (ES), and irrigation of alternate furrows with continuous flow (AC) and surge flow (AS). Performance indicators are water productivity of irrigation water (WP$_{irrig}$), the consumed fraction (CF), the beneficial water use fraction (BWUF) and the ratio actual to maximal yield (Y$_a$/Y$_{max}$). (Source:Horst et al. 2007)

Results (Fig. 10.14) show that adopting alternate furrows improves water use, particularly increasing both CF and WP$_{irrig}$ relative to supply to every furrow, thus indicating water saving. Results for surge-flow (both ES and AS) indicate a great increase in BWUF, i.e., less non-beneficial uses as percolation and runoff. However, the best yield indicator Y$_a$/Y$_{max}$ refers to the traditional EC method. This may be explained by difficulties faced by farmers in adopting the alternative methods. Under these circumstances, because farmers' incomes are directly related with the achieved yield, and because adopting surge-flow requires some investment, the intended water saving is only possible if appropriate incentives are given to farmers. In fact water saving is required in the area to combat the desertification problems in the Aral sea basin, but such savings are not an objective of the farmers.

A major factor for achieving high DU and water saving in basin irrigation is land levelling. An example of water saving is given in a study of basin irrigation in the upper Yellow River basin, in China (Pereira et al. 2007a). Figure 10.15 compares the infiltrated water depths along a 45 m long basin without and with precision levelling. The region, in Ningxia Province, is arid and improved irrigation is required for controlling excess water applications. Excess water applications have caused waterlogging and salinity problems, thus reduced yields and incomes.

Figure 10.15 shows that to achieve the target irrigation depth of 110 mm in non levelled fields farmers have to apply much more water, which percolates to the watertable. This is due to the fact that the current unevenness and slope of the field leads to uneven infiltration which produces percolation of the excess water. When

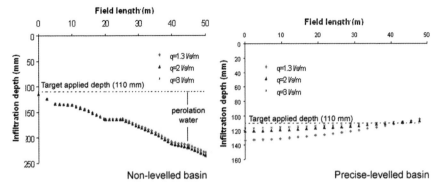

Fig. 10.15 Comparing infiltrated depths in basin irrigation of cereals for a non-levelled and a precise levelled basin adopting 3 inflow discharges: 1, 2 and 3 l/s/m. (Source: Pereira et al. 2007a)

precise zero-levelling is applied it becomes possible to attain a near uniform infiltration, highly reducing the non-beneficial percolation water. Results in Fig. 10.15 also show that for the levelled basins a better uniformity is achieved for the largest inflow discharge (3 l/s/m width). In this example the bottleneck resides on the lack of support to farmers since it is beyond their economic and social capabilities to adopt precise land levelling.

10.6.4 Demand Management: Improving Sprinkler Irrigation Systems

Main sprinkler systems (Keller and Bliesner 1990, Pereira and Trout 1999) are:

- *Set systems*: the sprinklers irrigate in a fixed position and can apply small to large water depths. Set systems include solid set or permanent systems as well as periodic-move systems, which are moved between irrigations, such as hand-move, wheel line laterals and hose-fed sprinklers. These systems are the least costly and the best adapted for small farms. A wide range of sprinklers can be selected for a variety of crops and soils as well as for environmental conditions.
- *Travelling guns*: a high pressure sprinkler continuously travels when irrigating a rectangular field. The high application rates and the characteristics of the moving system make travelling guns unsuitable for applying very small or large depths, or to irrigate heavy soils and sensitive crops. In addition, these systems have a high energy requirement (which is a very negative handicap under high energy prices) and may have low performance and high evaporation losses when operating under hot, arid and windy conditions.
- *Continuous move laterals*: the sprinklers operate while the lateral is moving in either a circular or a straight path. Large laterals are used, equipped with sprinklers or sprayers. The principal continuous-move systems are center-pivot and

linear move laterals. These systems are designed to apply small and frequent irrigation in large to very large fields. These systems are often automated.

In sprinkler irrigation, the irrigation uniformity essentially depends upon variables characterising the system such as the pressure, discharge, throw and application rate of the sprinklers, spacing between sprinklers and between the pipe laterals where sprinklers are mounted, and head losses along the pipe system network. These variables are set at the design phase and they are difficult to modify by the farmer. Thus distribution uniformity DU is the performance indicator that characterizes the irrigation system. The efficiency AE depends upon the same system variables as DU, and on management variables concerning system maintenance and, moreover, the duration and the frequency of the irrigation events. The AE indicator often follows DU and largely depends upon the irrigation scheduling applied. Thus, the irrigator can do little to improve the uniformity of irrigation and is constrained by the system characteristics for any improvements in irrigation performance.

Field evaluations can provide good information to farmers on how to improve management and introduce limited changes in the system, as well as useful feedback to designers and to those who manage quality control of design and services. Field evaluations allow the identification of problems in sprinkler systems under operation. In set sprinklers, the uniformity is often lower than that potentially attainable due to (i) excessive spacings between sprinklers and laterals, (ii) too much variation in pressure within the operating system, and (iii) insufficient or excessive pressure at the sprinklers. The main problems in travelling guns refer to (i) excessive distance between towpaths, (ii) inadequate pressure, (iii) asymmetric wet angle, and (iv) variable advance velocity. In lateral moving systems, more frequent causes for non-uniformity are (i) inadequate pressure distribution along the lateral, mainly when operating in non flat areas, (ii) application rates much above infiltration rates, (iii) use of an end gun sprinkler without an appropriate source of pressure, and (iv) excessive pressure in systems with sprayers. Wind effects are common to all systems but are more often a problem in travellers and lateral moving systems when emitters are on the top of the lateral. Problems reported above are generally due to poor design, or lack of design, where solutions are found exclusively to lower the investment costs. In addition, very often, the farmers have a very poor knowledge of their own systems and do not receive extension support for system selection or to maintain and manage the systems.

As for surface irrigation systems, the distribution uniformity plays a major role in sprinkler irrigation and is the main factor influencing system performance. The lower the DU, the larger is the difference between applied depths in the over-irrigated and the under-irrigated parts of the field, and thus the larger is the depth of water required to satisfy a given target application in the entire field. Consequently, the water use increases when DU decreases. As observed by Mantovani et al. (1995), when DU is low (40%) the farmers use nearly twice as much water as is applied when the DU is high (85%). However, if water is expensive farmers under-irrigate for low system uniformity and only fully irrigate when systems can achieve a high DU. Therefore, when looking for improvements of sprinkler systems to cope with water scarcity the most important factor to be considered is to attain higher distribution uniformity. This is in agreement with the trend to adopt a target DU for design (e.g., Keller and Bliesner 1990).

The improvement of sprinkler irrigation systems to cope with water scarcity, as summarised in Table 10.8, mainly concerns a variety of practices aimed at reducing the volumes of water applied, particularly to increase DU and AE, and to increase water productivity (Keller and Bliesner 1990, Pereira et al. 2002b). These practices and measures refer to:

a) *Optimising the overlapping of sprinkler jets*, through:

   1) Adopting sprinkler spacings in agreement with the characteristics of the sprinklers relative to the wetted diameter and the available pressure. The aim is to attain adequate performance and economic return of the crop.
   2) Similarly, choose the appropriate distance between travelling gun towpaths.
   3) Based on field evaluations, change sprinkler nozzles to make best use of the available pressure to improve overlapping.
   4) Adopt adequate alignment tools in continuous-move lateral systems to increase uniformity of water application.

b) *Minimising discharge variations within the operation system*

   1) Design to avoid excessive pipe head losses and control pressure variations to no more than 20% of the average pressure at the sprinklers in set systems, so as to attain appropriate uniformity.
   2) Adopt pressure regulators in sloping fields to avoid non-uniform discharges and runoff where excess pressure would produce excessive application rates.
   3) In continuous-move lateral systems, adopt booster pumps when an end gun sprinkler is used.
   4) Monitor and adjust the pumping equipment to ensure the upstream pressure matches that required by the sprinkler system.

c) *Minimising wind drift and evaporation water losses*:

   1) Avoiding irrigating during windy periods of the day and the hours when evaporation is high. However, this practice is constrained by system characteristics and by water delivery rules when a rigid rotation is applied.
   2) Use smaller spacings and select sprinklers with the lowest possible jet angles in windy areas to compensate for the jet disturbance caused by the wind. This option is easy to implement in new systems but requires higher investment costs.
   3) In case of continuous-move lateral systems, use suspended spray heads, or adopt the LEPA suspended emitter heads to minimise the wind effects. Because drop heads have higher application rates, special care is required to prevent this type of solution producing excessive runoff (and erosion).

Table 10.8 Improvements in sprinkler irrigation systems aimed at reducing volumes of water applied and increasing water productivity to cope with water scarcity

Objectives/Benefits	Techniques	Applicability
• *Optimising overlapping of sprinkler jets*	• Adopt or correct sprinkler spacings in set systems and towpath spacings of travelling guns for high DU	• Easy at design, more difficult once installed
	• Change sprinkler nozzles to suit available pressure	• Relatively easy and not expensive
	• Enhanced alignment of continuous lateral move systems	• Higher investment cost
• *Minimising discharge variations within the operation system*	• Design the pipe system for pressure variations < 20% of the average sprinkler pressure	• Easy to set at design
	• Pressure regulators in sloping fields	• Easy at design and after installation
	• Booster pumps for end gun sprinklers in moving laterals	• Easy at design and after installation
	• Monitor and adjust pumping equipment	• Needs technical support
• *Minimising evaporation and wind drift losses*	• Irrigation during non windy periods	• Limitations when delivery is rigid
	• Smaller spacings in windy areas	• Easy to set at design
	• Low jet angles for windy areas	• Easy to set at design
	• Suspended spray heads or LEPA piping and heads instead of sprinklers on the top of moving laterals	• Easy to set at design; costly after installation
	• Large sprinkler drops and application rates in windy areas	• Constrained by crops and soil infiltration
	• Orient set laterals and travelling gun towpaths perpendicular to prevailing wind direction	• Constrained by the design options
	• Avoid gun sprinklers under high winds	• Design decisions
• *Maximising infiltration and avoid runoff*	• Adopt application rates smaller than infiltration rate	• Easy for set and travelling gun systems
	• Soil management practices that favour infiltration	• Need technical support and investment
	• Furrow dams in sprinkled sloping fields (row crops)	• Easy for contour corrugated fields
	• Adapt speed and cycles of mobile systems to soil infiltration rates	• Limited to system characteristics
	• Improved spray heads and booms in moving laterals on heavy soils and sloping land	• Higher investment cost
• *Pre-condition for operation*	• Careful maintenance	• Easy to implement
• *Advanced irrigation and energy strategies*	• Automation and remote control, and precise water application	• Only for large farms with technological background
• *Improved fertiliser and water use efficiency*	• Adopt fertigation (fertilisers in water)	• Easy but requires advice and investment

4) Adopt sprinkler pressures that produce large drops and sprinklers having high discharge rates in windy areas. However, these options are limited by the sensitivity of crops and the soil infiltration characteristics.

5) Orient set laterals and travelling gun towpaths perpendicular to wind to compensate for the jet disturbance.

6) Avoid gun sprinklers under high winds and in heavy soils because under these conditions wind drift and runoff are unavoidable.

d) *Maximise infiltration* of applied water and avoid runoff:

1) In set and travelling gun systems, design systems for sprinkler application rates not exceeding the soil infiltration rate to avoid the runoff and erosion,

2) For heavy soils and sloping lands, where infiltration is lower, adopt soil management practices that favour infiltration as described in Section 10.5.4.

3) Adapt speed and cycles of continuous-move laterals to soil infiltration rates, i.e. minimise the time duration of water application in excess of infiltration rates. Good design and extension support are required.

4) Where relatively high sprinkler application rates are used, adopt furrow dams (Fig. 10.8) in sloping fields for row crops to store the non-infiltrated water in the ponds thus created and, therefore, avoid runoff. This practice, despite being costly, is highly effective in contour corrugated fields when slopes are not very high, but it is of limited benefit when cultivation is practiced down slope and/or slopes are quite steep.

5) In continuous-move lateral systems, use improved spray heads and sprayers on booms when irrigating low infiltration soils, particularly in undulating fields, to reduce the application rates and to control runoff and erosion. These design issues are more costly than the use of more common solutions but they are very effective in controlling the application rates.

e) *Adopt careful system maintenance* since it is a pre-condition for adequate functioning of the equipment and to achieve higher performance.

Other less specific measures contribute to improving the water use in sprinkler systems. However, some technologies apply only to large commercial farms having high technological support. This is the case for automation and remote control, which support enhanced management, including precise water application, real time irrigation scheduling and energy management strategies. It is also the case for fertigation - the application of fertilisers with the irrigation water – that improves the use of both the water and the fertilisers, and precise water application, when advanced information systems make it possible to apply differentiated water and fertiliser amounts allowing for differences in soil and crop conditions.

Sprinkler system design is a key factor for success (Fig. 10.16) and is a main area of concern because, as reported above, water savings are often hampered by poor design. New approaches are required not only to enhance design procedures but also to adopt specific approaches relative to design for management under scarcity. Then, mainly using models and expert systems tools, sprinklers and systems layout

Fig. 10.16 Excellent performance may be achieved with good equipment and careful design (courtesy by *Nelson*, http://www.nelsonirrigation.com/)

could be selected and simulated in response to specific environmental conditions and target performance. Field evaluation plays then a major role because, in addition to farmers advice, it produces relevant factual information that can be used in design.

Sprinkler irrigation is under continuous improvement by manufacturers of sprinkler heads, travellers, moving laterals, control and regulation devices, and automation equipment. Related information is nowadays available through a number of websites and many manufacturers provide there easy advice to users. This situation is very different from that of surface irrigation whose market is virtually non-existing in most countries. Progress is providing for upgraded performance but design became more difficult due to the variety of solutions offered by the market. This may be solved through an enhanced used of models, but these do not replace field observations and system evaluations.

10.6.5 Demand Management: Microirrigation Systems

Microirrigation, also called localized irrigation, applies water to individual plants or small groups of plants. Application rates are usually low to avoid water ponding and minimise the size of distribution tubing. The microirrigation systems in common use today (Keller and Bliesner 1990, Pereira and Trout 1999) can be classified in two general categories:

- *Drip irrigation*, where water is slowly applied through small emitter openings from plastic tubing. Drip tubing and emitters may be laid on the soil surface, buried, or suspended from trelisses.
- *Microspray irrigation*, also known as micro-sprinkling, where water is sprayed over the soil surface. Microspray systems are mainly used for widely spaced plants such as fruit trees but in many places of the world they are used for closed space crops in small plots.

A third type of localised irrigation, the bubbler systems, use small pipes and tubing to deliver a small stream of water to flood small basins adjacent to individual trees. Bubbler systems may be pressurised with flow emitters, or may operate under gravity pressure without emitters.

Microirrigation uniformity depends upon various system variables such as pressure variation and emitter discharge variation, emitter flow characteristics (pressure – discharge relationship, susceptibility to variations in temperature, emitter orifice size in relation to susceptibility to clogging), emitter material and manufacturing variability, emitter spacings, head losses in the lateral tubing, pressure variation due to field slopes, and filtering characteristics. With the exception of maintenance, the farmer can do very little to achieve good distribution uniformity. As for sprinkler systems, DU is the indicator for the system performance and essentially depends on decisions taken at design. The application efficiency mainly depends upon the same system variables as DU and on management variables related to the irrigation frequency and time duration. Therefore, the farmer may improve the application efficiency when adopting appropriate irrigation schedules, but performance are limited by the system constraints.

Field evaluations also play an important role in guiding farmers, creating information for design of new systems, and for quality control of design and services. Results of field evaluation show that micro-irrigation performance are often lower then expected. Pitts et al. (1996), referring to the evaluation of 174 micro-irrigation systems in USA, found an average DU of 70%, with seventy-five percent of cases having DU below the common target 85%. DU values lower than the target are observed world-wide. Low DU is mainly due to inappropriate control of pressure, discharge variation within the operating set, insufficient filtration and filters maintenance, lack of pressure regulators, poor selection of emitters, and poor information on manufacturing specifications and characteristics.

Uniformity in micro-irrigation affects the water saving capabilities of the systems, crop yields, and water productivity. For a long time it has been known that micro-irrigation design should base upon uniformity. Santos (1996) has shown that tomato yields did fall from near 102 ton/ha to 85 ton/ha when uniformity drops from 90% to only 60%, which was associated with an increase in water use from 470 mm to 500 mm. An extensive analysis by Ayars et al. (1999) shows the benefits for several crops of subsurface drip (Fig. 10.17). It can serve to maximise yields and considerably reduce water demand relatively to other methods.

Improvements in micro-irrigation systems to cope with water scarcity aim at achieving high uniformities of water distribution (DU) in the entire field (Keller and Bliesner 1990, Pereira et al. 2002b). This is seen as a pre-condition for efficient water use and productivity as well as to achieve water saving. They are summarised in Table 10.9 and concern:

a) *Use of a single drip line for a double crop row* to reduce water use (and system costs) when the soil has enough good lateral transmission of water to wet both rows.

b) *Use microsprayers in high infiltration soils* to avoid deep percolation that would be produced by drippers. It is easy to apply in orchards and low crops but not for tall crops. Then sprinkling would be more appropriate.

Fig. 10.17 Subsurface drip irrigation in Portugal: good design, installation and management provide for reduced demand and increased yields (courtesy by *TTape*, http://www.t-tape.com/)

c) *Adjust the duration of water application and timing to soil and crop characteristics* to control percolation and salt distribution and accumulation in the soil. For difficult conditions, support by extension services may be required.

d) *Control pressure and discharge variations within the operating set*:

1) Adopt pressure regulators in sloping areas to avoid pressure variations due to slope. Because micro-irrigation systems operate at low pressure, variations in elevation induce changes in pressure that are relatively high when compared with sprinkler systems. Pressure regulators may be installed in operating sets when these problems are identified through field evaluation.

2) Adopt self-compensating emitters in long and/or sloping laterals. This is an alternative to some cases where pressure regulators may be less efficient. Changing from turbulent emitters to self-compensating emitters in a system under operation is generally not feasible since it implies heavy costs.

d) *Use filters appropriate to the water quality and the emitter characteristics*, implying:

1) Selection of equipment and filtering sizes adequate to the problems to be solved, which is often more expensive than common equipment.

2) Select filter locations in agreement with the system layout and the management adopted for the sub-sets, mainly fertigation management.

3) Adopt carefully maintenance of the filters since they produce quite heavy head losses when they are not cleaned with appropriate frequency.

4) Apply chemicals periodically to clean the system and treat the clogged emitters.

f) *Adopt careful maintenance* since it is essential for achieving high performance.

g) *Automation*. Because micro-irrigation systems apply water with very high frequency, ease of operation requires that automation be used for appropriate implementation of irrigation and fertiliser management.

Table 10.9 Improvements in micro-irrigation systems aimed at reducing volumes of water applied and increasing the water productivity to cope with water scarcity

Techniques	Benefits	Applicability
• Single drip line for a double crop row	• Reduced water use	• For soils having good horizontal conductivity
• Microsprayers in high infiltration soils	• Avoid deep percolation which would be produced by drippers	• Orchards and low field crops
• Use drippers in low infiltration sloping soils	• Avoid runoff which could be produced by microsprayers	• No limitations
• Adjust application duration and timing to soils and crops	• Control of percolation and salt distribution and accumulation in the soil	• May require support of extension services
• Adopt pressure regulators in large sets and in sloping areas	• Provide for emitter uniformity by avoiding variations in pressure in the operating set	• To be implemented at the design stage
• Adopt self-compensating emitters in long and sloping laterals	• Provide for emitter uniformity by avoiding variations in pressure along each lateral	• Design limitations
• Use appropriate filtering and filters locations	• Control emitter clogging and consequent non-uniformity of emitter discharges	• To be defined at the design stage
• Frequent filter cleaning	• Control of emitter clogging	• Depends on the water quality
• Chemical treatment to combat emitter clogging	• Helps appropriate functioning of emitters	• To be implement out of crop season
• Careful maintenance	• Pre-condition to fully use the beneficial characteristics of the system	• According to system characteristics
• Automation	• Helps operation and the adoption of irrigation scheduling, mainly in real time	• Requires advice and investment
• Adopt fertigation (fertilisers with the irrigation water)	• Enhances water and fertiliser efficiency	• Requires advice/expertise
• Adopt chemigation	• Easy control of weeds and soil diseases	• Requires advice/expertise
• Improve system design for management under scarcity	• Provides for selecting emitters and system layout in agreement with environmental conditions, and helps to adopt non-optimal operation under conditions of limited water supply	• Requires technical support to farmers
• Field evaluation	• Identification of corrective measures to the system and, mainly to its management and maintenance	• Requires appropriate monitoring and extension services

h) *Adopt fertigation and chemigation* – i.e. application of fertilisers and herbicides or other soil chemicals with the irrigation water. These usually improve water use and plant responses to irrigation and fertiliser/chemical treatments.

The discussion above essentially refers to the need for good design and management when micro-irrigation systems are used in water scarce areas. Since these systems apply water only in a part of the field and near the crop roots, they are able to provide for higher production using less water than other irrigation systems, i.e. adopting micro-irrigation systems is a way to cope with water scarcity. However, these systems are much more expensive than surface and sprinkler systems despite the fact that costs tend to become lower over time. This implies that their adoption has to be economically feasible, i.e., that yields must be high enough to produce high farm incomes. To achieve this, systems have to be designed for achieving the best results without system operation water losses, but they cannot be designed just to save water. Progress in design is required to find solutions that may allow the systems to operate under optimal and non-optimal conditions when water is not sufficient to fully satisfy the crop requirements.

Concluding, microirrigation, in particular when only the crops root zone is wetted (Fig. 10.17), provides for water saving relative to other irrigation methods because soil evaporation is reduced. Good design and management lead to control percolation out of the root zone. In addition, particularly when fertilizers and agrochemicals are applied with the irrigation water, it leads to high water and land productivity.

10.6.6 Demand Management: Irrigation Scheduling

10.6.6.1 Irrigation Scheduling Techniques

Research has provided a large variety of tools to support improved irrigation scheduling, i.e. the timeliness of irrigation and the adequateness of volumes applied (Smith et al. 1996).

Irrigation scheduling techniques may be used with diverse objectives in the practice of farmers. More commonly, farmers seek to avoid any crop stress and maximise crop yields. When water is plenty, farmers tend to over-irrigate, both anticipating the timing of a need for irrigation and applying excessive water depths. Thus, the application of appropriate irrigation scheduling techniques permits them to optimise the timeliness and the volumes applied, thus controlling return flows, deep percolation, transport of fertilisers and agro-chemicals out of the root zone, and avoiding waterlogging in the parts of the field receiving excess water. Economic and environmental benefits are also obtained because better conditions are created for achieving the target yields, less water is used, the contamination of surface and ground-waters is controlled, the rising of saline water tables is avoided and, when the water source is the groundwater, the respective water levels are easily kept in the desirable depth range.

In water scarce regions, achieving such optimal management conditions is more important then under conditions of plenty since any excess in water use is a potential cause for deficit for other users or uses. Since water availability is usually limited, and certainly insufficient to achieve maximum yields, some kind of reduced demand scheduling has to be used. Reduced demand scheduling may be managed under very well controlled approaches as for deficit irrigation, or just to minimise the effects of water stress as for reduced irrigation when drought occurs. *Deficit irrigation* is an optimising strategy under which crops are deliberately allowed to sustain some degree of water deficit and yield reduction (English and Raja 1996). *Reduced irrigation* is a remedial strategy where crops are irrigated using the minimal amounts of water available during the more critical growth periods in such a way that some yield may be achieved and, for perennial crops, future yields are not compromised. Therefore, for deficit irrigation the timeliness and volumes of water applied must be well scheduled in such a way that the economic returns of the irrigated crops are optimised. Differently, for reduced irrigation only the water productivity may be optimal: farmers just try to crop and achieve the best possible yield by selecting the timings when the limited water resources are applied (e.g. in subsistence cropping).

Deficit irrigation is a sustainable strategy that should include in selected irrigation events the volumes required for leaching if poor quality water is used or when irrigating saline soils. Contrarily, reduced irrigation is not sustainable since yields do not respect any economic threshold and leaching requirements are not considered. In the long term, subsistence farming and increased soil salinity do not assist in maximising the yield and economic productivity of the water and controlling the impacts of irrigation on the environment.

Deficit irrigation under saline conditions is difficult, particularly in arid conditions. To make deficit irrigation and salinity control compatible may not be achievable. Then, reduced demand takes the form of *controlled saline irrigation* where leaching is achieved regularly through most seasons but not in years when water is scarce. Provisions for additional leaching have to be made as soon as more water is available. Several procedures may be used as outlined in Section 8.3.

Irrigation scheduling techniques and tools are quite varied and have different characteristics relative to their applicability and effectiveness for coping with water scarcity. Some of them are still only applicable in research or need further developments before they can be used in practice. Most of them require technical support by extension officers, extension programmes and/or the technological expertise of the farmers. However, there is great potential for their application in practice when appropriate programmes are implemented, including when oriented for effective support to small farmers. The problem is that in most countries these programmes do not exist because they are expensive, trained extension officers are lacking, there is not enough awareness of the issues of water saving in irrigation, or the institutional mechanisms developed for irrigation systems management are not adequate. Therefore, in general, large limitations occur for their use in the farmers practice.

The irrigation scheduling techniques are summarised in Table 10.10 with reference to the respective applicability and effectiveness.

Several techniques and tools refer to *soil water indicators*. They correspond to different levels of accuracy and sophistication and can be used in relation to the technological level of the farmers. The simplest approach is named soil appearance and feel and consists of assessing the soil water status by just feeling how dry the soil is. It can be quite effective to assess the timeliness of irrigation when

Table 10.10 Summary on irrigation scheduling techniques, tools, applicability, and effectiveness to cope with water scarcity

Techniques	Tools/procedures	Applicability	Effectiveness
Soil water indicators			
Soil appearance and feel	Hand probe, shovel	Field crops	Depending on farmers experience
Soil electrical resistance as depending on soil water content	Porous blocks, and electrode probes	Cash crops; calibration and advice are desirable	Depending upon selected irrigation thresholds
Soil water potential	Tensiometers, soil psychrometers, pressure transducers	Tree crops and horticultural crops. Calibration and advice to farmers are desirable	Very high when thresholds are well selected. Inappropriate for low soil water thresholds
Soil water content	Soil sampling, neutron probe, TDR and capacitative sensors	All crops. Require calibration and support by experts	Very high when thresholds are well selected
Soil water content and potential	Neutron probe or TDR and tensiometers	Research	Very high
Remotely sensed soil moisture	Thermal infrared scanner	Large areas	Limited
Crop indicators			
Appearance and feel	Observation of plant (leaf) signs of stress	Crops that give sign of stress prior to wilting	Depending on farmers experience
Leaf water content	Sampling	Mostly for research	Limited
Leaf water potential	Pressure chamber, psychrometers	Mostly for research	Very high
Stomatal resistance	Porometer	Mostly for research	High
Canopy temperature (and crop water stress index)	Infrared thermometers	Field crops in large fields. Needs expertise and calibration	Very high when thresholds are well selected
Changes in diameter of stems or fruits	Micrometric sensors	Tree crops. Needs expertise and calibration	Depending upon selected thresholds
Sap flow measurement	Electronic sensors	Mainly tree crops. Requires expertise and calibration	Very high when thresholds are well selected
Climatic indicators Pan evaporation (and rainfall)	Evaporation pans and evaporimeters	Generally included in regional irrigation scheduling programs	High but depends on the way information is provided to farmers
Crop evapotranspiration (and rainfall)	Weather stations and crop coefficients	Generally included in regional irrigation scheduling programs	High but depends the way information is provided to farmers
Remote sensing			
Remote sensing of crop evapotranspiration	Vegetation Index (NDVI) from several wavelengths including the thermal infrared	Regional irrigation programs with farmers advice in near real time	Now becoming operative for farmers advice. Needs spatially distributed water balance model to support information

the farmers perform observations in the entire root zone and not only on the soil surface. The method applies to non-frequent irrigations such as for surface irrigation.

Often used for tree and horticultural crops is the measurement of the soil water potential with tensiometers, soil psychrometers or pressure transducers. All these instruments are highly precise and can be used easily. However, they require expertise and external support to appropriately define the irrigation thresholds relative to the timeliness of irrigation. They are recommended when irrigation depths are constant throughout the irrigation season. They can be highly effective to support deficit irrigation when related irrigation thresholds are well defined

Assessing the soil water content allows to define both the timing and volume of irrigation. The simpler method is soil sampling for laboratory gravimetric analysis but can not provide information in real time. Another method easy to apply but less precise uses porous blocks and electrode probes to sense changes in the electrical resistance of the soil due to variations in soil moisture. It has been widely used for cash crops. Other more precise methods use neutron probes, TDR (time-domain reflectometry) sensors, or capacitative sensors and may provide information in real time. Neutron probes became less popular because of health concerns relative to the use of a neutrons source. All these equipments need calibration according to the soil characteristics and all need appropriate selection of irrigation thresholds. Thus, advisory support services are required. These methods, mainly TDR and capacitative sensors, are appropriate for scheduling deficit irrigation.

Irrigation scheduling techniques using *crop indicators* are useful for assessing the timeliness of irrigation events, thus crop stress indicators are used when irrigation depths are predefined, as it happens with microirrigation systems, or are kept constant during the irrigation season. The easiest technique is by observing the appearance of the crop, i.e., by detecting any particular stress behaviour such as changes in leaf rolling, leaf orientation or leaf colour. Its effectiveness depends upon the farmers skill in interpreting stress signs.

Techniques such as measuring the leaf water content, leaf water potential and stomatal resistance require sophisticated equipment and expertise for observations and interpretation of results. Thus, they are essentially used for research. Data may be of very good quality for scheduling irrigations.

Observations of the canopy temperature (and determination of the related crop water stress index) using hand held infrared thermometers are applicable to field crops in relatively large areas. This technique is based on the fact that the surface temperature of a well watered crop is some degrees lower than the air temperature while the surface temperature of a stressed crop is close to the air temperature. The use of this technique requires calibration of the water stress threshold, which is crop specific. It applies better to non frequent irrigation and is useful for deficit irrigation scheduling.

Observation of changes in diameter of stems or fruits, which are measured by using micrometric sensors, is becoming widely used for tree crops, including for deficit irrigation. However, it requires expertise and well selected thresholds. Various commercial applications are available. The measurement of sap flow through plant stems is a technique also becoming popular. It is mostly applied to trees and other ligneous plants using appropriate electronic sensors and is also useful to assess crop evapotranspiration (Paço et al. 2006). It requires calibration and well defined irrigation thresholds and needs external support.

The irrigation scheduling techniques using *climatic indicators* are widely used for tree, horticultural and field crops, and they are commonly applied with water balance techniques. They require local or regional weather data and some kind of water balance approach. The degree of accuracy depends upon the nature, quality and spatialisation of weather data, as well as the water balance approach used. Preferably, evapotranspiration or evaporation data should be used as input of water balance computational tools or models. Information may be processed in real time or, more often, using historical data. Techniques include:

(1) Evaporation (Ev) measurements with evaporation pans and evaporimeters or atmometers. Crop evapotranspiration (ET_c) is estimated using appropriate coefficients that relate Ev to ET_c. Pan evaporation data should preferably be averaged to 7 or 10 days. Its effectiveness depends upon the reliability of Ev measurements and accuracy on computing ET_c. The easiest way is to just broadcast how much millimetres evaporation occurred in the past days. More useful is to inform, crop by crop, how much has been the ET_c in a given "homogeneous" area.

(2) Assessment of crop evapotranspiration (ET_c) using weather data from climatic stations to estimate the reference evapotranspiration (ET_0) and crop coefficients (crop factors relating crop and reference evapotranspiration) to compute ET_c from ET_0 (see Allen et al. 1998, 2007a). Direct ET measurements with lysimeters or micrometeorological instrumentation are only used for research. The effectiveness of the crop evapotranspiration assessment method depends upon the accuracy of data collection and calculation procedures as well as the approach used to provide information to farmers. ET_0 and ET_c may also be estimated with limited weather data (Popova et al. 2006a) and using weather forecast messages (Cai et al. 2007).

(3) Remote sensing of vegetation from satellite observations. Indices such as the Normalized Difference Vegetation Index (NDVI) are computed from several wavelengths, including the thermal infrared, which allows to derive a crop coefficient and therefore estimate crop water requirements (e.g., Bastiaanssen and Harshadeep 2005, Garatuza-Payan and Watts 2005, Allen et al. 2007b). Generated data may be used at various scales for supporting irrigation advising programmes (Calera et al. 2005). The use of water balance simulation models operating in a GIS (geographical information systems) may be very helpful to further transmit information to farmers. This technique has a great potential for supporting water saving at project or basin scales.

Soil water balance techniques, which estimate the crop irrigation requirements, are very useful for irrigation scheduling because they allow the determination of the irrigation dates and volumes to be applied. Often models are used. When observed climate and crop data referred above are available these data may constitute an input to the model. Both real time and historical weather data (rainfall and ET_c) may be used. Models may produce typical irrigation calendars with a relatively large time step calculation or daily calendars adjusted to real time weather conditions. Soil water holding characteristics should be considered in addition to crop and climate data when performing the soil water balance to take into consideration the use of the stored water by the crop and, whenever possible, the capillary rise from the ground-water (e.g., Pereira et al. 2003d, Liu et al. 2006). Further sophistication in modelling includes the spatial distribution of data when operating the model in a GIS platform for supporting regional irrigation scheduling programmes (e.g., Fortes et al. 2005). Technologically advanced farmers compute themselves the water balance when they have access to rainfall and weather or ET_c data and have support to describe soil and crop conditions. Alternatively, they may accede to a model through an assisted Web page.

Irrigation scheduling techniques can only be applied when the farmers have control on the timing and duration of irrigation. However, the full effectiveness of irrigation management can only be achieved when the farmer is able to control the timing and the depth or volume of the irrigation. When surface irrigated areas are supplied from collective irrigation canal systems, farm irrigation scheduling depends upon the delivery schedule, e.g. discharge rate, duration of the deliveries and frequency of supplies, which are dictated by the operational policies used for the conveyance and distribution system (see Section 10.7). Discharge and duration impose constraints to the volume of application, and frequency determines the timing of irrigation. Surface irrigation delivery systems are often rigid and the time interval between successive deliveries may be too long. Then, farmers apply all water that is made available and over-irrigation is often practiced (Zardari and Cordery 2007). Under these conditions the water saving practices depend upon the off-farm conveyance and distribution. Therefore, the improvement of the on-farm irrigation systems should go together with the modernisation of the delivery system to allow more flexibility in selecting the appropriate inflow rates and supply times.

Pressurised conveyance and distribution systems are often managed on demand, i.e. farmers are in control of irrigation timings and duration. However, to certain limits imposed by economic and technical reasons, discharges available are limited. Therefore, irrigators supplied by these systems are free to select and adopt the irrigation schedules they consider more appropriate to their crops and farming practices. When rigid delivery rules are enforced by the system managers during periods of drought or limited water supply, farmers have to adapt irrigation timings and duration in accordance with the periods when water is delivered, but they keep some degree of freedom to schedule irrigations. In alternative, if water is measured at the farm hydrants, system managers may impose restrictions on volumes and price penalties for excess water use. Then farmers have to optimise the respective schedules aiming at reducing the demand. Pressurised systems are therefore far more favourable than surface systems for adopting rational irrigation scheduling practices, including when supply restrictions are imposed.

10.6.6.2 Reduced Demand Irrigation Scheduling

Reduced demand irrigation scheduling implies the adoption of some irrigation scheduling technique, empirical or scientific. The information above, summarised in Table 10.10, needs to be added by that referring to how irrigation scheduling techniques may be used for reducing the demand for irrigation water, as summarised in Table 10.11. However, it is appropriate to discuss the constraints of the irrigation methods relative to the application of irrigation scheduling.

In surface irrigation, irrigation depths are generally large to very large, 80–120 mm, depending upon the system characteristics as reported earlier (10.6.3) Larger values correspond to deep rooting crops in heavy soils. Smaller depths (30–50 mm) are only feasible when the field is precisely levelled and some degree of automation is used. Therefore, the rule is the adoption of large irrigation depths and large time intervals between irrigations.

In sprinkler irrigation, depths depend upon the system. Set systems may apply very small depths (5–10 mm) with very high frequency when permanent and automated systems are available, up to large and infrequent irrigations when systems are movable. Generally medium depths, around 30–40 mm, with several days interval are common. Travelling rain-guns generally apply 20–40 mm at each event. Lateral moving systems are designed for small up to medium depths, 5–30 mm, and to high irrigation frequency. Micro-irrigation systems apply from very small depths, such as few millimetres several times in the day, up to 15–20 mm for intervals of 2 or 3 days. These systems may be used for larger intervals when applied for supplemental irrigation in orchards and vineyards.

The irrigation scheduling methods reported above may be used to define irrigation timings and depths summarized in Table 10.11 where the irrigation frequency, irrigation depths and the applicability of the methods are referred.

Crop appearance and soil moisture feeling methods are less appropriate when aiming to reduce the demand for water. They base upon farmers knowledge about the periods when water stress affects crop yields. The method is inefficient when a new crop is introduced. Differently, soil water observations are very useful to support farmers' decisions using a target irrigation threshold, generally called soil water management allowed depletion (MAD). The MAD changes from crop to crop and with crop stages, and depends upon the water availability for irrigation. Thus, advisory support is required. The MAD values have to be defined in accordance with the irrigation method and in agreement to the constraints imposed by the irrigation system as mentioned above. The technology involved is easy to be implemented in large farms and is feasible for small farms when regional or project irrigation management programs are implemented and observations are performed in selected demonstration farm fields.

Crop stress indicators are generally used for cash crops but, as for canopy temperature indicators, they may be also used for field crops. Crop stress indicators provide for the decisions on the irrigation timings. These have to be selected not only in relation to the crop and crop stage but also taking into account the constraints imposed by the irrigation system. Thresholds are lowered when water is limited and may be

Table 10.11 Irrigation management for coping with water scarcity when farmers are in control of irrigation timings and volumes

Managerial information/target	Irrigation frequency	Irrigation depth	Applicability
Crop appearance			
Crop observation and knowledge of crop critical stages by farmers	• Irrigation at planting if required to crop emergence • First irrigation delayed for root development • Low irrigation frequency, avoiding heavy stress at critical crop growth stages	• Surface irrigation: large depths dictated by system conditions, • Sprinkling: depth selected to refill the root zone under constraint of system characteristics	• Field crops • Surface irrigation and low frequency sprinkler systems
Observed soil water			
Management allowed soil water depletion (MAD). The MAD threshold changes with crop stages and water availability	• Irrigation when the MAD is attained • Low irrigation frequency for surface irrigation • High irrigation frequency for micro-irrigation and lateral move systems	• Surface irrigation: depths dictated by system conditions • Sprinkling and micro-irrigation: depths according to system characteristics	• For most crops and cropping conditions • For most irrigation methods
Observed crop stress indicators			
Crop stress thresholds specific to crops and sensed data. Thresholds are lower when limited water is available	• Irrigation timings when thresholds are attained • Low irrigation frequency for surface irrigation • Higher frequency for micro-irrigation and lateral move systems	• Surface irrigation: depths dictated by system conditions • Sprinkling and micro-irrigation: depths according to system characteristics	• Most crops, more often cash crops using sprinkler and micro-irrigation • Commercial farms with technological expertise
Climate information			
Daily or several-day information on evaporation and/or crop ET to estimate several day crop water consumption	• Irrigation dates decided when cumulated ET equals the target depth to be applied • Frequency according to crop and irrigation method but lower when water is limited	• Surface irrigation: large depths dictated by system conditions • Sprinkling and micro-irrigation: depths according to crop and system characteristics	• Most crops and irrigation methods • Easy for climates where rainfall is negligible during irrigation season • Regional advisory services

Table 10.11 (continued)

Managerial information/target	Irrigation frequency	Irrigation depth	Applicability
Water balance information including for real time scheduling			
Weather data (ET and rainfall) to compute irrigation needs. Actual irrigation and soil water data to update the water balance for real time. Targets: an economic acceptable yield reduction or MAD	• Irrigation dates decided when the cumulated soil water depletion equals the target depth to be applied • Frequency in accordance with crop and irrigation method, but lower when water is limited	• Surface irrigation: large depths dictated by system conditions • Sprinkling and micro-irrigation: smaller depths in accordance with crop and system characteristics	• For every crop, cropping system and irrigation method • Commercial farms • Irrigation advisory services • Use of information systems including through the Web

optimised when deficit irrigation is used. Crop stress indicators require expertise and commonly support by irrigation advisors.

Information on weather data, such as daily or several-day information on ET, including when computed from weather forecasts and remote sensing, is very useful to estimate crop water consumption, mainly using a model. The irrigation amounts and dates may then be easily estimated in agreement with the characteristic of the irrigation system used. The irrigation frequency may be lower when water supply is limited, thus allowing some crop water stress. It can also be used by small farmers when the information is broadcasted or provided to their community, however not expressed in mm but in the units used by farmers.

Water balance information resulting from weather data input to a water balance model to compute the irrigation requirements is probably the most often used approach. The water balance may be performed using simple hand written check books, computer spread sheets, or simulation models. Irrigation data observed by the farmer (irrigation date and estimated volume applied) is used to update the water balance including when farmers are connected to a central via web. Irrigation thresholds and water depth volumes may be easily selected considering the constraints imposed by the crop, the system and the water availability. The above referred soil MAD is part of input data. The method is best applied when daily crop ET and rainfall are available in real time and/or resulting from remote sensing. Simulation models apply various farms when models run in a GIS database.

The brief methodological review presented above shows that the implementation of irrigation scheduling practices require the involvement of water and agriculture management institutions, farmers organisations, water user associations, and NGO's to stimulate farmers and the providers of the technologies to make use of them. In other words, to just say that farmers have to adopt improved scheduling methods is not appropriate. Farmers do it when they understand the benefits, are informed, receive appropriate support and the communities are involved. Thus, before adopting new technologies for irrigation scheduling it is necessary that problems to be solved in a given location be clearly identified, objectives be set, tools and means be made available by the government, the market or any other institution. Then farmers apply an improved schedule that will increase water productivity, control environmental impacts, save water and increase agricultural incomes.

10.6.6.3 Deficit Irrigation

Deficit irrigation, as mentioned before, is an optimising strategy under which crops are deliberately allowed to sustain some degree of water deficit and yield reduction (English and Raja 1996). Deficit irrigation is a sustainable issue to cope with water scarcity because the optimisation approach leads to economical viability and the allowed water deficits favour water saving, help controlling percolation and runoff return flows and reduces the losses of fertilisers and agro-chemicals. In addition it shall provide for application of rational leaching requirements to cope with salinity. The adoption of deficit irrigation implies appropriate knowledge of crop ET, of crop responses to water deficits, including the identification of critical crop growth periods, and of the economic impacts of yield reduction strategies. Therefore, some degree of technological development is required to support the application of deficit irrigation scheduling techniques.

Generally, deficit irrigation schedules are built upon validated simulation models or on extensive field trials. Field trials essentially concern the evaluation of the impacts on yields resulting from different irrigation depths and timings. Economic data have also to be considered. Simulation models must include appropriate yield-water functions to assess the yield impacts of water deficits and should be calibrated using field trials data (e.g. Popova et al. 2006a). Economical data are often lacking and extrapolations from other studies are difficult because costs and benefits change with farming systems, labour, production factors and water prices, while benefits vary with the price of the harvestable yields and local markets. Summarising, there is a great uncertainty and risk in the decision process relative to deficit irrigation. More research is therefore required and better information has to be produced to support farmers' decisions.

Two examples of deficit irrigation studies in Central Tunisia are shown here, one relative to a cereal crop, winter wheat, and the second to a cash crop, tomato (Zairi et al. 2003). Selected results for the wheat crop are given in Fig. 10.18 relative to average and drought demand conditions. They show that a decreased supplemental irrigation leads to reducing the gross margin per unit surface cropped (GM/ha) but to an increase of the gross margin per unit water applied (GM/m^3). This indicates that, both gross margins being positive, it is feasible to adopt deficit irrigation of wheat under drought conditions despite lower yields and incomes.

Differently, for a summer cash crop as tomato, results in Fig. 10.19 show that the GM/m^3 increases for average demand conditions when less water is applied, behaving similarly to the wheat. But when the climatic demand increases such as under drought, GM/m^3 increases only for small values of the water deficit. This indicates that when the crop relies essentially on irrigation and the rainfall contribution is small the economic threshold is high, i.e. larger water deficits are not economically

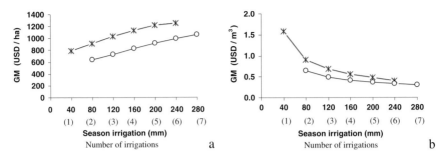

Fig. 10.18 Gross margins per unit surface (**a**) and per unit volume of water applied (**b**) for alternative deficit irrigation strategies for the wheat crop in Central Tunisia under average (*), and very high (o) climatic demand conditions (Zairi et al. 2003)

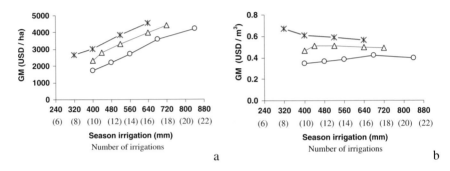

Fig. 10.19 Gross margins per unit surface (**a**) and per unit volume of water applied (**b**) for alternative deficit irrigation strategies of tomato in Siliana, Tunisia, for average (*), high (Δ) and very high (o) demand conditions (Zairi et al. 2003)

viable. It must also be noted that an appropriate adoption of deficit irrigation requires that the irrigation systems' performance is good. When the distribution uniformity is low deficit irrigation easy leads to under-irrigation in the part of the field receiving less water resulting that the crop is heavily stressed and impacts on yields are high.

10.7 Supply Management

The importance of supply management strategies to cope with water scarcity is well identified in literature and observed in practice, particularly for surface irrigation systems which constitute the majority of large and medium water systems in water scarce regions (Renault et al. 2007).

In the improvement of farm irrigation systems was analysed in the previous sections it has been shown that the role of farmers in reducing the demand is limited both by the farm system constraints and by their capabilities to be in control of the discharge rate, duration and frequency of irrigation. These limitations are more important relative to deficit irrigation because farmers require some flexibility in the deliveries to decide the optimal irrigation timings and depths, and deliveries need to be reliable, dependable along the irrigation season and equitable among upstream and tail end users. Therefore, the adoption of reduced demand strategies largely requires improved quality of supply management (Goussard 1996, Hatcho 1998).

Supply management is generally considered under the perspective of enhancing reservoir and conveyance capabilities to provide higher reliability and flexibility of deliveries required for improved demand management. This is true not only for irrigation systems as mentioned above but also for non-agricultural water uses as analysed in previous sections. Therefore, supply management includes:

- *Increased storage capacities*, including large reservoirs with capacities for inter-annual regulation, and small reservoirs, namely for supplemental irrigation,
- *Improved conveyance and distribution systems*, including compensation and intermediate reservoirs for improving the flexibility of deliveries and avoiding system operation water losses during periods of low demand, as well as improved regulation and control, including automation and remote control, to favour that deliveries match the demand,
- *Good maintenance* of reservoir and conveyance systems, which is a pre-condition for water saving and the reliability of deliveries, and
- The development of *new sources of water supplies* to cope with extreme conditions of water scarcity, as it is the case for droughts.

Supply management is also considered under the perspective of systems operation, particularly related to delivery scheduling, thus including:

- *Hydrometeorological networks*, data bases and information systems to produce appropriate information for effective implementation and exploration of real time operation of any water systems,
- *Agrometeorological irrigation information systems* including tools for farmers to accede to information, often comprising GIS, to support local or regional irrigation management programs, as well as decision support systems serving the reservoir operation, the water system management, and the users to select crop patterns, irrigation scheduling and irrigation systems, and
- *Planning for droughts* to establish allocation and delivery policies and drought operation rules.

Supply management for irrigation also refers to the farm water conservation, as indicated in Table 10.12. In dry semiarid and arid zones, water harvesting plays a central role. In other water scarcity areas different issues may be considered.

The management of irrigation and multi-purpose water supply systems to cope with water scarcity includes the utilisation of a variety of managerial techniques and tools that are summarised in Table 10.13 and described below. Particular references are made to irrigation systems because these systems are the most frequent and larger water systems in water scarce regions. However, outlined issues also apply to municipal and industrial uses with due adaptations to the respective operational conditions since these systems are generally pressurised, have centralised control, and operate regularly throughout the year.

Table 10.12 Supply management for water conservation in agriculture

Objective	Technology
Increased storage	Small reservoirs for runoff storage
	Groundwater recharge from excess runoff
Increase water yield	Water harvesting
	Vegetation management to control runoff
Increased use of rainfall	Spate irrigation
	Micro-catchments, land forming
	Terracing
	Conservation tillage
Add to available supplies	Unconventional water systems
	Reservoirs, conveyance and intra-basin transfers

Additional emergency supply for drought mitigation concerns the exceptional use of waters during periods when the normal sources for the storage, conveyance and distribution are insufficient to meet the demand. These periods are limited in time such as the duration of a drought, but the operation of these additional sources has to be planned in advance to be effective and to prevent negative impacts on health and on the environment.

1) *New sources of surface water*, including the use of the dead storage in reservoirs and short distance water transfers from nearby systems and or sub-systems, generally associated with negotiations of water rights among farmers and non-agricultural users, including for nature, recreation, municipal and industrial uses. Since most of surface water sources in water scarce areas are already developed and water rights assigned, the use of additional waters requires appropriate planning and institutional framework.

2) *Increased groundwater pumping*, changing from the exploration of the perennial yield to the mining yield. A continuous mining of the groundwater is not sustainable but a controlled use of the mining groundwater yield during periods of water scarcity is sustainable when appropriate planning, management and monitoring are adopted, as discussed in Chapter 7.

3) *Transfer of water rights* between users, where those having the right for a given fraction of stored volumes sell temporarily these rights to other users. This applies to societies where individual water rights are well recognised by law, thus being of difficult application in other societies.

4) *Use of low quality water and reuse of wastewater* for irrigation of agricultural crops, and landscape, including lawns and golf courses, in addition or in alternative to water of good quality. Aspects relative to the safe use of these non-conventional waters are dealt in Chapter 8.

5) *Develop conjunctive use* of the surface waters that are mobilised through the existing water systems and the waters from emergency sources such as groundwater mining and non-conventional waters mentioned above. Adopting a conjunctive use approach provides for rational and sound water resources allocation and maximises the use of the available rainfall in addition to the diverted water

Table 10.13 Management of irrigation and multi-purpose supply systems to cope with water scarcity

Management techniques	Benefits	Applicability
Add to supply for drought mitigation		
• New sources of surface water, short distance water transfers	• Increase local water availability	• Difficult, needs integrated planning
• Increased groundwater pumping	• Adds to normal water sources	• Needs control/monitoring
• Transfer of water rights	• Re-allocation of available water	• In accordance with existing water laws
• Use/reuse of low quality water for irrigation and landscape	• Alternative sources of water	• Depending on crops and uses. Monitoring required
• Conjunctive use	• Maximises use of available rainfall and water resource	• Needs appropriate planning and management
• Reinforce the use of other non-conventional waters	• Prevents extreme scarcity	• Where non-conventional waters are already in use
Improved reservoirs operation		
• Information systems, including remote sensing, GIS, models	• Information for optimised operation and management	• Non limited but expensive for small schemes
• Hydrological forecasting and drought watch systems	• Improved assessment of supplies	• At large scale, large projects or regional level
• Upgrading monitoring	• Improved use of operation tools	• General but involving costs
• Application of optimisation, risk, and decision models	• Optimised management rules; and water allocation	• Non limited but expensive for small schemes
Conveyance and distribution systems		
• Canal lining	• Avoidance of seepage	• Limited by costs
• Improved regulation and control	• Higher flexibility, better service and reduced operation losses	• Needs investment and technology
• Automation and remote control in canal management	• Improved delivery management and low operation losses	• High technological requirements
• Low pressure pipe distributors	• Reduced spills and leaks, higher flexibility, easier water metering	• Limited by costs but easy to implement
• Change from supply oriented to demand oriented delivery schedules	• Favours farmers to apply water saving irrigation management	• Needs communication tools
• Intermediate storage (in canal, reservoirs, farm ponds)	• Increased flexibility and reduced operation losses	• Requires investment and management tools

Table 10.13 (continued)

Management techniques	Benefits	Applicability
Delivery schedules and rules		
• Involve farmers in delivery schedules planning to cope with limited supply	• Allows farmers to adopt best management practices	• Needs appropriate institutional arrangements
• Adopt demand delivery scheduling in pressurised systems	• Higher flexibility for water saving at farm	• Only constrained by the system characteristics
• Water prices in relation to volumes of water diverted and times for use	• Induce farmers to save water and to irrigate by night (automation)	• Requires appropriate water metering and water pricing
• Information systems	• Provides for optimised operation maintenance and management	• General but primarily to large schemes
• Application of optimisation methods to schedule deliveries	• Increased reliability and equity, and reduced farm demand	• Needs feed-back information from farmers
Maintenance and management		
• Effective systems maintenance	• Avoids spills and leaks and improves operation conditions	• Requires planning and trained staff
• Water metering	• Data for operation and billing	• Requires equipment
• Monitoring system functioning	• Identification of critical areas and system losses	• Requires planning and staff
• Assessment of system performance (physical, environmental and service)	• Provides follow-up on water saving programmes	• As above
• Personnel training	• Allows to implement more demanding technologies	• General, not highly costly when well planned
• Information to farmers and other users	• Knowledge on the system constraints and saving issues	• General

resources. Nevertheless, conjunctive use requires appropriate management of the water systems and good technological operative conditions.

6) *Reinforcing the use of non-conventional waters* such as rainfall harvesting for domestic uses – drinking water cisterns – and fog collection, referred in Section 8.5. These issues apply to areas where water collection systems are already available and in use since during drought they may not mobilise the required water quantities.

Improving reservoirs operation aims at controlling operational losses and better allocation the available water. In few cases, measures may include changes in equipment that control releases but most of improvements concern management. Several aspects relative to the improved use of surface waters are discussed in Chapter 6, particularly relative to small systems and reservoirs. Issues include:

1) *The use of information systems and modern technologies* such as remote sensing, GIS, and models that provide for the state variables relative to the reserves in storage and to the uses and demand. Information systems are essential to appropriately explore reservoir decision tools mentioned below and to create information relative to users decisions in agreement to the water availability.

2) *Hydrological forecasting and drought watch systems* that allow for real time, or near real time prediction of reservoir inflows, evaporation and seepage, thus to better estimate the time evolution of storage that dictate the operation rules. Reservoir losses are discussed in Section 6.2. These management tools may also be useful to estimate how the demand will evolve when appropriate feedback is developed with canal managers in case of irrigation uses. Then, the support by information systems is particularly useful.

3) *Upgrading monitoring of reservoir inflows and releases* to better support the adoption of information tools mentioned above.

4) *Application of optimisation, risk, and decision models* to reservoir operation, to define optimised system management rules and to decide the allocation of water resources among the different users – municipal, irrigation, industrial, recreational, energy and nature. A large panoply of such tools has been made available by research, many are in use, but their adoption under water extreme water scarcity still is low due to the difficulty in gathering the information required to optimise decisions.

5) *Changes in reservoir water release equipment* for more accurate control of volumes supplied, easy adjustment of discharges in the course of the time, and adopting automation when decision models are effectively operating in real time.

Improving conveyance and distribution systems refers both to equipment and management software. An enormous amount of research has been recently devoted to these subjects and a very large number of papers and books refer to these matters, particularly for irrigation systems. Approaches that help coping with water scarcity are generally oriented to control seepage and operational losses, to provide for higher flexibility in water deliveries to irrigated farms, and to improve the levels of service by matching supply to demand, increasing the reliability of supplies and enhancing the dependability of deliveries along the operation season and the equity of the distribution in the areas served. Main issues relate to:

1) *Canal lining* to avoid seepage. However, canal lining is only fully effective when canal management is improved, maintenance is carefully and timely performed, and other canal structures are also improved for enhanced conveyance and distribution service. Otherwise, investment costs may not be justified and resulting water costs may be excessive for farmers if the water service remains at low performance levels.

2) *Improved regulation and control* of canal and pipeline systems, including local or centralised automation, generally permits higher delivery flexibility and

improved conditions at farm level to adopt water saving irrigation practices as discussed in Section 10.6. Adopting appropriate regulation and control provides for reduced operation losses and for easy maintenance since water levels vary much less during operation. In general, reliability, adequacy and equity of deliveries are enhanced, while dependability depends upon the policies for reservoir management. Increasing the levels of service gives better opportunity to farmers to adopt improved farm irrigation, and to include practices that lead to water conservation and saving, as well as to control environmental impacts, especially in saline environments. Appropriate regulation and control is probably the most important issue for irrigation and multipurpose water systems. However, regulation and control systems may be expensive and require technological capabilities to be fully explored despite local control automation may be easy to adopt and apply.

3) *Automation and remote control in canal management* is a step further in technological advancement in regulation and control. Remote control allows a better operation of the system particularly to take into consideration the users demand in real time. The appropriateness of these technologies to cope with water scarcity relate to improved water service, more easier application of irrigation scheduling and the farmers ability to improve the water use leading to higher water productivity and water saving. Because high technology is required and the demand has to be known in real time, these technologies adapt better to systems serving large and commercial farms.

4) *Adopting low-pressure pipe distributors* in surface irrigation systems instead of open channels and ditches is an effective solution to reduce spills and leaks, to achieve higher flexibility and service performance, and to easily adopt water metering. The investment costs may be compensated by lower operational costs when compared with open channel distributors. Benefits at farm level are once again related with the flexibility in the deliveries. This technology has no particular technological requirements but requires updated design.

5) *Changing from supply oriented to demand oriented delivery schedules* is the desirable orientation of system management when regulation and control are reliable enough. In fact, demand oriented delivery schedules assume that managers give priority to satisfy the demand rather than optimising the supply service. Thus, it makes possible that farmers apply improved irrigation schedules to save water and increase water productivity. It requires that regulation and control be modernised and some kind of communication between farmers and managers is adopted. This communication may be performed through direct contact between farmers and canal operation personnel, by phone or via computer.

6) *Intermediate storage* in canal reaches, small reservoirs linked with selected canal nodes, or farm ponds are often used to increase the flexibility of the system to respond to variations in demand, to reduce operation losses during periods of reduced water use such as the night-time and holidays, and to permit the use of farm irrigation systems having discharge and duration requirements different of those provided by the delivery schedule practiced. The latter is

the case of farms adopting micro-irrigation or sprinkling where distribution systems adopt delivery schedules designed for surface irrigation.

7) *Involving farmers in decisions to change delivery schedules dictated by limited supply*. During periods of limited water supply the delivery schedules have to be modified in order to satisfy the priority uses and to enforce restrictions for other uses. When farmers are involved in the decision process that leads to modify the delivery schedules, or to fix the respective water quotas, the farmers may negotiate these emergency measures in order to adopt the best practices that accommodate with water constraints to be enforced. This involvement may be difficult in systems serving a large number of small farms but the involvement of farmers in this decision is important for the effective adoption of emergency water saving programmes.

8) *Adopting demand delivery scheduling in pressurised systems* because this is generally the most appropriate for the required flexibility to the farm use of sprinkler and micro-irrigation systems. As analysed above (Section 10.6) these systems may be managed for a variety of irrigation depths and frequency, thus for adopting water saving at farm too. When water supply is limited and restrictions to the demand have to be enforced the respective decisions should better involve the farmers. However, adopting rigid schedules in pressurised systems generally do not lead to easy adoption of saving at farm level.

9) *Adopting water prices that induce farmers to irrigate by night* is a policy particularly appropriate for pressurised systems but that can also be used for open canal systems when some kind of metering is adopted. It consists in differentiating the day-time and night-time water prices to induce night irrigation, which gives larger flexibility to the system operation, improves the service performance and reduces operational losses due to excess water flowing in the systems at night, when the demand is lower. In addition to metering, automation of farm systems is also necessary.

10) *Adopting water prices that induce farmers to save water* may be an appropriate policy for pressurised systems where metering is available. It can also be used for open canal systems but its application is then quite difficult. It consists in water prices that vary in accordance with the water use, which increase after a given volume is diverted to the farm. The price structure ideally varies with the type of crop following policies on cropping patterns, and with the available supply, both affecting the minimum volume and the rate for increasing the prices. Water metering is essential for a fair application of this policy. Systems where water costs are associated with the land surface cropped may adopt alternative pricing policies differentiated by crop and water volume but its enforcement requires appropriate field surveys to check the areas declared by the farmers. Water pricing must take into account the feasibility of the water user activities and not lead them to abandon them due to exaggerated costs.

11) *Information systems* may play an important role when decision models are used for systems operation or to help farmers selecting the respective management to cope with water scarcity. Information systems are particularly useful to identify

the state variables of the system, inclusive in real time, thus providing for better matching deliveries to demand. Information systems are even more useful in multipurpose systems to support decisions on allocation of water by user sectors, especially when the management of the reservoir and the conveyance and distribution system are linked.

12) *Optimisation tools for water allocation and to schedule deliveries* are techniques that complement those modelling and decision tools mentioned before. Difficulties in application are due to insufficient economic information to adequately perform optimisation, and to the required feed-back information from the users. However, when it is possible to be applied, it may support achieving higher reliability, adequacy and equity in deliveries and the enforcement of water saving.

Systems maintenance and management are essential to cope with water scarcity. When these are adequate, they provide for controlling water wastes, seepage and water spills and provide for water saving. When maintenance and management are poor, not only system losses are high but the water service is poor, less reliable and non-dependable, tail end users receive the poorest service and incentives for the users to save water are lacking. Issues relate to:

1) *Effective systems maintenance* which is required not only to avoid seepage, water spills and leaks but for adequate operation of the hydraulic structures, regulation and control, and good service to users. Maintenance needs trained personnel and equipment as well as planning. Particular attention should be paid to periods when water supply is limited, when all available water is insufficient to meet the demand.

2) *Water metering* flow depth and discharges in surface systems and pressure and discharges in pressurised systems – is required to support operation and, at outlets and hydrants, for billing the users. Data from metering also provides basic data on system variables useful for management. The adoption of information and decision support tools mentioned above is not possible without metering at critical nodes of the system and, for more advanced technological levels, without metering the water deliveries to users.

3) *Monitoring* system functioning and system performance is required to identify the critical reaches of the conduits and canals and service areas and to provide the follow-up of maintenance programmes, improvements in equipment, implementation of upgraded management tools, as well as the quality of service provided. Monitoring allows to quantify system operation losses, priority areas for improvement and to evaluate upgrading programmes. Data produced are an essential input to information systems and decision support tools in addition to metering. As for metering, monitoring does not produce water saving but is essential for their effective implementation.

4) *The assessment of the system performance* physical, environmental, economic and service performance – provides for the evaluation of the actual functioning of the systems complementing monitoring and metering. Actual indicators

are useful for planning modernisation, rehabilitation and betterment of the systems and are generally useful to base decisions required for implementing such programmes, consequently for planning for water saving and conservation.

5) *Training of personnel in operation, maintenance and management* is required to enhance the quality of service, for technological upgrading of the systems and to carry out water saving programmes. Training is also necessary to develop skills required to contact with the public and for the communication with users.

6) *Users and farmers information* is of paramount importance to increase the awareness of the value of the water and the importance of water saving. Information is required to involve farmers and other users in water saving and conservation programmes. The need for increased awareness, education and training is discussed in Chapter 12. Information is considered herein as comprising training and participation in the decision making process, as it is desirable and achievable in case of farmers. On the one hand, there is the need for training and assistance to farmers to improve on-farm irrigation systems and scheduling, and to implement water conservation and demand reducing practices. On the other hand, measures have to be adopted to implement effective farmers' participation in decisions relative to supply management. This includes the information to farmers on characteristics and limitations of the water supply systems.

In the irrigation domain, the improvement of conveyance and distribution systems must go together with improvements in farm irrigation as well as in drainage. An example referring to an irrigation system located in the Upper Yellow River basin, China is presented to illustrate this approach. It refers to the Huinong irrigation system, located in an arid region, but where too much water is allocated and diverted to irrigation. Upstream diversion facilities and the conveyance and distribution canal system are in the need for modernization, as well as the drainage system (Pereira et al. 2003b). Field studies allowed the calibration and validation of various models for irrigation management (Pereira et al. 2007a, Gonçalves et al. 2007). Using a decision support system and field data it was possible to foresee the evolution of water use both off- and on-farm assuming an immediate reduction in the water diversion to decrease seepage, runoff and percolation, thus to control waterlogging and salinity and to improve crop conditions. Progressive improvements in the canal system refer to control and regulation structures, canal lining, management of diversion structures and changes in delivery schedules. At farm, improvements refer to basin irrigation and scheduling. In addition, crop patterns and the rice cropped area may also change. Results for water use are presented in Fig. 10.20, which show a reduction of more than 50% is achievable, mainly due to reducing the N-BWU relative to the canal system. Figure 10.21 shows that the economic water productivity referring to the total irrigation water and to the farm shall highly increase.

To be noted, however, that more stringent improvements may not be economically viable since improvements after the first 5 years show low impacts (Fig. 10.20) and the gross margin per unit volume of water use increases at a very low rate after main improvements are considered (Fig. 10.21). This relates with ratios between production costs and yield value.

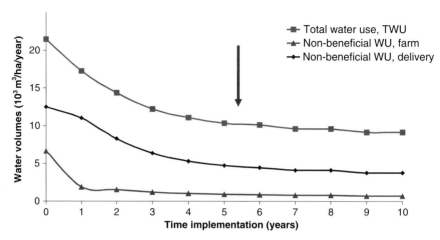

Fig. 10.20 Foreseen dynamics of total water use and non-beneficial water use at farm and system levels along the process of implementation of improvements in irrigation and drainage systems, Huinong irrigation district

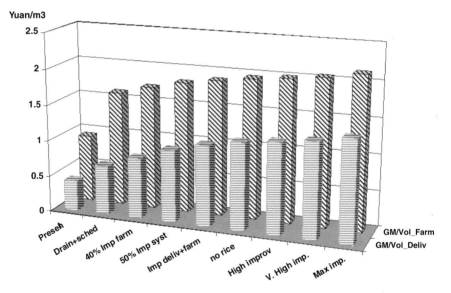

Fig. 10.21 Foreseen evolution of the economic water productivity due to the progressive irrigation and drainage improvements at farm and system level, Huinong irrigation district

10.8 Concluding Remarks

This chapter outlines a large number of practices and managerial tools for water conservation and saving to cope with water scarcity. First, these practices and tools were presented in very general terms in relation with the different water scarcity regimes (Section 9.4): natural aridity and drought, and man-made desertification and water shortage. In this Chapter, these water conservation and saving issues are dealt with more detail relative to the main water use sectors. However, in this analysis there is no focus on the source for water scarcity.

Most of water conservation and saving practices and tools are known from professionals in the respective domains. Many of them are already practiced by many communities world-wide. Thus the main idea herein is to have them outlined in a publication for the general public, not focusing a specific group of professionals. It is important that professionals in every water use sector know and understand how other sectors may find solutions to cope with water scarcity. This is why those issues are analysed in relation to the most common difficulties for their implementation. This is particularly the case for irrigated agriculture where a large panoply of practices and tools are available but whose application is yet disappointing. As explained above, there are reasons for that unfavourable situation. When professionals and managers in other areas could understand why progress is not made they may favour to find more appropriate conditions.

Issues outlined do not focus the most stringent water scarcity conditions. Practices and tools must be known and selected in accordance with the problems to be solved. The practices and managerial tools have been outlined as a guide for further selecting, evaluating and implementing those more adequate to a given region or country. The adoption of water conservation and saving is not only a technological problem but involves many other considerations relative to the social behaviour of the urban and rural communities, the economic constraints, or the legal and institutional framework that may favour the adoption of some measures and not other.

Moreover, adopting water conservation and saving practices and tools requires that a better knowledge of the water scarcity regimes be developed, the water be given a social, economic and environmental value in agreement with the society, and education for water be effectively installed. For these reasons, water conservation and saving constitute just one chapter in a wider book.

Chapter 11
Social, Economic, Cultural, Legal and Institutional Constraints and Issues

Abstract The social, economic, cultural, legal and institutional constraints and issues relative to water management to cope with (high) water scarcity are discussed as oriented to local communities, urban developments, rural areas, user groups, and administrative, public and private organisations which are concerned with water supply.

11.1 Local Communities

In regions of (high) water scarcity small communities have usually developed, over time, informal systems for water sharing, use and disposal. These systems may not be efficient or equitable but some degree of acceptance usually prevails. Problems are likely to occur in such communities when the population size changes or when new ideas are introduced or developed for water use. Cooperation is needed between individuals – in fact is vital for the survival of the system and for harmony within the community. There is usually, or there needs to be, an agreed set of expectations and operating rules. It is likely these will not have any formality, will not be written down, but will be widely accepted. Friction and upheaval is likely to occur if any newcomers to the community do not accept or adhere to the accepted practices. Hence in summary, cooperation and thoughtfulness towards others is required.

Communities in regions of water scarcity are particularly vulnerable to actions of nearby communities. For example a community which catches surface water from the few runoff events that occur during a year are likely to be adversely affected by any developments upstream either of capture of the water resource by someone else or of release of wastewater. Similarly the community itself may adversely affect downstream users. Hence care, sensitivity and cooperation are needed at all times. Similarly where a local community is dependent on groundwater, any actions it takes, or actions of others who access the same aquifer will affect all users of that aquifer. Again cooperation is required. However as the scarce water resource comes under increasing pressure, cooperation and dependence on the care and sensitivity of all involved is usually insufficient and it becomes necessary to introduce more formal agreements and even to develop institutional and legal frameworks for exploitation of the resource. Without such institutional arrangements upstream and/or economically more powerful communities will tend to dominate and jeopardize the survival of other communities. Where water is abundant such considerations may not be necessary but in regions of water scarcity, unless the population is very stable and continues traditional water use practices, problems are almost certain to occur. Populations almost always increase and so innovative approaches for water supply are usually needed. Without introduction of some kind of institutional arrangements conflicts are likely to be bitter and protracted.

One means of avoiding or overcoming disputes where the resource is not able to meet everyone's expectations is to innovate. There are many opportunities for local communities to either use the available water more effectively, and therefore need to consume less, or to develop additional resources. Most local communities already utilize the obvious and traditional sources of water such as the local, perhaps ephemeral, stream or groundwater. However there are often other neglected resources, which may appear small, but which in fact may be capable of increasing the water availability by a considerable amount. Most innovative water capture or exploitation opportunities for communities in regions of water scarcity are very site specific. From the list that follows there may be none applicable to some local communities but in other areas it may be that more than one of these ideas could be utilized.

Some innovations that could be utilized may be

- Re-design of roofs to allow easy capture of roof runoff
- Development of a receptacle-making industry for the manufacture of storage tanks to hold roof water or water from other sources while minimizing seepage and evaporation losses and discouraging disease vectors. For example in northeast Thailand the development of widespread use of ferro-cement jars and tanks with capacities from a few hundred, up to a few thousand litres has revolutionized social patterns in this region which has a 9–10 month annual dry period. Prior to development of roof catches and water jars families drifted to city-centres late in the dry season causing large social disruption. The necessity for this annual movement has been largely removed by the availability of water storage facilities for each household. These storage jars/tanks are constructed using local materials and skills, and in addition to securing a supply of potable water their manufacture is adding to the local employment base.
- Import water from a region of water abundance. This solution usually has high capital demands but may be useable in some regions (as discussed in Section 8.5). It usually involves transfer of water from one basin to another and so environmental considerations may also be quite important.
- Customary uses of water may be quite wasteful. To reduce the waste may, in effect, increase the available water resource quite markedly. For example, as referred in Section 10.1.6, where potable water is disposed from households to the sewer system and other potable water is used for urban cleansing and irrigation, some of the disposed water, with minimal treatment, could be used for the cleansing and irrigation activities, reducing the demand on the potable supply.

In water scarce regions no water should be "thrown away". All wastewater, usually after only minor treatment, can be reuse for a range of non-drinking purposes.

- In some communities large amounts of water are traditionally used for (ritual) bathing. Minor changes in the location, shape and operation of bathing facilities may enable reuse of this water and avoid a need for additional water supplies.
- Water reuse has been mentioned above and dealt with in Section 8.2. Many possibilities are available for water reuse. The technology is usually quite simple but large capital costs may be involved to make possible the re-routing of large volumes. Pumping costs may also be involved as systems have usually developed using gravitational flows, and reuse would usually involve redirecting flow from a point of disposal, at the lowest elevation of the system.
- Development of a culture of care for water resources. In most communities, both in developed and developing countries, there is only very limited appreciation of the links between general everyday behaviour and water resources. Simple activities like careless disposal of wastes, or cleansing of paved surfaces with water under pressure from a hose, rather than by sweeping, are not recognised by most perpetrators as being wasteful of precious scarce water resources. There is a huge need for education to alert the public to the effects of their actions and to encourage more appropriate, water-conservation behaviour.

11.2 Urban Centres

11.2.1 General

For small urban centres it may be possible to draw the water needs from the stream which passes through the town or from the aquifer below the town. However as urban developments increase in size (increased industry or larger population or both) there is often a need to obtain access to more water than is available from the local source. It is likely that both the stream and the groundwater below the urbanizing area will have become contaminated by wastewater and other effluents of (careless) human activity. Obtaining addition water may involve transfer of water from another basin. Usually when water supply is sourced from another basin the wastewater is not transported back to that other basin but is disposed in the basin in which the urban centre lies. This means the downstream flow characteristics of both streams will be changed – one having its flows decreased, the other increased but with poor quality. In the past it has been common for the source basin flows to be reduced practically to zero, meaning that the environment of that river valley is changed completely. In the 21st century changes of this magnitude are usually unacceptable since they have large impacts on the local population and drastic impacts on the ecology of the stream. Similarly the stream and aquifer to which the water is added, usually in the form of wastewater, suffer changes of increased flows, changed timing of flows and a deterioration in general water quality.

11.2.2 Environmental Consequences

Exploitation of groundwater from certain types of geological formations can lead to problems of land subsidence. There are many documented cases around the world (e.g. USGS 2000) of large subsidence causing changes in surface gradients, leading to changes in flow paths for drainage and flood flows and even the development of wetlands on residential land, with the need to abandon expensive housing developments. Similarly exploitation of a resource, of either surface water or groundwater, will usually impact on already existing uses of those streams or aquifers. Careful investigation and planning is needed before new water sources are developed to avoid very costly unforeseen consequences.

11.2.3 Water Reuse

Urban centres usually provide ideal opportunities to practice water reuse. Wastewater of a range of qualities is available and is usually already concentrated, or will need to be concentrated as development continues, at a few locations in the wastewater collection system (sewer). In any region where water is scarce all forms of water reuse need to be considered, from use of that water in untreated form for irrigation, to full treatment for domestic (drinking) water. Water reuse usually provides several advantages over collection of "new" water.

- The water has already been collected and is concentrated at one or a few locations;
- The quality of the wastewater is usually technically easy to improve. Bacterial and solids contamination are easy to remove, but dissolve salts of any kind are very difficult to remove;
- The cost of reuse is usually smaller than the cost of collection and treatment of "new" water.

11.2.4 Reduce Evaporation

In all regions of water scarcity it is important to avoid water loss by evaporation. Wherever possible water storages should be covered, irrigation should be conducted at night and in windless conditions. Without care well over 50% of all collected water can be lost by evaporation.

11.2.5 Water Conservation Education

In most urban areas it is common to see large quantities of water being wasted due to thoughtlessness or carelessness. Many urban dwellers take their water supply for granted and have little appreciation of its value. Almost all urban dwellers worldwide need continual reminders of the importance of saving water. This is particularly true in areas of water scarcity. As discussed in the next chapter there is need to continually provide education to all urban dwellers, beginning with small children, concerning the need to conserve our precious water resources. People need to be shown that small changes in behaviour, such as properly closing taps, minimizing water for bathing, etc., can reduce consumption by up to 50%.

11.2.6 Pricing of Urban Water

In most urban areas where there is a piped supply to households the water is supplied free, or the users pay a fixed amount. World Bank has reduced its funding of water supply schemes because most governments are unwilling to charge users enough to cover the real costs of operating the supply system. As mentioned earlier (Chapters 5 and 10), in regions of water scarcity it makes sense to charge at least the real cost of supply. In some cases a sensible management strategy would be to use the water supply charges to add a small amount to tax income. It is easily possible to arrange tariff structures so that essential water costs very little but profligate use incurs large penalties. Around the world experience has shown that water price combined with education can have a beneficial effect on consumption. (Spulber and Sabbaghi 1998).

11.2.7 Institutional Framework

Without widely accepted legal and institutional arrangements concerning water supply there is likely to be either hardship for the urban population or considerable unnecessary wastage of the scarce resource. There needs to be a body or bodies with clear responsibility for development, maintenance and management of the supply system. The manager could be a public or private body, but it needs legal backing for collection and ownership of water and to be able to regulate and charge for water consumption. Without such agreed or accepted legal backing it is almost impossible to have a fair and efficient supply system. In a region of water scarcity efficient (usually meaning minimum) water use and avoidance of leakages, theft and other losses is essential. These ideal conditions can usually only be achieved when a fair, widely accepted (by the water users) legal and institutional framework of operation is in place.

11.2.8 Research

The science of optimal water use is not yet perfect and there are many areas where there are opportunities for making more efficient use of limited water resources. Irrigation offers many opportunities to use less water to obtain the same or greater crop production. Evaporation reduction is always a challenge. Not only is research needed for these fundamental issues but there are many opportunities to save water in local situations by encouraging small changes in current practices. Research is needed both to discover where savings are possible and to find ways to encourage changing of some traditional or even culturally based practices. Research is needed by social scientists in addition to water scientists and engineers.

In most urban centres there are many opportunities to reduce use of scarce water resources. As can be seen from the above some of these opportunities involve changes in peoples' thinking and then of their behaviour. Changes in thinking and attitudes may be needed by the city leaders and politicians, by the water managers and by all members of the community at large.

There is a large opportunity to develop water reuse. Already in most large urban areas the cost of treating water for reuse is less than the cost of collecting and treating "new" water. However research is needed to develop methods which will guarantee the safety of reclaimed water and to persuade potential users of this water that it is in fact safe.

Continued research effort is needed for treatment of water to potable quality. Costs of desalination are falling but energy sources will limit widespread use of this technology. Other low cost methods need to be found for removing dissolved salts (particularly toxic salts) and bacterial contaminants. For regions of water scarcity where settlements are small there is a need to develop methods that can be used at the household level.

11.3 Rural Areas

11.3.1 Water for Households and Irrigation

In most regions of the world there are enough water resources to sustain small populations. The difficulty of translating this potential into reality lies primarily in the complexities of managing the water. In most water-scarce regions supply is usually abundant during the few days of the year when there is rain or snow. How can this very short term, relative abundance be managed to provide a steady supply until the next occurrence of rain?

In general, management of scarce water resources is primarily concerned with storage of water and protection from contamination. This applies whether the immediate supply source is groundwater or a surface reservoir of some kind. In essence management is about directing the short period, relatively abundant water into some form of storage, maintaining the water in the storage, and then extracting it at a rate sufficient to meet the needs, but not at a high enough rate to exhaust the storage before the next inflow occurs.

11.3.2 Capture of Available Water

When rain occurs there may or may not be runoff from the soil surface. If there is runoff, infrastructure should be in place to capture this water. Use can be made of small dams across stream beds or of excavated ponds into which surface flow can be directed. As noted earlier (Section 6.3) it is important that any constructed reservoirs should be as deep as possible, with as large a volume to surface area ratio as possible. Evaporation continues to remove water from the surface of any reservoir and so it is vital that the reservoir be much deeper than the depth of water that will be removed by evaporation before the next inflow. In arid regions with high temperatures the evaporation can remove up to 4 m of water per year. Water usage from arid zone reservoirs may be of the order of only 20% of the reservoir capacity. Evaporation and minor seepage takes the remainder. It may be worthwhile excavating a hole in the bottom of any arid zone reservoir to provide a source of drinking water for emergency situations when the interval between inflows is longer than expected. Such an excavated hole needs to be located where it will not become filled with silt from each inflow, and it needs to be cleaned of accumulated silt regularly.

A significant source of water in water scarce regions can be roof runoff or runoff from other impervious surfaces such as roads or school playgrounds. Some effort is usually required to modify roofs or impervious land surfaces to develop capture and collection systems. It may mean that traditional roof systems or roofing materials may need to be changed to facilitate water capture. Similarly water storage ideas may need to be revised. For example in the north-east of Thailand a cottage-industry has developed around the manufacture of ferro-cement water jars which range in size from 10 l up to 3,000 l. These have the advantage that they can have small openings and so very little of the captured water is lost by evaporation and insect pests can be kept out. A roof area of 20 m^2 can provide up to 350 l from a 20 mm rainfall event.

Some ingenuity is needed to adapt current rainwater drainage systems and to make use of locally available, low-cost materials to permit capture and storage of this water. However the effort of the investigation and persuasion of people to modify their traditional practices needs to be balanced against the hardships of inadequate water and the cost of collecting that water from a distance. Roof water is collected and can easily be stored adjacent to the point of use and impervious land surfaces are usually very near to points of water use because the impervious soil surface is usually the product of the passage of many feet.

11.3.3 Education for Evaporation Reduction

Evaporation is relentlessly driven by solar radiation and the passage of dry air over water surfaces. Everyone living in an area of water scarcity needs to understand this. Until people understand the basic principles of evaporation they are unlikely to take any action to minimize evaporation of the precious water. The keys to evaporation reduction are the minimization of the water surface that is in contact with the air and the protection of the water from solar radiation. Water should be stored on the surface in deep reservoirs rather than in extensive, shallow basins. Total elimination of evaporation is only possible if the water is totally enclosed, as in a roofed-over man-made tank. Enclosed tanks can only have limited capacity. Ponds or reservoirs in the land surface should be located to maximize the depth to surface area ratio.

Similarly outside water use should, as far as possible, be restricted to times when evaporation potential is lowest – at night and when there is no wind.

These principles are unlikely to be known or practiced by the population unless they are presented to them frequently. As discussed in Chapter 12 the education process needs to begin with small children and continue throughout life. Schools need to have consciousness of the means of avoiding water loss by evaporation as a basic part of their curriculum. However there is a need to ensure that those for whom school is not available, and the post-school population are also continually reminded of the fact that their survival may depend on their attention to water saving and particularly to reduction of evaporation. In regions of high evaporation there is little point in capturing water in shallow lakes and ponds. However deepening of the ponds, even over only a small part of their bed makes good sense.

11.3.4 Water Harvesting

Water harvesting is commonly practiced in areas of water scarcity (see Sections 6.4 and 8.5). However there is always need to enlist and stimulate the thinking and ideas of the local population to find better ways to maximize the beneficial use of all water that is precipitated. A climate of discussion and interchange of ideas needs to be encouraged so that maximum benefits can be achieved for the community from the small amount of rain that falls. Professional, expert help can be enlisted to develop better water-use and capture methods but the local people need to be enlisted as the greatest source of ideas. There may be long-established traditional practices of water use that may need some change. It will usually only be possible to effect changes in such practices by first getting to understand the culture and traditions that surround them and then developing a sensitive, locally adapted educational program. Actions to direct or coerce changes in traditional practices are unlikely to be successful or will lead to resentment, long term distrust, social disruption and worse.

11.3.5 Avoidance of Environmental Damage

It is often thought that manipulation of water resources in water scarce regions will not have any environmental effects. In fact the opposite is more likely to be true. Any changes or management of the scarce resource is almost certain to affect other parts of the environment, most of them adversely. Any water captured for any purpose will be a deficit for the local environment. The effect may be so small as to be negligible. However in some regions capture of water for household use in a region of scarce water will reduce the water available to vegetation downslope of the point of capture. Disposal of wastewater can have the opposite effect, encouraging growth of vegetation that is uncharacteristic of the region. More widespread and more noticeable effects must be expected from larger enterprises. The changes may be subtle and slow developing but they can be devastating for the local populations. For example disposal of nitrogen-rich water may cause eutrophication of a pond a little further downslope. Development of or changes to water supply and wastewater disposal schemes need considerable imaginative environmental planning to avoid unexpected harmful outcomes which may outweigh any benefits of the development. Examples abound.

11.3.6 Irrigation Performances

Irrigation is discussed in some detail in Section 10.6 and performance is discussed in Chapter 9. In regions of water scarcity irrigation should only be practiced where high value returns can be expected and where loss of water by evaporation and seepage below the root zone can be avoided.

Since water is an extremely scarce resource it should only be used for practices which produce large or life preserving benefits. While it may not be possible to assign a monetary value to in-house water use, most of this water use is essential for the survival of the population. Similarly the monetary returns from irrigation may or may not be high but the important question to consider is whether or not the irrigation water use is vital to the survival of the local community. If water use of all types is given a utility value by the local population it should be possible to determine whether or not irrigation is in the best interests of the community. Overlying the making of such decisions are questions of ownership of water, which has to do with the constitutional, legislative and cultural traditions of the region, which are discussed below.

If irrigation is to be practiced it is essential that the crop return per unit of water used is maximized. Large efforts need to be made to minimize unproductive use of water. In other regions there may be different objectives but in areas of water scarcity these are essential aims. The amount applied needs to be such that little or none is percolated beyond the root zone. Wherever possible irrigation water should be applied in the evening, allowing all ponded water to infiltrate before the sun rises and encourages evaporation. In dry regions sprinklers should, wherever possible, only be used on windless nights. Atomisation of water into small droplets which have large surface area encourages rapid vapourisation in dry air. In low humidity, windy conditions, up to 50% of sprinkler or spray water may not reach the soil surface, and even some of that will evaporate before infiltrating. If daytime use of sprinklers is the only option for water application, it may be better to save the water for some alternative beneficial use. Differences between the proportion of sprinkler water penetrating the root zone in daytime and night-time can be as high as 50%. Where water is highly scarce, a non-beneficial use of water of this magnitude could be considered a crime against the local people.

Modern changes to drip irrigation and sub-surface water application have distinct advantages over sprinkler, flood and furrow methods of applying water in regions of extreme evaporation potential. Changing water application methods is very expensive, so there is a need to encourage use of drip and subsurface methods for all new irrigation developments in places of very high evaporation.

Research and development aimed at reducing water use in irrigation, or rather, increasing crop yield per unit of water applied, is of great importance. In all irrigation methods a large fraction of the applied water may be consumed by evaporation. If this could be reduced, without reducing crop productivity, much water could be available for other uses.

11.3.7 Water Rights

The issue of ownership is always important in regions of scarcity. For maximum development and harmonious relationships there must be clear and widely accepted agreement on sovereignty. In one sense water is owned by everyone. It is a natural resource. However no practical scheme has yet been devised to administer such an ownership system. In some states all water is owned by the head of state, in others it is owned by those on whose land it is found. All ownership or administration systems are human inventions and as such reflect the innate greed and selfishness of the species. No system is perfect. There are always some who are advantaged and some who are disadvantaged. This reflects the history of civilization. Some systems of ownership or tenancy or management or rights to use of water have remained in place for a long time, others have changed continually. Change must be expected. As circumstances change existing systems will be seen to be inadequate for current circumstances and changes will occur, either by discussion and negotiation or by violence and dominance, but changes will occur. Unfortunately no political system or water management policy coming from the political system is perfect. There is a continual need for water managers to improve the efficiency of water use either within the framework of water governance, or perhaps by working to have the water governance or ownership system changed.

At a less political level, the rules governing water resources allocation and use have huge effects on the efficiency of water use. Stable systems encourage investment in efficient water use and water application methods. Year to year leasing of water does not encourage any investment and leads to opportunistic use of water with minimal preparation or planning. Even at the small village level there are considerable advantages in having in place agreed rules concerning who has access to water, how much can they use, for what purposes can it be used, what is to be done with return water and what is to be its quality, etc. As systems become larger the need for agreed rules become much more important. But the rules need to be widely accepted. Disagreement over any parts of the rules inevitably leads to instability and a distinct reduction in the benefits that will be obtainable per unit of water.

11.4 User Groups

User is defined in the broad sense of a consumer of water. Householders who receive a supply, irrigators who extract water for their farm by pumping from a river or reservoir or who are supplied water via a pipeline are all included in this term "user". Sometimes users are dispersed residents, commercial enterprises and farmers. Sometimes the users band together to form a group which extracts or collects or buys water which is then distributed among the group members.

For any water user group or association or cooperative to be successful it must have some set of rules which govern its operation. These rules may be formal or informal, written or traditional but there must be some accepted mode of operation. Preferably rules should be written since unwritten rules tend to change little-by-little and almost inevitably lead to disputes. Written rules are also often disputed and need to be adjudicated in some manner. However unwritten, informal rules suffer the problem that each user has his own idea of what the rules mean, and over time those ideas change. When a disagreement occurs it becomes very difficult to find agreement on the meaning of the rules since by this time each interested party has a different understanding of what the rules are. Disputes over unwritten rules, where most parties are intent on preserving or (perhaps subconsciously) increasing their benefits become very difficult, even impossible to resolve and inevitably mean that relationships between at least some of the parties become strained or are terminated. When this happens everyone loses because the social disruption, disputation and perception of being wronged prevent further cooperative activities. Then managing the water resource has priorities which focus on social sensitivities and vested interests rather than on optimal use of a scarce resource to benefit the whole community.

It sometimes occurs that a user group is effectively owner of the resource and carries out total management of the resource, from capital investment, to maintenance, to revenue collection and distribution, to protection of water quality, etc.

Institutional frameworks of user groups need to be different from place to place to fit in with the local political, legislative and cultural environment.

Development of an autocratic regime where a single person has complete authority to determine water distribution is unlikely to be acceptable in a country where there is a long standing democratic tradition and vice versa. In some situations it may be considered there is no need to develop formal rules since there is currently harmony and acceptance of the informal agreements or cooperation. However in any situation of water scarcity it is inevitable the demands for water will exceed available supply and dissatisfaction and disputes will inevitably arise. Better to have an accepted, defensible, formal system in place as early as possible.

The structure of user groups and the agreements and rules they develop for allocation and use of the scarce water resource will need to reflect the interests and needs of potential users, or customers, as they could be considered. For example what is the nature of each potential users claim to be able to appropriate a share of the water? Are they traditionally the owners of the water or users of water that belongs to everyone? In the local tradition who is the owner of the water – the monarch, the head of state, parliament, the local mayor or village headman, the landowner or the whole population? This issue must be settled before any formal arrangements can be possible. On what terms will users be able to take water – by purchase? – to the highest bidder in the market? in proportion to size of landholding? – as a land owner or also as a land renter? as a shareholder in a cooperative to which a volume of water is assigned? as a paying customer of an agent charged with managing a part of the resource? There are myriad possible arrangements, many of which either alone or in combinations could provide for efficient overall management of the resource. But as stated earlier, some formal arrangement, which is acceptable to potential users and fits in with the political and cultural context is likely to provide the most stability for the users. It will also enable the environment within which the water use must occur, and the environment from which the water is extracted, and to which any wastewater is to be returned, to be considered early in the process, and not left to a "patch-up" requirement after environmental problems have become obvious, which almost always leads to less than efficient compromises both for the water users and the environment.

User groups will usually develop self help programs for their own benefit. Of high priority for user groups should be educational programs to assist users to develop efficient water use practices. Independent of whatever charges are leveed for water use an education program should aim to convey a sense of the great value of a water resource and the value of efforts to protect and maintain the resource. For users to pay the real, full costs of managing the water resource is probably the surest way to convey this sense of value, but where there is no tradition for users to pay there is still a very important need for all to understand the true nature of the value of water.

User groups also need to support and encourage research activities to enable them to make greater use of their available limited resource. They should encourage the search for new, untapped potential sources but they should particularly be encouraging use of less water for each task and an awareness of the benefits of maintaining a high quality of all water, particularly of any water released back into the natural environment.

11.5 Administration of Water Use – Public and Private Organizations

11.5.1 Types of Administrative Structures

Structures can be privately organized, or they can be related to government activity, or it could be there is no structure. It is common to have no structure at all when a land area is being newly settled. While the available water resource far exceeds the needs of the potential users there may be no need for any administrative structure or any real management of the resource. The resource is large and the users have little effect on it.

When water use begins to approach the available supply, either during a brief dry period or perhaps during a drought, there are usually calls from among the users, particularly from those at the fringe of the resource or those who have only simple water collection methods, for some organizational system to be developed to ensure that they are not cut off from any supply during a period of scarcity. Since water is rarely in abundant supply, and since its occurrence in time is random, following the

Fig. 11.1 Where water has to be collected out of the village, women constitute an essential users group that should receive particular attention (courtesy by Angela Moreno)

laws of chance, there will inevitably be periods when no water is available unless action is taken during wetter periods to store some water to provide supply during dry periods.

If storage schemes are to be implemented (surface reservoirs, recharge of groundwater aquifers), should the system be that every user looks after himself, or should cooperation and community action be encouraged? In some small parts of the world, water supply is a totally private matter. Each household (Fig. 11.1) and other enterprise looks after its own needs. In other areas there are private companies which collect water and distribute it to those who pay. In the majority of regions water administration is a government (either local or national) activity and the governments have set up structures to manage the collection, storage and supply of water. In many cases, governments found that participatory management is the most adequate solution and the management responsibility is given to water user associations, as often happens quite successfully with irrigation systems (Groenfeldt and Svendsen 2000). In yet other areas governments have decided they do not need to be involved in such activities and have sold their water infrastructure to private companies which aim to make a profit from water management. In many of these cases the water company is a monopoly and has no competitors because they own the complete (and interconnected) collection, storage and distribution system. In some of these cases government regulators are appointed to ensure the companies become neither takers of immense profits nor inefficient wasters of both the water and the funds of the users. One method used to attempt to maintain accountability of monopoly companies is for the government (therefore the people) to own a substantial portion of the shares.

In addition to the several methods of administering collection, storage and distribution of water there can be many different modes of determining how much of the available water each user may acquire. In some systems (which are not very common in areas of water scarcity, though they are not unknown) users are free to take as much as they wish provided they pay for it. For example in many large cities this is the system adopted. The supply organization undertakes to supply water in whatever quantities the user wishes and the supplier recovers all costs plus a reasonable profit margin. For this system to succeed the supply manager needs to plan many years ahead, constructing collection and storage systems to meet all needs. The manager needs access to large amounts of capital. He also needs to forecast trends in population and lifestyles (per capita water use) and to be able to influence water use activities (fashions) by using education, pricing and perhaps coercion and restrictions on supply to ensure his objectives are met. In many parts of the world it has been realized that access to unlimited supply by large populations in large cities is not necessary for human well-being and is usually very damaging to the environment of the surrounding country.

Hence in many places the aim of water supply managers has changed from supplying whatever is requested, to managing demand by use of the methods mentioned above, and as discussed in Chapter 10 for different uses of the water. Methods used to manage demand include making large usage expensive, using persuasion backed by environmental arguments, and where necessary restricting supply by only making water available in the distribution system for part of each day (McMahon 1993).

11.5.2 *Equity and Rights*

Who has a right of access to water? – to how much? – how often? In regions of water scarcity sustainability of the water resource and of the human population will only be achieved when the population regards the system as fair. Inequities can be maintained for the advantage of some over others for some time but eventually inequitable systems will be swept aside. The problem in this equity concept is how is fairness defined? What is regarded as fair by one group may be intolerable to another. Water rights reflect human rights. Human rights have some absoluteness about them, but they can never be separated from cultural considerations. Who has access to land and water ownership? Who determines where wastewater can be disposed and by whom? These are issues that will change continually over time and from location to location. Is the water I have, available for me to use and dispose as I choose or am I a steward, expected to preserve and protect my water against any possible damage or loss? Is there an agreed set of laws in my region which govern how I and my neighbours manage and use the water? Is the water a marketable commodity available to the highest bidder with the price governed only by supply and demand? Can we allow an essential life-support resource to be controlled totally by market forces? There are myriad questions of this type. They must be answered within the context of local cultural, political and economic systems, but they must be answered. The more of these questions that are openly considered before a water management/administration system is set up in any area of water scarcity the greater the chances of the long term success of the scheme.

Chapter 12
Education

Abstract Education is a fundamental aspect of coping with water scarcity. The focus of the discussion is to consider how attitudes can be changed and how an educational program may be established. Education is meant in the widest sense, aimed at children and youths, women, with their role in the family and the community, farmers and industrial water users, managers, operational and maintenance personnel, educators, and other professionals.

12.1 Need to Change Attitudes to Water

12.1.1 Current Attitudes

In most regions of water scarcity the population has developed its strategies for survival as best it can, using its ingenuity and available skills. However as populations increase and the water resource available per person declines, and as global climate change threatens increased uncertainty concerning the size of the limited available resource, there is a need for populations to have new attitudes to water. The general prevailing attitudes could be summarized as follows:

- Water is a resource provided free by "nature"
- There will always be water available for my needs

Unfortunately these attitudes are reinforced for the majority of the world population who live in urban environments by the reasonably efficient public water supply to which they have access. In the developed world it is taken for granted that high quality, abundant water is always available at several points within every housing unit. In a number of developing countries, even in regions of water scarcity it is taken for granted that safe quality water is delivered to all housing units at least for a few hours every day. However there are at least 1.1 billion people without access to safe water and 2.6 billion people who lack access to improved sanitation, which represented 42% of the world's population.

The regular household supplies, for which the charges, if any, are quite modest, have insulated the majority of the population from any need to consider conservation of water. In these regions where a large part of the populace is satisfied with, and taking for granted, continuation of their water supply, there is little demand on suppliers (often government agents) from customers to improve the efficiency of their water gathering/supply operations, nor on the customers to think about water conservation.

An unfortunate consequence of these worldwide circumstances is there is much waste of water by those who have access to an organized supply system, but availability of water to those who must gather their own supplies continually declines in both quantity and quality. As outlined by the WHO (2004):

- In 2002, 1.1 billion people lacked access to improved water sources. They represented 17% of the global population.
- Over half of the world's population has access to improved water through household connections or yard tap.
- Of the 1.1 billion without improved water sources, nearly two thirds live in Asia.
- In sub-Saharan Africa, 42% of the population is still without improved water.

In order to meet the water supply target, until 2015 an additional 260,000 people per day need to gain access to improved water sources; however, between 2002 and 2015, the world's population is expected to increase every year by 74.8 million people, which makes that challenge difficult to match.

It can be confidently stated that in most regions of the world there is large waste of this precious natural resource due to:

- *Lack of care.* If water is always available and if it costs almost nothing a very complacent attitude is almost certain to develop. Even for community minded people there is no encouragement or incentive to be careful with water.
- *Lack of thought.* Unless water users are educated about the scarcity of supplies and the potential for contaminating water, in rural as well as urban environments, the majority are unlikely to independently arrive at the conclusion that water supplies are being endangered by their activities. To combat this attitude deficiency there is a need for a major community educational program.
- *Lack of realistic pricing.* In most parts of the world water supplies are subsidized from general tax revenue. If all users met the actual cost of the water they used, including the costs of disposal of the waste water, there would be some incentive for improving interest in water conservation. In regions of real water scarcity it makes sense to increase charges for water by adding a tax raising component to the water supply charges. Even if all operational and capital costs are covered by the charges, the water is still quite inexpensive. Water from organized supply systems will usually cost much less than for households to collect (and pay with their labour) and treat water themselves. However charges may not be enough to provide incentive for individuals to be water conservation minded. Clever pricing structures can be a means of drawing the real cost of water (including all the environmental costs) to the attention of everyone in the community. The charge per litre for small usage can be very low, but increase rapidly to encourage large users to consume less. Experience has shown that subsidizing water supply

encourages waste and eventually leads to the destruction of the scheme (statement made by World Bank officials on many occasions, Siregar 2003, World Bank 2006). It is rare that long term water supply costs, such as for operation, maintenance and debt servicing can be continually met from general tax revenue. History shows that where no charges are levied on users the whole scheme progressively runs down and then an unpaid capital debt remains, which prevents reinvestment to refurbish or replace the lost system (as has occurred with the attempts to privatize the Dar-es-Salaam water supply in 2008). Recent experience with increasing prices per volume used has shown that consumption can be price sensitive, even in relatively affluent urban areas (e.g., in Sydney, use has declined 30% after a 60% price increase over 3 years), when it is accompanied be strong public information about the need to conserve water. However Gaudin (2006) has shown that residential water demand is little affected by price alone, but Berbel Vecino and Gutiérrez Martín (2004) and Chohin-Kuper et al. (2003) show that for irrigation there is sensitivity to price. It seems that irrigators tend not to respond to increasing water costs until those costs begin to have serious effects on the irrigators business viability. Obviously at this stage the irrigator must re-evaluate their water use practices or go out of business.

- *Lack of maintenance.* Water supply schemes need many kinds of maintenance activities. As the infrastructure ages the demand for maintenance increases. Realistic allowance is needed in budgets to cover the ever increasing maintenance costs. Most water supply schemes are largely out of sight (buried pipes) and are taken for granted. The most modern, best maintained schemes lose about 10% of their water by leakage and general losses. In poorly maintained schemes the losses exceed the water reaching customers. In addition to the revenue foregone by this lack of maintenance, in regions of water scarcity the squandering of more than half the resource is inexcusable – it condemns the customers to a more difficult than necessary lifestyle. Similarly in regions of rising water tables and water logging, leakage from the supply system increases the problems. Commitment to good maintenance is imperative as a key strategy for coping with water scarcity.

- *Lack of sense of ownership.* In most cultures there are proverbs or popular sayings along the lines "if I don't own it I have no responsibility to care for it" and "if it is free it cannot be worth worrying about". Such attitudes ensure a short life for any project. It is important that the local population regards their water supply scheme as their own. If they have a sense of ownership they will treat it well and insist that others do the same. If the scheme is seen as being owned by a remote company or the government they are likely to feel no responsibility and to see the owner as the bearer of all costs of maintenance and repairs, even if these are the result of lack of care, or even malicious actions by customers. A sense of ownership can be encouraged by requiring customers to have a real ownership stake in the enterprise and by constant educational efforts. Can this sense of ownership and responsibility be developed if the water supply system is privately owned? The motivation for private ownership is often stated to be to free up capital for other infrastructure needs but there may be a need to reconsider these aims. Since water supply is the life support infrastructure and encouraging its adequate funding and maintenance is a continual challenge, perhaps it is not a good idea to encourage its private ownership and operation.

12.1.2 How Can Attitudes be Changed?

The immediate response to a question such as this is "education". But how can education be used to bring to every member of the community a sense of responsibility for the protection and efficient use of its water resources? Water use is ubiquitous but cultural attitudes to water and its usage vary greatly from region to region.

At first there needs to be some agreement on what are the ideal practices to be followed in relation to water management. Some rules or framework of agreed principles need to be put together which (at least) encompass the issues shown in the Box 12.1.

Box 12.1 Questions that need to be answered for successful water management

- Who owns water? – rainwater, water flowing over the surface, water flowing in surface channels, groundwater.
- Who can make use of the water?
- Activities to be avoided near the community water sources – near the wells, near the pond, in the catchment. Who will supervise compliance? What will be their powers?
- How will water be collected and distributed? Who will undertake this activity? How much will they be paid? How much will be charged for supplied water? Will the price vary with quality? How to ensure everyone, irrespective of position or wealth, has reasonable access to the supply?
- After use, how will the wastewater be disposed? Who will be responsible for disposal and for environmental effects of the disposal?
- How much of the wastewater will be recycled? What uses will be made of recycled water?
- During periods of acute shortage (e.g. during drought), how will rationing of the scarce supply be managed to ensure continuation of minimum essential supplies and equity in the community?

Once agreed principles are assembled there needs to be an investigation (research) of how ideas and attitudes among community members can be changed most effectively. This is a most important step since there are large links between water and culture and tradition, and every small community has developed its own idiosyncrasies and "best" practices. The outcome of this research should be a plan for educating and changing attitudes to water and water use that will facilitate the sustainability of the community. One educational program will not suit all communities in a region, so some sensitivity and flexibility will be needed in both the development and delivery of appropriate programs. Note that education to bring about attitude change cannot be considered a task to be completed in the short term. A permanent, ongoing educational program will be needed. More details are considered later in this chapter.

12.1.3 Aims of an Educational Program

The aim of any water education/public relations/information/propaganda program needs to be to develop both in the community as a whole, and in each individual, attitudes which will ensure supply of water of appropriate quality and quantity to sustain the community. Attitudes needed are those which foster conservation and sustainability of existing water resources and encouragement of innovative and practical ideas to allow more effective use of the available resource. Incentives are needed to encourage individuals to seek out and develop additional resources that are economic but do not interfere with resource use by others, nor degrade or endanger the environment or bring about unacceptable environmental changes. This thinking would need to encompass water collection and harvesting, storage, wastewater disposal and reuse, water quality issues of all types and accessibility to water.

The role of education in the above aims is to inspire everyone in the community to be thinking about water in constructive ways. If each person can be encouraged to consider how they, personally, can save water and preserve its quality, how they can contribute ideas to increase the community water resource and how they can contribute to the community effort to increase and preserve the available water resource, then the education system may be considered to have been a success. However once some success is achieved the effort must not be diminished since communities need continuing education and ongoing consciousness raising in order to maximize the ongoing support from (and benefit to) the community for water resource protection and preservation.

If community members ideas are solicited, and the best ones are given some attention and implemented, a community spirit for maximizing beneficial uses of the resource and minimizing its abuse will develop, for the benefit of all. Education and dissemination of important public information can play large roles in bringing about such an ethos.

A very important issue that should be considered together with education is that of public participation. The more the local people can be encouraged to involve themselves in the development of rules and regulations and development of policies concerning water use and protection the better. Public participation is often seen as being slow and difficult, even obstructive to advancement, but when the public (the local people) are not involved in the decision making they have no sense of ownership and most of the problems outlined above become the norm. Public officials and engineers often object to public participation. They consider it to cause

Fig. 12.1 A drawing by
S. Khasabov aimed at
supporting new attitudes
relative to water and
desertification
(UNCCD, 1997; reproduced
by permission of UNCCD)

delays and be inefficient. However top-down decision making rarely builds public confidence and almost never builds a sense of ownership, which, as outlined above, is important for developing public assistance and support in development and sustainability of public infrastructure. Public participation in planning, development of operation and maintenance programs and in development of pricing and revenue collection and enforcement schemes would seem to be of vital importance for the sustainability of community water health (Fig. 12.1).

12.2 Education and Training

12.2.1 Aims of Water Education and Training

As stated above one of the aims of any education program in a region of water scarcity must be to develop attitudes to water that will lead the population of the region to sustainably obtain the maximum possible benefits from the limited water resource and to minimize or eliminate wastage of the resource.

12.2.2 Overall Water Education

Education on water issues is not just needed by the technocrats, managers and operators of water supply schemes. Since everyone in the population uses water and has a vital interest in any supply scheme, educational programs need to be aimed towards the whole population. Everyone has opportunity to misuse water and by undertaking irresponsible activities each inhabitant can have devastating effects on the whole scheme and on everyone dependent on it. Everyone in the community needs to be targeted

Fig. 12.2 A wall painted message in Creole language, Cape Verde: "Water, source of life"

- To raise their awareness of water (particularly scarcity) issues,
- To develop attitudes of care and responsibility for water (Fig. 12.2)
- To foster a sense of ownership of the resource.
- To foster protection of the resource and of individuals from water-borne disease by inculcating sensible attitudes to hygiene, washing and separation of wastes from the supply system, and to protect the water from invasion by undesirable insects and other disease vectors.

Any educational program needs a many faceted approach and needs to be inclusive of all of the population, with appropriate messages (Fig. 12.2) and aims for each part of the population, from the very young to the aged.

12.2.3 Educating Children and Youths

Children have an involvement with water from the very beginning of life. Drinking and bathing are an important part of their everyday activities. Education and training on water and its importance to every one of us needs to begin long before children start school. Their carers need to be instilling appropriate attitudes to water as early as possible. To do this the carers (usually, but not always, mothers) need to be educated on water issues. The instruction and attitude formation developed around the home needs to be reinforced by the popular entertainment media the children encounter, which may be the village storyteller, children's storybooks, children's radio and television. A good example is the education kit produced by the UNCCD (http://www.unesco.org/mab/ekocd/index_cartoon.html) oriented for education of youths on desertification (Fig. 12.3). Books and radio and TV programs are needed which have the specific aim of fostering good attitudes to water. These can take the form of direct instructional material (documentary type) but they must also be supplemented by good attitudes to water being built in to fiction and popular radio and TV series. A huge amount of community attitude formation occurs as people (including children) identify with popular entertainment characters and their activities. Children are very impressionable and open to teaching on all manner of topics. Development of attitudes of care for our vital water resources is a topic to which children's (entertainment) TV could make a large contribution. It is noticeable that in some regions anti smoking campaigns have been greatly assisted by teaching

Fig. 12.3 Educational
cartoon (with permission of
UNCCD)

very small children that smoking is unhealthy. Children have an ability to influence
the actions of their adult relatives by their innocent questioning of adult behaviour.

Education of very small children on water issues must be a long term strategy
and therefore once begun must continue – forever. The benefits of educating the
very young on water issues can be that for the rest of their lives these people will
be aware of water issues and are likely to be open to consider new thinking on
how water can be used more effectively, less wastefully and with less environmental
impacts.

Schools. All school curricular should include instruction on water issues appropriate to the region and to the age of the children. In regions of water scarcity
the instruction should deal with minimization of water use and protecting water
from contamination. Instruction should begin at the school entry level and should
continue through all years of school attendance. At secondary school level syllabi
could consider innovative ideas on water collection, water treatment, wastewater
treatment and water reuse. The need for implementation of some of these ideas will
be obvious to children in water scarce regions. The contributions they can make
to developing new ideas for conserving water should be fostered and welcomed by
their communities.

Tertiary level. Here the focus of water education programs needs to be on environmental issues, policy changes, legal and economic frameworks and details of
technologies. Traditionally water issues have been considered in the engineering
faculty but this is very restrictive. As can be seen from the list of issues, above,
there is a need to enlist the thinking of all educational disciplines to encourage
the broadest possible development of ideas on how to cope with water scarcity. As
outlined elsewhere in this book, strategies for coping with water scarcity must not
be limited to development of technological solutions, but must involve development
of legal systems, the harmonious integration of new water management ideas with
local cultural values and therefore needs the attention of many disciplines.

Population in general. Alongside all the direct teaching of educational institutions there is need for ongoing education and reminders for the whole population.
Advertising can contribute to this need but it is expensive and not very effective.
More important is the introduction of water scarcity issues into popular culture –
popular radio and TV series, pop music, cartoons, popular fiction writing, newspapers and popular magazines. In regions of water scarcity the need to reinforce

the message of water conservation and protection is so important there should be consideration given to providing government incentives, even subsidies for these activities. Perhaps media that reach a wide audience with water conservation messages could receive favourable taxation treatment.

12.2.4 The Role of Women

In many regions, particularly those where water is scarce, the role of women is very important:

- Women have a large influence in shaping the attitudes of the young, the future majority of the population
- Women often have a large role (with children) in water collection, where there is no direct household supply (Fig. 11.1)
- Women largely control in-house water use.

Educational programs of all the types listed above (formal, popular culture, advertising, etc.) are needed which target women. These programs should be aimed at attitude formation and modification.

Since women provide much of the labour in water collection and disposal they have considerable incentive to find and develop less labour demanding methods to conduct these activities. It is therefore imperative that the women be at least as well educated on water issues as any other members of the community. They need to be encouraged to contribute their ideas on ways to use water most effectively and therefore to maximise the benefits that can be gained from their labour. They are also best placed and have the greatest incentive to develop methods of water collection and distribution that minimise labour requirements and free them to make other contributions to community needs.

12.2.5 Farmers, Households and Industrial Water Users

Here there are usually very large opportunities to save water. University courses on irrigation and industrial water use put large emphasis on optimal water use. Unfortunately much less effort is made to transfer these ideas to the individuals who are directly involved in using the water, and these individuals are unlikely to have had any opportunity to participate in University courses. As a result, in many regions of water scarcity the farmers and industrial water users consume very large quantities of water for quite small productivity, for example in Pakistan's "warabandi" system (Bandaragoda 1998). Some of the responsibility for the resulting waste lies with inefficient water laws, water allocation arrangements and water pricing structures. Anecdotal evidence suggests that water supplied free is rarely valued by the user. There need to be large incentives to encourage farmers to make optimal use of water. Even incentives by themselves are not sufficient. Programs of education for farmers are needed. These need to be run by well trained extension workers who have time to work with the farmers to show them how to make better use of their water. Alongside this dispatch of extension advisers to the farms there need to be well publicized, easily accessible demonstration farms where optimal water use is practiced, where farmers can see for themselves the potential benefits of good water-use practices. Farmers will respond quickly when they see with their own eyes that the returns they could reap for their labour could be increased many times, especially, as will usually be the case, where the physical effort and time required may be no more than for their current practices.

Similar incentives are needed for industrial water users. Incentives can be in the form of subsidies for purchase of more water-use efficient plant, encouragement of water reuse by removing taxes from recycling plants, by increasing the price of input raw water, or combinations of these. The need is to maximize returns per unit of water, to minimize the unproductive loss of community water and to minimize environmental damage.

Households need to be educated to use water sparingly – for showers, bathing, clothes washing and toilet flushing. Similarly outdoor household water use should be confined to keeping vegetation healthy – by irrigating only in the evening. Washing of outdoor surfaces should be minimized and in fact is often unnecessary; a broom can be more effective than a pressurized jet of water.

Research is needed on how to effectively communicate water saving ideas to farmers, industrial operators and households. In every culture some different approaches will be needed. The aim should be to discover how to raise the consciousness of everyone in the community to the reality that water is scarce and very precious. Everyone in the community needs to be encouraged not just to stop wasting water, but to be looking for ideas on how to get more value from every drop of water.

12.2.6 Managers, Operational and Maintenance Personnel

This is the group that is probably most sensitive to water pricing. As a manager seeks to increase company profits he or she will quickly respond to any large cost item. Few enterprises, except irrigation, really need to consume large volumes of water. Perhaps they have traditionally operated in a particular manner, without recognizing that they were using far more water than was really necessary. Realistic pricing of water quickly attracts managers' attention to possibilities for consuming less. In many industries recycling can be achieved at very little cost, but even where the cost is higher, the water saved may be of more value (here is an incentive for realistic pricing of water supplies) than the cost of purchase and operation of a recycling plant. The initial outlay to develop a recycling system may be high and this is where government funded financial incentives can be effective.

Leakage of precious water is a serious matter in a region of water scarcity. Well motivated maintenance staff can keep leakage and other unproductive losses of water to a minimum. Plumbing staff need to be educated on the value of preventing loss of water by attending to details such as leaking valves, automatic flushing systems which flush more frequently than necessary, leaking delivery pipes, etc.

12.2.7 Educators, Agronomists and Engineers

An integral part of the education of all teachers, agricultural specialists and engineers should be to instil healthy attitudes to the value of water and the need to optimize the benefits of water use. These professionals have large influence on others who deal with water in their everyday work and so they are in a position to influence water use practices and attitudes to water use. The need is to raise their awareness of the precious nature of water in water scarce regions and to encourage them to pass on these ideas as they go about their professional activities. Bringing the professionals to the point where they take some responsibility for developing beneficial attitudes to scarce water resources should not be taken for granted, or be assumed to be a simple task. Their background, their experience of government activity, their political leaning relative to the incumbent powers, all influence their willingness to assist the community in this way. After their own, or a relative, friend or colleague's bad experience in attempting to change attitudes to water with an employer or in relations with other agencies, some people are likely to be disaffected and take the attitude "why should I bother?" While this position may be viewed as unprofessional, it is often reality. Therefore programs and leadership are needed to inspire professionals to take their own leadership roles in promoting sound community attitudes to water scarcity. It is unlikely to occur spontaneously. Effort is required at community leadership levels to inspire others to assist in the development of sound community attitudes to coping with water scarcity.

12.3 Need for New Developments and Research

12.3.1 New Technologies

There are many areas of water resources management where new technologies are needed. Some areas of need for developments are possibly unattainable, for example the need to significantly reduce evaporation from water surfaces. However there are many areas where research could lead to technologies that could make more water available and hence increase the quality of life of large populations. For example can we develop new technologies that will

- Allow increased agricultural and industrial production with less water use?
- Prevent the need for disposal of any "waste" water?
- Enable desalination of brackish and sea water for costs similar to current costs of collecting and treating fresh water?

Education lies behind the possibilities for success in these and all other aspects of water saving. Education is needed to bring the whole community to understand water scarcity issues – not just to observe that water is scarce, but to understand why it is scarce and to be determined to attempt to find solutions to the resulting inadequate availability of water. This understanding has the potential to lead to much innovative thinking among the people, on how to reduce waste, how to avoid contamination, how to use less to achieve more. The best developments will not necessarily be thought up by the water professionals, or even by researchers, but when the whole community is looking for ideas, amazing developments can occur. Not only will the community be contributing ideas but they are likely to be receptive to the need for funding to test the better ideas that are proposed. Testing and developing new ideas usually requires considerable funding and months to years of examination and development of pilot projects to demonstrate the practicality of proposals. To carry such activities through there is a need for widespread community support both in spirit and in resources. Without a solid educational program such support is unlikely to occur and without an ongoing commitment to the educational/public relations effort such innovative, but in many individual cases, ultimately unproductive investigations, are unlikely to be sustained. The supporting community needs to understand that many good ideas, when thoroughly tested, may not produce the benefits expected, but at the same time it needs to continue to solicit and test good ideas.

Research by social scientists is needed on how to develop support in the community for the assignment of resources to investigation of new ideas and approaches to water scarcity issues. Initially enthusiasm may be high but as some avenues of research inevitably produce little of benefit there will be a need to find ways to sustain the community interest and willingness to pay.

In parallel with investigation and development of new technologies there is a need to develop management philosophies that can be used most efficiently to cope with water scarcity within the already existing cultural and traditional environments of the region. Management philosophies need to be developed to

- Encourage the population to understand that the scarce water resource belongs to them
- Make the very best use of the available water resource
- Encourage within the population an attitude of care for water
- Encourage innovation and development of ideas and setting up of pilot studies of the most promising ones
- Find means of funding all of the research and development activities.

12.4 Development of Public Awareness of Water Scarcity Issues

It is probable that this is really the major educational challenge. History shows it is never possible to bring the whole population to support even the most excellent idea or philosophy.

In order to reach the maximum number of people and to have some chance of recruiting them to the "cause" of coping well with water scarcity there is a need for a many-faceted approach. As a minimum there needs to be very positive leadership at the political level, and also among the opinion makers in the society. All communities have their opinion leaders. In the developed countries these tend to be popular radio or TV show hosts. Sometimes a newspaper columnist can be very influential. In other countries popular TV series, often called "soap operas" can have a huge influence on public opinion. In other areas religious leaders have considerable influence. Advertising and general public information are also important, although these tend to have a more supportive role and are rarely in the forefront of developing community attitudes. However they can be very important in conveying the truth to the people, as truth is not usually obtained reliably from popular entertainment sources.

12.4.1 Media in General

Broadcasters and column writers who have a large popular following can be extremely influential in shaping public opinion and in raising awareness of important issues. In regions of water scarcity an effort is needed to recruit these people, especially the ones who have a large public following, to the need for adoption of strategies for coping with water scarcity. Not only does their interest need to be recruited, but further effort is required to keep them informed of developments and of local needs. These publicists are only likely to give small amounts of on-air time to a particular cause (e.g. coping with water scarcity) if they are continually reminded of it and if the information they receive is attractively presented and easily assimilated.

Similarly other writers and artists (particularly cartoonists) need to be recruited to the cause. It is important that water scarcity issues become the subject of cartoons, comics and jokes in the community. Authors of these works need to be targeted for special attention concerning coping with water scarcity. Some kind of incentives to discuss "coping with water scarcity", may also need to be provided.

Documentaries for radio, TV and cinemas need to be prepared. While these do not have large public impact they can be very useful in conveying truthful, factual information and providing support to the popular opinion makers.

12.4.2 Advertising

Advertising is probably not very effective, can be very expensive, but it has an important place in getting messages to the public. It provides an opportunity to make sure scientifically correct information is circulated. Advertising cannot be done "on the cheap". It is expensive, and needs to be put in the hands of a successful, culturally aware advertising agency which can attract public attention. Paid advertising can really only draw attention to an idea. It cannot provide all the information needed to shape public opinion. However it should be an important part of the overall "awareness arousing" strategy. In locations to which electronic media do not reach and where education levels are not high there may be a need to have a small team of "information providers" visit on a regular basis – perhaps 3 monthly – to raise awareness. This may need to be undertaken in a variety of ways including street theatre, fun visits to schools, public meetings (unlikely to be well attended), free public film shows and even high density use of posters.

No mention has been made above concerning the financing of these activities. It is likely that the only sources of funding are likely to be government, water supply agencies or charitable NGOs. There needs to be ongoing awareness raising of the need for funding of public education on water issues by international agencies such as those associated with the UN and the associations of relevant professional specialists.

12.4.3 General Public Information

Usually, provision of information important for life is considered a government responsibility. All the activities discussed above could be considered to be public education, also a government responsibility. There is a need to initiate and sustain a campaign to raise public awareness that coping with water scarcity is a responsibility of every member of the population. We all desire to continue living, and water is needed for life. Given some commitment and determination it should not be too difficult to obtain funding and assemble a group of dedicated, determined people to promote attitudes of care for water in the community. The group is needed to coordinate activities and particularly to recruit opinion makers to the cause. They need to be free to use all possible means to achieve their aims. They need to devise approaches that fit in with the local culture, customs and even current fashions. If there is no tradition of reading newspapers, or if many in the population are illiterate then written messages or advertising will have little effect. If TV or radio are not available or have little appeal then these media will similarly not provide a means of shaping opinions. The existing local communication systems should be used, and if these are not appropriate for transmitting opinion changing material on coping with water scarcity the group may need to devise some new form of communication method (– a considerable challenge!). This may seem an extreme proposal, but we need to remember the issue is survival! If communities do not learn to cope with water scarcity in an adequate, locally appropriate and equitable, sustainable manner they will go out of existence!

Therefore there is a need to find ways to capture and retain, in the long term, public interest in, and willingness to contribute to finding ways to cope with water scarcity.

12.5 Conclusion

Education to develop routine attitudes to coping with water scarcity is needed at all levels of society. Commitment is required of the local community to encourage the development and ongoing promotion of education about coping with water scarcity to the whole community. Commitment is stressed because to sustain beneficial attitudes to water the educational programs must be continuous, not just one-off or short term activities. Commitment is also needed to provide the funding needed to support ongoing programs. Educational activities need to be aimed at all levels of the community and therefore a single activity cannot be successful. Education will not be successful if it is only conducted in the classroom. It must be integrated into all forms of popular culture and entertainment and become the responsibility of every member of the community.

Bibliography

Abernethy C L (ed) (2001) Intersectoral Management of River Basins. IWMI, Colombo

Abraham E, Tomasini D, Maccagno P (eds) (2003) Desertificación. Indicadores y Puntos de Referencia en America Latina y el Caribe. Zeta Editores, Mendoza, Argentina

Aertgeerts R, Angelakis A (eds) (2003) State of the ART REPORT: Health Risks in Aquifer Recharge Using Reclaimed Water. World Health Organization, Geneva, and WHO Regional Office for Europe, Copenhagen, Denmark

Agnew C, Anderson E (1992) Water Resources in the Arid Realm. Routledge, London

Aiken G R, Kuniansky E L (eds) (2002) Artificial Recharge (Workshop Proceedings, Sacramento, April 2–4) U.S. Geological Survey, Open-File Report 02-89. http://water.usgs.gov/ogw/pubs/ofr0289/

Allen R G, Pruitt W O, Businger J A, Fritschen L J, Jensen M E, Quinn FH (1996) Evaporation and transpiration. In: Wootton T P, et al. (eds) ASCE Handbook of Hydrology. American Soc. Civil Engineers, New York, pp. 125–252

Allen R G, Willardson L S, Frederiksen H D (1997) Water use definitions and their use for assessing the impacts of water conservation. In: de Jager J M, Vermes L P, Ragab R (eds) Sustainable Irrigation in Areas of Water Scarcity and Drought (Proc. ICID Workshop, Oxford). British Nat. Com. ICID, Oxford, pp. 72–81

Allen R G, Pereira L S, Raes D, Smith M (1998) Crop Evapotranspiration. Guidelines for Computing Crop Water Requirements. FAO Irrig Drain Pap 56, Rome

Allen R G, Wright J L, Pruitt W O, Pereira L S, Jensen M E (2007a) Water requirements. In: Hoffman G J, Evans R G, Jensen M E, Martin D L, Elliot R L (eds) Design and Operation of Farm Irrigation Systems (2nd Edition). ASABE, St. Joseph, MI, pp. 208–288

Allen R G, Tasumi M, Morse A, Trezza R, Wright J L, Bastiaanssen W, Kramber W, Lorite I, Robinson C W (2007b) Satellite-based energy balance for mapping evapotranspiration with internalized calibration (METRIC). – Aplications. J Irrig Drain Engng 133:395–406

Alley W M (1984) The palmer drought severity index: Limitations and assumptions. J Climate Appl Meteo 23:1100–1109

Alley W M, Reilly T E, Franke O L (1999) Sustainability of Ground-Water Resources. U.S. Geological Survey Circular 1186, Denver. http://pubs.usgs.gov/circ/circ1186/

Al-Marshudi A S (2001) Traditional irrigated agriculture in Oman: Operation and management of the aflaj system. Water Internat 26:259–264

Al-Nakshabandi G A, Saqqar M M, Shatanawi M R, Fayyad M, Al-Horani H (1997) Some environmental problems associated with use of treated wastewater for irrigation in Jordan. Agric Water Manag 34:81–94

AMS (2004) Meteorological drought. Bulletin American Meteorological Society 85 http://www.ametsoc.org/POLICY/droughtstatementfinal0304.html

Antón D, Delgado C D (2000) Sequía en un Mondo de Agua. Piriguazu Ed., Montevidéo, Uruguai, and Centro Interamericano de Recursos del Agua, Univ. Autónoma del Estado de Mexico, Toluca, Mexico

Aoki C, Memon M A, Mabuchi H (2006) Water and Wastewater Reuse. An Environmentally Sound Approach for Sustainable Urban Water Management. UNEP, IETC, Osaka

Arar A (1988) Background to treatment and use of sewage effluent. In: Pescod M B, Arar A (eds) Treatment and Use of Sewage Effluent for Irrigation. Butterworths, London, pp. 10–20

Archer S, Stokes C (2000) Stress, disturbance and change in rangeland ecosystems. In: Arnalds O, Archer S (eds) Rangeland Desertification. Kluwer, Dordrecht, pp. 17–38

Arnell N (ed) (1989) Human Influences on Hydrological Behaviour: an International Literature Survey. IHP III, Project 6.1, UNESCO, Paris

Arreguín-Cortés F I (1994) Efficient use of water in cities and industry. In: Garduño H, Arreguín-Cortés F (eds) Efficient Water Use. UNESCO Regional Office, Montevideo, Uruguay, pp. 61–91

Asano T, Burton F L, Tchobanoglous G (2006) Water Reuse: Issues, Technologies and Applications, Metcalf & Eddy, Inc., McGraw-Hill Book Co., New York

Ayars J E, Phene C J, Hutmacher R B, Davis K R, Schoneman R A, Vail S S, Mead R M (1999) Subsurface drip irrigation of row crops: a review of 15 years research at the water management research laboratory. Agric Water Manag 42:1–27

Ayers R S, Westcot D W (1985) Water Quality for Agriculture. FAO Irrigation and Drainage Paper 29 Rev. 1, FAO, Rome

Balabanis P, Peter D, Ghazi A, Tsogas M (eds) (1999) Mediterranean Desertification. Research Results and Policy Implications (Proc. Int. Conf., Crete, 1996). Directorate General for Research, European Commission, EUR 19303, Luxembourg

Bandaragoda D J (1998) Design and practice of water allocation rules: lessons from warabandi in Pakistan's Pujab. Research Report No. 17. IWMI, Colombo, Sri Lanka

Bastiaanssen W G M, Harshadeep N R (2005) Managing scarce water resources in Asia: The nature of the problem and can remote sensing help? Irrig Drain Syst 19:269–284

Berbel Vecino J, Gutiérrez Martín C (2004) Sustainability of European Irrigated Agriculture under Water Framework Directive and Agenda 2000. WADI: EVK1-CT-2000-00057. Directorate-General for Research, EUR 21220 EN, European Commission, Brussels

Berghuber K, Vogl C R (2005) Descripción y análisis de los puquios como tecnología adaptada para la irrigación en Nasca, Perú. Zonas Áridas 9:40–55

Berney O, Charman J, Kostov L, Minetti L, Stoutesdijk J, Tricoli D (2001) Small Dams and Weirs in Earth and Gabion Materials. AGL/MISC/32/2001, FAO, Rome

Biswas A K (1980) The environment and water development in the Third World. ASCE, J Water Resour Plann Mgmt Div 105:319–332

Blaikie P, Brookfield H (1987) Land Degradation and Society. London: Routledge

Bonaccorso B, Bordi I, Cancelliere A, Rossi G, Sutera A (2003) Spatial variability of drought: An analysis of the SPI in Sicily. Water Resour Manag 17:273–296

Bordi I, Fraedrich K, Gerstengarbe F-W, Werner P C, Sutera A (2004) Potential predictability of dry and wet periods: Sicily and Elbe-Basin (Germany). Theor Appl Climatol 77:125–138

Bordi I, Fraedrich K, Petitta M, Sutera A (2006) Large-scale assessment of drought variability based on NCEP/NCAR and ERA-40 Re-Analyses. Water Resour Manag 20:899–915

Bordi I, Sutera A (2007) Drought monitoring and forecasting at large scale. In: Rossi G, Vega T, Bonaccorso B (eds) Methods and Tools for Drought Analysis and Management. Springer, Dordrecht, pp. 3–27

Bos M G, Burton M A, Molden D J (2005) Irrigation and Drainage Performance Assessment. Practical Guidelines. CABI Publish., Wallingford

Botterill L C, Wilhite D A (eds) (2005) From Disaster Response to Risk Management. Australia's National Drought Policy. Springer, Dordrecht

Briassoulis H, Junti M, Wilson G (2003) Mediterranean Desertification. Framing the Policy Context. Directorate General for Research, European Commission, EUR 20731, Brussels

Burman R, Pochop L O (1994) Evaporation, Evapotranspiration and Climatic Data. Elsevier, Amsterdam

Burt C M, Clemmens A J, Strelkoff T S, Solomon K H, Bliesner R D, Hardy L A, Howell T A, Eisenhauer D E (1997) Irrigation performance measures: efficiency and uniformity. J Irrig Drain Engng 123:423–442

Cacciamani C, Morgillo A, Marchesi S, Pavan V (2007) Monitoring and forecasting drought on a regional scale: Emilia-Romagna Region. In: Rossi G, Vega T, Bonaccorso B (eds) Methods and Tools for Drought Analysis and Management. Springer, Dordrecht, pp. 29–48

Cai J B, Liu Y, Lei T W, Pereira L S (2007) Estimating reference evapotranspiration with the FAO Penman-Monteith equation using daily weather forecast messages. Agric For Meteo 145:22–35

Calera A, Jochum A M, Cuesta A, Montoro A, López P (2005) Irrigation management from space: Towards user-friendly products. Irrig Drain Syst 19:337–353

Cancelliere A, Di Mauro G, Bonaccorso B, Rossi G (2007) Drought forecasting using the Standardized Precipitation Index. Water Resour Manag 21:801–819

Chohin-Kuper A, Rieu T, Montginoul M (2003) Water policy reforms: pricing water, cost recovery, water demand and impact on agriculture. Lessons from the Mediterranean experience. Water Pricing Seminar, Agencia Catalana del Agua & World Bank Institute

Choukr-Allah R, Malcolm C V, Hamdy A (eds) (1996) Halophytes and Biosaline Agriculture. Marcel Dekker Inc., London

Clemmens A J, Molden D J (2007) Water uses and productivity of irrigation systems. Irrig Sci 25:247–261

Conover W J (1980) Practical Non Parametric Statistics. J Wiley & Sons, New York

Cordery I (1999) Long range forecasting of low rainfall. Int J Climatol 19:463–470

Cordery I (2003) A case for increased collection of water resources data. Australian J Water Resour 6:95–103

Cordery I (2004) Arid zone surface runoff in Australia – a significant resource. Aust J Water Resources 7:115–121

Cordery I, Cloke P S (2005) Monitoring for modelling reality and sound economics. Headwater2005, 6th Int Conference on Headwater Control, Bergen, Norway

Cordery I, Curtis B R (1985) Drought – a problem of definition. Hydrology & Water Resources Symposium, Inst Engrs Aust, Sydney

Cordery I, McCall M (2000) A model for forecasting drought from teleconnections. Water Resour Res 36:763–768

Correia F N (ed) (2004) Desertificação em Portugal. Incidência no Ordenamento do Território e no Desenvolvimento Urbano. DGOTDU, Lisbon (2 vol)

Courault D, Seguin B, Olioso A (2005) Review on estimation of evapotranspiration from remote sensing data: From empirical to numerical modeling approaches. Irrig Drain Syst 19:223–249

Critchley W, Reij C P, Turner S D (1992) Soil and Water Conservation in Sub-Saharan Africa. IFAD, Rome

Critchley W, Siegert K (1991) Water Harvesting. FAO AGL/MISC/17/91, FAO, Rome

Davis S N, DeWiest, R J M (1966) Hydrogeology, J. Wiley and Sons, New York

Dalton P, Raine S, Broadfoot K (2001) Best management practices for maximising whole farm irrigation efficiency in the cotton industry. Final Report for CRDC Project NEC2C. National Centre for Engineering in Agriculture Publication 179707/2, USQ, Toowoomba

Davis S N, DeWiest R J M (1966) Hydrogeology, J. Wiley and Sons, New York

Deloncle A, Berk R, D'Andrea F, Ghil M (2007) Weather regime prediction using statistical learning. J Atmos Sci 64:1619–1635

Demissie M, Stout G E (eds) (1988) The State-of-the-Art of Hydrology and Hydrogeology in the Arid and Semi-Arid Areas of Africa (Proc. Sahel Forum, Ouagadougou, Burkina Fasso). UNESCO, Paris

Depeweg H (1999) Off-farm conveyance and distribution systems. In: van Lier H N, Pereira L S, Steiner F R (eds) CIGR Handbook of Agricultural Engineering, vol. I: Land and Water Engineering, ASAE, St. Joseph, MI, pp. 484–506

Dracup J A, Lee K S, Paulson E D (1980) On the definition of droughts. Water Resour Res 16:297–302

Droogers P, Seckler D, Makin I (2001) Estimating the potential of rain-fed agriculture. Working Paper 20. IWMI, Colombo, Sri Lanka

Duarte E A, Reis I B, Fragoso R, Martins M, Laranjeira J (2005) How the planning and efficient management can answer to the new environmental challenges? A detergent industry practical experience. In: III International Conference on the Efficient Use and Management of Urban Water. Int. Water Assoc (IWA), pp. 78–84

Edwards D (2000) SPI defined. In: http://ulysses.atmos.colostate.edu/SPI.html

Edwards C A, Lal R, Madden P, Miller R H, House G (eds) (1990) Sustainable Agircultural Systems. Soil and Water Conservation Society, Ankeny, and St. Lucie Press, Delray Beach, FL

El Amami H, Zairi A, Pereira L S, Machado T, Slatni A, Rodrigues P N (2001) Deficit irrigation of cereals and horticultural crops. 2. Economic analysis. Agricultural Enginnering International (www.agen.tamu.edu/cigr/), Vol. III, Manuscript LW 00 007b

El-Sayed Y M (2007) The rising potential of competitive solar desalination. Desalination 216:314–324

Engelman R, Leroy P (1993) Sustaining Water: Population and the Future of Renewable Water Supplies. Population and Environment Program, Washington DC

English M, Raja S N (1996) Perspectives on deficit irrigation. Agric Water Manag 32:1–14

Ennabli N (1993) Les aménagements hydrauliques et hydro-agricoles en Tunisie. INAT, Tunis

Enne G, Peter D, Pottier D (eds) (2001) Desertification Convention: Data and Information Requirements for Interdisciplinary Research (Proc. Int. Workshop, Alghero, 1999). Directorate General for Research, European Commission, EUR 19496, Luxembourg

Enne G, Peter D, Zanolla C, Zucca C (eds) (2004) The MEDRAP Concerted Action to Support the Northern Mediterranean Action Programme to Combat Desertification, Nucleo Ricerca Desertificazione, University of Sassari, Italy

Enne G, Zanolla C, Peter D (eds) (2000) Desertification in Europe: Mitigation Strategies, Land-Use Planning (Proc. Adv. Study Course, Alghero, 1999). Directorate General for Research, European Commission, EUR 19390, Luxembourg

Enne G, Zucca C (2000) Desertification Indicators for the European Mediterranean Region. Nucleo Ricerca Desertificazione, University of Sassari, and ANPA, Ministero dell'Ambiente, Roma

Falkland A C, Brunel J P (1993) Review of hydrology and water resources in humid tropical islands. In: Bonell M, Hufschmidt M M, Gladwell J S (eds) Hydrology and Water Management in the Humid Tropics, Cambridge University Press, pp. 135–163

FAO (1990) An International Action Programme on Water and Sustainable Agricultural Development. FAO, Rome

FAO (1994) Water Harvesting for Improved Agricultural Production. FAO Water Report 3, FAO, Rome

FAO (1997) Seawater Intrusion in Coastal Aquifers- Guidelines for Study, Monitoring and Control. Water Reports 11, FAO, Rome

FAO (1999) The Future of Our Land: Facing the Challenge. Guidelines for Integrated Planning for Sustainable Management of Land Resources. FAO and UNEP, Rome

FAO (2000) Manual on Integrated Soil Management and Conservation Practices. FAO Land and Water Bulletin 8, Rome

FAO (2002) Land–Water Linkages in Rural Watersheds. FAO Land and Water Bulletin 9, Rome

FogQuest (2004) 3rd International Conference on Fog, Fog Collection and Dew (11–15 October 2004 Cape Town, South Africa). FogQuest, Kamloops BC, Canada

Fortes P S, Platonov A E, Pereira L S (2005) GISAREG – A GIS based irrigation scheduling simulation model to support improved water use. Agric Water Manage 77:159–179

Foster S, Loucks D P (eds) (2006) Non-renewable Groundwater Resources. A Guidebook on Socially-Sustainable Management for Water-Policy Makers. IHP-VI Series on Groundwater No. 10, UNESCO, Paris

Frederick K D (1993) Balancing Water Demands with Supplies: The Role of Management in a World of Increasing Scarcity. Technical Paper No. 189, The World Bank, Washington, DC

FUNASA (2007) Técnicas agrícolas para contenção de solo e água. UFCG, Campina Grande

Gale I (ed) (2005) Strategies for Managed Aquifer Recharge (MAR) in Semi-arid Areas. UNESCO IHP/2005/GW/MAR, Paris

Galloway D L, Jones D R, Ingebritsen S E (eds) (1999) Land Subsidence in the United States. U.S. Geological Survey Circular 1182, 177p

Gandhidasan P, Abualhamayel H I (2007) Fog collection as a source of fresh water supply in the Kingdom of Saudi Arabia. Water Environ J 21:19–25

Garatuza-Payan J, Watts C J (2005) The use of remote sensing for estimating ET of irrigated wheat and cotton in Northwest Mexico. Irrig Drain Syst 19:301–320

Garduño H, Arreguín-Cortés F (eds) (1994) Efficient Water Use. UNESCO Regional Office ROSTLAC, Montevideo, Uruguay

Garen D C (1993) Revised surface-water supply index for western United States. J Water Resour Plann Manage 119:437–454

Gaudin S (2006) Effect of price information on residential water demand. Appl Econom 38:383–393

Geeson N A, Brandt C J, Thornes J B (eds) (2002) Mediterranean Desertification. A Mosaic of Processes and Responses. John Wiley & Sons, Chicheste

Giordano M, Villholth K G (eds) (2007) The Agricultural Groundwater Revolution. Opportunities and Threats to Development. CABI Publishing, Wallingford

Gleick P (ed) (2002) The World's Water 2002–2003. Island Press, Washington

Gonçalves J M, Pereira L S, Fang S X, Dong B (2007) Modelling and multicriteria analysis of water saving scenarios for an irrigation district in the Upper Yellow River Basin. Agric Water Manage 94:93–108

Goosen M F A, Shayya W H (eds) (2000) Water Management, Purification & Conservation in Arid Climates. Technomics Publ. Co., Lancaster, Pennsylvania

Gordon N D, McMahon T A, Finlayson B L (1992) Stream Hydrology. An Introduction to Ecologists. John Wiley & Sons, Chichester

Goussard J (1996) Interaction between water delivery and irrigation scheduling. In: M. Smith et al. (eds) Irrigation Scheduling: From Theory to Practice. FAO Water Reports 8, ICID and FAO, Rome, pp. 263–272

Grigg N S, Vlachos E C (eds) (1990) Drought Water Management (Proc. Nat. Workshop, Washington DC). Colorado State University, Fort Collins

Groenfeldt D, Svendsen M (eds) (2000) Case Studies in Participatory Irrigation Management. World Bank Institute, Washington DC

Guerrero-Salazar P, Yevjevich V (1975) Analysis of Drought Characteristics by the Theory of Runs. Hydrology Paper n 80, Colorado State University, Fort Collins

Gupta S K, Sharma S K, Tyagi N K (eds) (1998) Salinity Management in Agriculture. Central Soil Salinity Research Institute, Karnal, India

Guttman N B (1998) Comparing the Palmer drought index and the standardised precipitation index. J Am Water Resour Assoc 34:113–121

Haan C T (1977) Statistical Methods in Hydrology. Iowa State University Press, Ames

Haan C T, Barfield B J, Hayes J C (1994) Design Hydrology and Sedimentology for Small Catchments. Academic Press, San Diego, CA

Hamdy A, El-Gamal F, Lamaddalena N, Bogliotti C, Guelloubi R (eds) (2005) Non-Conventional Water Use. Options Méditerranéennes, Série B, N. 53, CIHEAM, Paris

Hamdy A, Karajeh F (eds) (1999) Marginal Water Management for Sustainable Agriculture. (Proc. Advanced Short Course, Aleppo, Syria), ICARDA, Aleppo and CIHEAM/IAM-B, Istituto Agronomico Mediterraneo, Bari

Hastenrath S, Greischar L (1993) Further work on the prediction of northeast Brazil rainfall anomalies. J Climate 6:743–758

Hatcho N (1998) Demand management by irrigation delivery scheduling. In: Pereira L S, Gowing J W (eds) Water and the Environment: Innovation Issues in Irrigation and Drainage, E &FN Spon, London, pp. 239–246

Hayes M J (2006) Drought Indices. http://drought.unl.edu/whatis/indices.htm

Hayes M J, Svoboda M, Le Comte D, Redmond K T, Pasteris P (2005) Drought monitoring: new tools for the 21st century. In: Wilhite D A (ed) Drought and Water Crisis. Science, Technology, and Management Issues, Taylor & Francis, Bocca Raton, pp. 53–69

Heath R C (1983) Basic Ground-water Hydrology. USGS, Water Supply Paper 2220. http://pubs.er.usgs.gov/usgspubs/wsp/wsp2220

Heathcote R L (1969) Drought in Australia: a problem of perception. The Geograph Rev 59:174–194

Heim R R (2002) A review of twentieth-century drought indices used in the United States. Bull Amer Meteor Soc 83:149–1165

Helsel D R, Hirsch R M (1992) Statistical Methods in Water Resources. Elsevier, Amsterdam

Hennessy J (1993) Water Management in the 21st Century. Keynote Address, 15th ICID Congress, The Hague, ICID, New Delhi

Hirsch R M (1982) A comparison of four streamflow record extension techniques. Water Resour Res 18:1081–1088

Hoffman G J, Evans R G, Jensen M E, Martin D L, Elliot R L (eds) (2007) Design and Operation of Farm Irrigation Systems (2nd Edition). ASABE, St. Joseph, MI

Hoffman G J, Shalhevet J (2007) Controlling salinity. In: Hoffman G J, Evans R G, Jensen M E, Martin D L, Elliot R L (eds) Design and Operation of Farm Irrigation Systems (2nd Edition). ASABE, St. Joseph, MI, pp. 160–207

Horst M G, Shamutalov Sh Sh, Gonçalves J M, Pereira L S (2007) Assessing impacts of surge-flow irrigation on water saving and productivity of cotton. Agric Water Manage 87:115–127

Hufschmidt M M, Kindler J (1991) Approaches to Integrated Water Resources Management in Humid Tropical and Arid and Semiarid Zones in Developing Countries. UNESCO Technical Documents in Hydrology, UNESCO, Paris

Huisman L, Olsthoorn T N (1983) Artificial Groundwater Recharge. Pitman, Boston, 320p

Iacovides I (1994) Artificial Ground Water Recharge. Mission to Libya. UNDP Project LIB/002 Water Resources Development, UNDP, New York

Iacovides I (2001) Information Systems for Water Resources Management. WMO Bulletin 50(1), Hydrology in the Service of Water Management, WMO, Geneva

ICRC (1998) Forum: War and Water. International Committee of the Red Cross, Geneva

IGRAC (2004) Global Groundwater Regions. http://igrac.nitg.tno.nl/pics/region.pdf lvOct05

IPCC (2007) Climate Change Impacts, Adaptation and Vulnerability: Summary for Policymakers. http://www.ipcc.ch/SPM6avr07.pdf

Irrigation Association (2005a) Landscape irrigation scheduling and water management. Report by Water Management Committee. In: McCabe J, Ossa J, Allen R G, Carleton B, Carruthers B, Corcos C, Howell T A, Marlow R, Mecham B, Spofford T L (eds), http://www.irrigation.org/

Irrigation Association (2005b) Turf and landscape irrigation: best management practices. Report by Water Management Committee. In: McCabe J, Ossa J, Allen R G, Carleton B, Carruthers B, Corcos C, Howell T A, Marlow R, Mecham B, Spofford T L (eds), http://www.irrigation.org/

Jensen M E (1996) Irrigated agriculture at the crossroads. In: Pereira L S, Feddes R A, Gilley J R, Lesaffre B (eds) Sustainability of Irrigated Agriculture. Kluwer Academic Publishers, Dordrecht, pp. 19–33

Jensen M E (1993) The Impacts of Irrigation and Drainage on the Environment. 5th Gulhati Memorial Lecture (15th ICID Congress, The Hague) ICID, New Delhi

Jensen M E (2007) Beyond irrigation efficiency. Irrig Sci 25:233–245

Jones K R, Berney O, Carr D P, Barrett E C (1981) Arid Zone Hydrology. FAO Irrig Drain Pap 37, Rome

Kaiser H M, Drennen T E (1993) Agricultural Dimensions of Global Climate Change. St Lucie Press, Delray Beach, Florida

Kalma J D, Franks S W (2000) Rainfall in Arid and Semi-Arid Regions. The University of Newcastle, Callaghan, Australia

Kandiah A (ed) (1990) Water, Soil and Crop Management Relating to the Use of Saline Water. FAO AGL/MISC/16/90, FAO, Rome

Kanwar R S (1996) Agrochemicals and Water Management. In: Pereira L S, Feddes R A, Gilley J R, Lesaffre B (eds) Sustainability of Irrigated Agriculture. Kluwer Academic Publishers, Dordrecht, pp. 373–393

Karl T R (1986) The sensitivity of the Palmer Drought Severity Index and Palmer's Z index to their calibration coefficients including potential evapotranspiration. J Climate Appl Meteo 25:77–86

Karl T R, Quinlan F, Ezell D Z (1987) Drought termination and amelioration: Its climatological probability. J Climate Appl Meteo 26:1198–1209

Keller J, Bliesner R D (1990) Sprinkler and Trickle Irrigation. Van Nostrand Reinhold, New York

Kepner W G, Rubio J L, Mouat D A, Pedrazzini F (eds) (2006) Desertification in the Mediterranean Region: a Security Issue. NATO Sc.Com., AK/Nato Publishing Unit, Springer-Verlag, Dordrecht

Khawaji A D, Kutubkhanah I K, Wie J-M (2008) Advances in seawater desalination technologies. Desalination 221:47–69

Khouri J, Amer A, Salih A (eds) (1995) Rainfall Water Management in the Arab Region. UNESCO/ROSTAS, Cairo

Kijne J W, Barker R, Molden D (eds) (2003) Water Productivity in Agriculture: Limits and Opportunities for Improvement, IWMI and CABI Publ., Wallingford

Kim T W, Valdés J B, Nijssen B, Roncayolo D (2006) Quantification of linkages between large-scale climatic patterns and precipitation in the Colorado River Basin. J Hydrology 321:173–186

Kite G W (1988) Frequency and Risk Analysis in Hydrology. Water Resources Publications, Fort Collins, Colorado

Kondrashov D, Ide K, Ghil M (2004) Weather regimes and preferred transition paths in a three-level quasigeostrophic model. J Atmos Sci 61:568–587

Kosmas C, Kirkby M, Geeson N (eds) (1999) The Medalus project: Mediterranean desertification and land use. Manual on key indicators of desertification and mapping environmentally sensitive areas to desertification. Office for Official Publications of the European Communities, Luxembourg

Kottegoda N T (1980) Stochastic Water Resources Technology. The Macmillan Press, London

Kovalevsky V S, Kruseman G P, Rushton K R (eds) (2004) Groundwater Studies: An International Guide for Hydrogeological Investigations. IHP-VI, Series on Groundwater No. 3, UNESCO, Paris

Kozlovsky E A (ed) (1984) Hydrogeological Principles of Groundwater Protection, UNESCO and UNEP, Moscow

Kundzewicz Z W, Robson A (eds) (2000) Detecting Trend and other Changes in Hydrological Data, WCDMP-45 (WMO/TD-No. 1013), UNESCO and WMO, Geneva

Kundzewicz Z W, Robson A J (2004) Change detection in hydrological records – a review of the methodology. Hydrol Sci J 49:7–19

Labat D (2006) Oscillations in land surface hydrological cycle. Earth Planet Sci Lett 242:143–154

Lamaddalena N, Sagardoy J A (2000) Performance Analysis of On-Demand Pressurized Irrigation Systems. FAO Irrigation and Drainage Paper 59, FAO, Rome

Lazarova V (2000) Wastewater disinfection: assessment of available technologies for water reclamation. In: Goosen M F A, Shayya W H (eds) Water Management, Purification and Conservation in Arid Climates. Technomic Publ. Co., Lancaster, Penn., pp. 171–198

Lieth H, Al Masoom A (1993) Towards the Rational Use of High Salinity Tolerant Plants, Kluwer Academic Publ., Dordrecht

Lin X (1999) Flash Floods in Arid and Semi-Arid Zones. IHP-V, Technical Documents in Hydrology, n 23, UNESCO, Paris

Lindemann J H (2004) Wind and solar powered seawater desalination. Applied solutions for the Mediterranean, the Middle East and the Gulf Countries. Desalination 168:73–80

Liniger H, Critchley W (eds) (2007) Where the Land Is Greener – Case Studies and Analysis of Soil and Water Conservation Initiatives Worldwide. WOCAT, Bern. http://www.wocat.org/

Liu Y, Pereira L S, Fernando R M (2006) Fluxes through the bottom boundary of the root zone in silty soils: Parametric approaches to estimate groundwater contribution and percolation. Agric Water Manage 84:27–40

Llerena C A, Invar M, Benavides M A (eds) (2004) Conservación y Abandono de Andenes. Univ. Nacional Agraria La Molina, Lima

Lohani V K, Loganathan G V, Mostaghimi S (1998) Long-term analysis and short-term forecasting of dry spells by the Palmer drought severity index. Nordic Hydrol 29:21–40

Lohani V K, Loganathan G V (1997) An early warning system for drought management using the Palmer drought index. J Am Water Resour Assoc 33:1375–1386

Louro V (ed) (2004) Desertificação. Sinais, Dinâmica e Sociedade. Instituto Piaget, Lisboa

Ma W, Zhao Y Q, Wang L (2007) The pretreatment with enhanced coagulation and a UF membrane for seawater desalination with reverse osmosis. Desalination 203:256–259

Maidment D R (ed) (1993) Handbook of Hydrology. McGraw-Hill, New York

Manel K (ed) (2001) On-Site Wastewater Treatment. (Proc. Ninth Nat. Symp. On Individual and Small Community Sewage Systems, Fort Worth, Texas). ASAE, St. Joseph, MI

Mantovani E C, Villalobos F J, Orgaz F, Fereres E (1995) Modelling the effects of sprinkler irrigation uniformity on crop yield. Agric Water Manage 27:243–257

Mao Z, Dong B, Pereira L S (2004) Assessment and water saving issues for Ningxia paddies, upper Yellow River Basin. Paddy Water Environ 2:99–110

Mara D, Cairncross S (1989) Guidelines for the Safe Use of Wastewater and Excreta in Agriculture and Aquaculture. WHO and UNEP, Geneva

Margat J, Saad K F (1984) Deep-lying aquifers: water mines under the desert? Nat Resour 20(2): 7–13

Margeta J (1987) Water Resources Development of Small Mediterranean Islands and Isolated Coastal Areas. MAP Technical Reports, Nr. 12, Split, Croatia

Margeta J, Iacovides I, Azzopardi E (1997) Integrated Approach to Development, Management and Use of Water Resources. PAP/RAC – MAP, Split, Croatia

Margeta J, Iacovides I, Sevener M, Azzopardi E (1999) Guidelines for Integrated Coastal Urban Water System Planning in the Mediterranean. PAP/RAC, Split, Croatia

Martin de Santa Olalla F (ed) (2001) Agricultura y Desertificación. Ediciones Mundi-Prensa, Madrid

Martinéz T, Palerm J, Pereira L S, Castro M (eds) (2009) Riegos Ancestrales de Iberoamerica, Editorial Mundiprensa, Barcelona, and Colegio de Postgraduados, Montecillo, Texcoco, Mexico

Martínez Beltrán J, Koo-Oshima S (eds) (2006) Water Desalination for Agricultural Applications. Land and Water Discussion Paper 5, FAO, Rome

Mathioulakis E, Belessiotis V, Delyannis E (2007) Desalination by using alternative energy: Review and state-of-the-art. Desalination 203:346–365

McCabe G J, Legates D R (1995) Relationships between 700 hPa height anomalies and 1 April snowpack accumulations in the western USA. Int J Climatol 15:517–530

McKee T B, Doesken N J, Kleist J (1993) The relationship of drought frequency and duration to time scales. In: 8th Conference on Applied Climatology. Am. Meteor. Soc., Boston, pp. 179–184

McKee T B, Doesken N J, Kleist J (1995) Drought monitoring with multiple time scales. In: 9th Conference on Applied Climatology, Am. Meteor. Soc., Boston, pp. 233–236

McMahon T A (1993) Hydrologic design for water use. In: Maidment D R (ed) Handbook of Hydrology, McGraw-Hill, New York

Minhas P S (1996) Saline water management for irrigation in India. Agric Water Manage 38:1–24

Minhas P S, Tyagi N K (1998) Guidelines for Irrigation with Saline and Alkali Waters. Central Soil Salinity Research Institute, Karnal, India

Mishra A K, Desai V R (2006) Drought forecasting using feed-forward recursive neural network. Ecol Modell 198:127–138

Missaoui H (1996) Soil and water conservation in Tunisia. In: Pereira L S, Feddes R A, Gilley J R, Lesaffre B (eds) Sustainability of Irrigated Agriculture. Kluwer Academic Publishers, Dordrecht, pp. 121–135

Molden D (2007) Water for Food, Water for Life: A Comprehensive Assessment of Water Management in Agriculture. Earthscan, London, and IWMI, Colombo

Molden D, de Fraiture C (2004) Investing in Water for Food, Ecosystems and Livelihoods. Comprehensive Assessment of Water Management in Agriculture, IWMI, Colombo, Sri Lanka

Molden D, Sakthivadivel R, Habib Z (2001) Basin-level use and productivity of water: Examples from South Asia. Research Report 49. IWMI, Colombo, Sri Lanka

Molden D, Murray-Rust H, Sakthivadivel R, Makin I (2003) A water-productivity framework for understanding and action. In: Kijne J W, Barker R, Molden D (eds) Water Productivity in Agriculture: Limits and Opportunities for Improvement, IWMI and CABI Publ., Wallingford, pp. 1–18

Moreira E E, Paulo A A, Pereira L S, Mexia J T (2006) Analysis of SPI drought class transitions using loglinear models. J Hydrol 331:349–359

Moreira E E, Paulo A A, Pereira L S (2007) Is drought occurrence and severity increasing due to climate change? Analysing drought class transitions with loglinear models. In: Rossi G, Vega T, Bonaccorso B (eds) Methods and Tools for Drought Analysis and Management. Springer, Dordrecht, pp. 67–81

Moreira E E, Coelho C A, Paulo A A, Pereira L S, Mexia J T (2008) SPI-based drought category prediction using loglinear models. J Hydrol 354,116–130

Mourits L J M, Salih A M A, Sherif M M (1997) Wadi Hydrology and Groundwater Protection (Proc. UNESCO/NWRC/ACSAD Workshops) IHP-V, Technical Documents in Hydrology, No. 1, UNESCO Cairo Office, Cairo

NAS (1974) More Water for Arid Lands. Promising Technologies and Research Opportunities. National Academy of Sciences, Washington, DC

NDMC (2005) National Drought Monitoring Center. http://www.drought.unl.edu/

NDMC (2006) What is drought? National Drought Mitigation Center. http://drought.unl.edu/whatis/what.htm

Nicholls N, Lavery B (1992) Australian rainfall trends during the twentieth century. Int J Climatol 12:153–163

Nichols N, Coughlan M J, Monnik K (2005) The challenge of climate prediction in mitigating drought impacts. In: Wilhite D A (ed) Drought and Water Crisis. Science, Technology, and Management Issues, Taylor & Francis, Bocca Raton, pp. 33–51

Nicholson S E (1993) An overview of African rainfall fluctuations in the last decade. J Climate 6:1463–1466

NRC (1991) Toward Sustainability: A Plan for Collaborative Research on Agricultural and Natural Resource Management. National Research Council, National Academy Press, Washington, DC

NRC (1996) A New Era for Irrigation. National Research Council, National Academy Press, Washington, DC, 203pp

NSSTC (2003) NASA working to take the guesswork out of long-term drought prediction. http://www.msfc.nasa.gov/news/NSSTC/news/releases/2003/N03-008.html

NWRC (2000) Wadi Hydrology (Proc. International Conference UNESCO and NWRC, Sharm El-Sheik, Egypt). National Water Research Center, Cairo

Okun D A (2000) The conservation of water through wastewater reclamation and urban nonpotable reuse. In: Goosen M F A, Shayya W H (eds) Water Management, Purification & Conservation in Arid Climates. Technomics Publ. Co., Lancaster, Pennsylvania, pp. 29–58

Oron G, Campos C, Gillerman L, Salgot M (1999) Wastewater treatment, renovation and reuse for agricultural irrigation in small communities. Agric Water Manage 38:223–234

Ortega-Guerrero A Rudolph D L, Cherry J A (1999) Analysis of long-term land subsidence near Mexico City: Field investigations and predictive modeling. Water Resour Res 35:3327–3342

Otchet A (2000) Black and blue, Libya's liquid legacy. The Courier, UNESCO

Oweis T, Hachum A, Kijne J (1999) Water Harvesting and Supplemental Irrigation for Improved Water Use Efficiency in Dry Areas. ICARDA and IWMI, SWIM paper 7, IWMI, Colombo, Sri Lanka

Oweis T, Prinz D, Hachum A (2004) Indigenous Water Harvesting Systems in West Asia and North Africa. ICARDA, Aleppo, Syria

Paço T A, Ferreira M I, Conceição N (2006) Peach orchard evapotranspiration in a sandy soil: Comparison between eddy covariance measurements and estimates by the FAO 56 approach. Agric Water Manage 85:305–313

Palmer W (1965) Meteorological Drought. U.S. Weather Bureau, Res. Paper N0 45, Washington, DC

Paulo A A, Pereira L S, Matias P G (2003) Analysis of local and regional droughts in southern Portugal using the theory of runs and the Standardized Precipitation Index. In: Rossi G, Cancelliere A, Pereira L S, Oweis T, Shatanawi M, Zairi A (eds) Tools for Drought Mitigation in Mediterranean Regions, Kluwer, Dordrecht, pp. 55–78

Paulo A A, Pereira L S (2006) Drought concepts and characterization: Comparing drought indices applied at local and regional Scales. Water Internat 31:37–49

Paulo A A, Pereira L S (2007) Prediction of SPI drought class transitions using Markov chains. Water Resour Manage 21:1813–1827

Paulo A A, Pereira L S (2008) Stochastic prediction of drought class transitions. Water Resour Manage. doi: 10.1007/s11269-007-9225-5

Pereira L S (1999) Higher performances through combined improvements in irrigation methods and scheduling: a discussion. Agric Water Manage 40:153–169

Pereira L S, Trout T J (1999) Irrigation methods. In: van Lier H N, Pereira L S, Steiner F R (eds) CIGR Handbook of Agricultural Engineering, vol. I: Land and Water Engineering, ASAE, St. Joseph, MI, pp. 297–379

Pereira L S, Dargouth M S, Klostermeyer W C, Lotti C, Mancel J, Omolukum A O, Oulad-Cherif B (eds) (1990) The Role of Irrigation in Mitigating the Effects of Drought (Trans. 14th ICID Congress, Rio de Janeiro). ICID, New Delhi

Pereira L S, van den Broek B, Kabat P, Allen R G (eds) (1995) Crop-Water Simulation Models in Practice. Wageningen Pers, Wageningen, DC

Pereira L S, Feddes R A, Gilley J R, Lesaffre B (eds) (1996) Sustainability of Irrigated Agriculture. Kluwer Academic Publishers, Dordrecht

Pereira L S, Cordery I, Iacovides I (2002a) Coping with Water Scarcity. UNESCO IHP VI, Technical Documents in Hydrology No. 58, UNESCO, Paris

Pereira L S, Oweis T, Zairi A, (2002b) Irrigation management under water scarcity. Agric Water Manage 57:175–206

Pereira L S, Cai L G, Hann M J (2003a) Farm water and soil management for improved water use in the North China Plain. Irrig Drain 52:299–317

Pereira L S, Cai L G, Musy A, Minhas P S (eds) (2003b) Water Savings in the Yellow River Basin. Issues and Decision Support Tools in Irrigation. China Agriculture Press, Beijing

Pereira L S, Teodoro P R, Rodrigues P N, Teixeira J L (2003c) Irrigation scheduling simulation: the model ISAREG. In: Rossi G, Cancelliere A, Pereira L S, Oweis T, Shatanawi M, Zairi A (eds) Tools for Drought Mitigation in Mediterranean Regions. Kluwer, Dordrecht, pp. 161–180

Pereira L S, Calejo M J, Lamaddalena N, Douieb A, Bounoua R (2003d) Design and performance analysis of low pressure irrigation distribution systems. Irrig Drain Syst 17:305–324

Pereira L S, Dukhovny V A, Horst M G (eds) (2005) Irrigation Management for Combating Desertification in the Aral Sea. Assessment and Tools. Vita Color, Tashkent (Russian version in paperback, English version in CD-ROM)

Pereira L S, Louro V, Rosário L, Almeida A (2006) Desertification, territory and people, a holistic approach in the Portuguese context. In: Kepner W G, Rubio J L, Mouat D A, Pedrazzini F (eds) Desertification in the Mediterranean Region: a Security Issue. NATO Sc.Com., AK/NATO Publishing Unit, Springer-Verlag, Dordrecht, pp. 269–289

Pereira L S, Gonçalves J M, Dong B, Mao Z, Fang S X (2007a) Assessing basin irrigation and scheduling strategies for saving irrigation water and controlling salinity in the Upper Yellow River Basin, China. Agric Water Manage 93:109–122

Pereira L S, Rosa R D, Paulo A A (2007b) Testing a modification of the Palmer drought severity index for Mediterranean environments. In: Rossi G, Vega T, Bonaccorso B (eds) Methods and Tools for Drought Analysis and Management. Springer, Dordrecht, pp. 149–167

Perez-Trejo F (1994) Desertification and Land Degradation in the European Mediterranean. Office for Official Publications of the European Communities, Luxembourg

Pescod M B (1992) Wastewater Treatment and Use in Agriculture. FAO Irrigation and Drainage Paper 47, FAO, Rome

Pessarakli M (ed) (1999) Handbook of Plant and Crop Stress (2nd ed.). Marcel Dekker, New York

Pewe T L (1990) Land subsidence and earth-fissure formation caused by groundwater withdrawal in Arizona; A review. In: Higgins C G, Coates D R (eds) Groundwater Geomorphology: The Role of Subsurface Water in Earth-Surface Processes and Landforms. Boulder, Colorado, Geological Society of America Special Paper 252, pp. 218–233

Pilgrim D H, Cordery I (1993) Flood runoff. In: Maidment D R (ed) Handbook of Hydrology, McGray-Hill, New York, pp. 9.1–9.42

Pitts D, Peterson K, Gilbert G, Fastenau R (1996) Field assessment of irrigation system performance. Appl Engng Agric 12:307–313

Plusquellec H, Ochs W (2003) Water Conservation: Irrigation. Technical Note F.2, The World Bank, Washington, DC

Popova Z, Eneva S, Pereira L S (2006a) Model validation, crop coefficients and yield response factors for maize irrigation scheduling based on long-term experiments. Biosyst Engng 95: 139–149

Popova Z, Kercheva M, Pereira L S (2006b) Validation of the FAO methodology for computing ETo with missing climatic data. Application to South Bulgaria. Irrig Drain 55:201–215

Prinz D (1996) Water harvesting – past and future. In: Pereira L S, Feddes R A, Gilley J R, Lesaffre B (eds) Sustainability of Irrigated Agriculture. Kluwer Acad. Publ., Dordrecht, pp. 137–168

Ragab R, Pearce G, Kim J C, Nairizi S, Hamdy A (eds) (2001) Wastewater Reuse Management (Proc. International Workshop, Seoul, Korea, ICID and CIGR), Korean Nat. Com. on Irrigation and Drainage, Seoul, Korea

Rangeley W R (1990) Irrigation at a Crossroads. 4th Gulhati Memorial Lecture (14th ICID Congress, Rio de Janeiro), ICID, New Delhi

Raziei T, Saghafian B, Paulo A A, Pereira L S, Bordi I (2008) Spatial patterns and temporal variability of drought in Western Iran. Water Resour Manage. doi 10.1007/s11269-008-9282-4

Refsgaard J C, Alley W M, Vuglinsky V S (1989) Methodology for Distinguishing between Man's Influence and Climatic Effects on the Hydrological Cycle. IHP III, Project 6.3, UNESCO, Paris

Reij C, Critchley W (1996) Sustainability of soil and water conservation in Sub-Saharan Africa. In: Pereira L S, Feddes R A, Gilley J R, Lesaffre B (eds) Sustainability of Irrigated Agriculture. Kluwer Acad. Publ., Dordrecht, pp. 107–119

Renault D, Facon T, Wahaj R (2007) Modernizing irrigation management – the MASSCOTE approach. Mapping System and Services for Canal Operation Techniques. Irrigation and Drainage Paper 63, FAO, Rome

Reynolds J E, Stafford Smith D M (eds) (2002) Global Desertification. Do Humans Cause Deserts? Dahlem University Press, Berlin

Rhoades J D (1990) Strategies to facilitate the use of saline water for irrigation. In: Kandiah, A. (ed.) Water, Soil and Crop Management Relating to the Use of Saline Water. FAO AGL/MISC/16/90, FAO, Rome, pp. 125–136

Rhoades J D, Chanduvi F, Lesch S (1999) Soil Salinity Assessment. Methods and Interpretation of Electrical Conductivity Measurements. FAO Irrigation and Drainage Paper 57, FAO, Rome

Rhoades J D, Kandiah A, Mashali A M (1992) The Use of Saline Waters for Crop Production. FAO Irrigation and Drainage Paper 48, FAO, Rome

Robertson A W, Ghil M (1999) Large-scale weather regimes and local climate over the western United States. J Climate 12:1796–1813

Rodier J A (1993) Paramètres charactéristiques des fortes crues dans les régions tropicales sèches. I-Coefficient de ruissellement. Hydrologie Continentale 8(2):139–160

Rodier J A (1994) Paramêtres charactéristiques des fortes crues dans les régions tropicales sèches. II-Fonction de transfer (temps de base, coefficient de pointe, temps charactéristique de base). Hydrologie Continentale 9(1):33–68

Rosegrant M W, Cai X, Cline S A (2002) Global Water Outlook to 2025. Averting an Impending Crisis. A 2020 Vision for Food, Agriculture, and the Environment Initiative. International Food Policy Research Institute, Washington, D.C., and International Water Management Institute. Colombo, Sri Lanka

Rosenzweig C, Hillel D (1998) Climate Change and the Global Harvest. Potential Impacts of the Greenhouse Effect on Agriculture. Oxford Univ Press, New York

Rossi G, Cancelliere A, Pereira L S, Oweis T, Shatanawi M, Zairi A (eds) (2003) Tools for Drought Mitigation in Mediterranean Regions. Kluwer, Dordrecht, 257p

Rossi G (2003) Requisites for a drought watch system. In: Rossi G, Cancellieri A, Pereira L S, Oweis T, Shatanawi M, Zairi A (eds) Tools for Drought Mitigation Mediterranean Regions, Kluwer, Dordrecht, pp. 147–157

Rossi G, Vega T, Bonaccorso B (eds) (2007a) Methods and Tools for Drought Analysis and Management. Springer, Dordrecht

Rossi G, Castiglione L, Bonaccorso B (2007b) Guidelines for planning and implementing drought mitigation measures. In: Rossi G, Vega T, Bonaccorso B (eds) Methods and Tools for Drought Analysis and Management. Springer, Dordrecht, pp. 325–347

Roxo M J, Mourão J M (eds) (1998) Desertificação. Instituto Mediterrânico, Universidade Nova de Lisboa, Lisboa

Ruiz J E, Cordery I, Sharma A (2005) Integrating ocean subsurface temperatures in statistical ENSO forecasts. J Climate 18:3571–3586

Ruiz J E, Cordery I, Sharma A (2006) Impact of mid-Pacific Ocean thermocline on the prediction of Australian rainfall. J Hydrol 317:104–122

Ruiz J E, Cordery I, Sharma A (2007) Forecasting streamflows in Australia using the tropical Indo-Pacific thermocline as predictor. J Hydrol 341:156–164

Rushton K R (1986) Surface water-groundwater interaction in irrigation schemes. Int. Assn Hydrol. Sc., Publ. No. 156, pp. 17–27

Saad K F, Khouri J, Al-Drouby A, Gedeon R, Salih A (1995) Groundwater Protection in the Arab Region. ACSAD and UNESCO. Paris and Cairo

Salas J D, Harboe R, Marco-Segura J (eds) (1993) Stochastic Hydrology and its Use in Water Resources Systems Simulation and Optimization, Kluwer, Dordrecht

Salas J D (1993) Analysis and modeling of hydrologic time series. In: Maidment D R (ed) Handbook of Hydrology. Ch. 19, McGraw-Hill, New York

Salvatore M, Pozzi F, Ataman E, Huddleston B, Bloise M (2005) Mapping Global Urban and Rural Population Distributions. Environment and Natural Resources Working Paper 24, FAO, Rome

Samra J S, Singh G, Daggar J C (2006) Drought Management Strategies in India. Indian Council of Agricultural Research, New Delhi

Sanderson M, Markova O, Nullet D, Ohmura A, Kuznetsova Z, Mather J R, Borzenkova I I, Budyko M I (1990) UNESCO Sourcebook in Climatology for Hydrologists and Water Resource Engineers UNESCO, Paris

Santos M A (1983) Regional droughts: a stochastic characterisation. J Hydrol 66:183–211

Santos F L (1996) Quality and maximum profit of industrial tomato as affected by distribution uniformity of drip irrigation system. Irrig Drain Syst 10:281–294

Santos J A, Corte-Real J, Leite S M (2005)Weather regimes and their connection to the winter rainfall in Portugal. Int J Climatol 25:33–50

Santos J, Corte-Real J, Leite S (2006) Atmospheric large-scale dynamics during the 2004/2005 winter drought in Portugal. Int J Climatol. doi 10.1002/joc.1425

Schemenauer R S, Cereceda P (1994b) Fog collection's role in water planning for developing countries. Natural Resources Forum, 18, 91-100, United Nations, New York

Schemenauer R S, Bridgman H (eds) (1998) Fog and Fog Collection (Proc. First Int. Conf., Vancouver, Canada), Conference on Fog and Fog Collection, North York, Ontario

Schemenauer R S, Cereceda P (1994a) A proposed standard fog collector for use in high elevation regions. J Appl Meteo 33:1313–1322

Seckler D, Upali A, Molden D, de Silva R, Barker R (1998) World Water Demand and Supply, 1990 to 2025: Scenarios and Issues. Research Report 19, IWMI. Colombo, Sri Lanka

Semiat R (2000) Desalination. Present and future. Water Internat 25:54–65

Shalhevet J (1994) Using water of marginal quality for crop production: major issues. Agric Water Manage 25:233–265

Sharma K D (2001) Rainwater harvesting and recycling. In: Goosen M F A, Shayya W H (eds) Water Management, Purification and Conservation in Arid Climates. Technomic Publ. Co., Lancaster, PA, pp. 59–86

Shaw E M (1983) Hydrology in Practice. Van Norstrand Reinhold, UK

Shiklomanov I A, Rodda J C (2003) World Water Resources at the Beginning of the 21st Century, UNESCO Int. Hydrology Series, Cambridge University Press (extended summary http://webworld.unesco.org/water/ihp/db/shiklomanov/summary/html/summary.html#6.%20 Anthropoge)

Shiklomanov I A (1998) World Water Resources. A New Appraisal and Assessment for the 21st Century. UNESCO, Paris

Shiklomanov I A (2000) Appraisal and assessment of world water resources. Water Internat 25: 11–32

Shin H S, Salas J D (2000) Regional drought analysis based on neural networks. J Hydrol Engng 5:145–155

Silva M A, et al. (2006) Barragem Subterrânea. EMBRAPA, Petrolina, Brasil

Simmers I (ed) (1997) Recharge of Phreatic Aquifers in Semi- Arid Areas. International Association of Hydrogeologists. A A Balkema, Rotterdam and Brookfield

Siregar P R (2003) World Bank and ADB's role in privatizing water in Asia Region. In: Asia Pacific Conference on Debt and Privatization of Water and Power Service (Jubilee South/APMDD, Bangkok, 8–12 December), Asian Dev Bank

Smakhtin V, Revenga C, Döll P (2004) Taking into Account Environmental Water Requirements in Global-scale Water Resources Assessments. Comprehensive Assessment Research Report 2, IWMI, Colombo, Sri Lanka

Smith M, Pereira L S, Berengena J, Itier B, Goussard J, Ragab R, Tollefson L, Van Hoffwegen P (eds) (1996) Irrigation Scheduling: From Theory to Practice. FAO Water Report 8, FAO, Rome

Sovocool K, Rosales J (2001) Turf and landscape irrigation. The X factor. Irrigation Business and Technology Sep-Oct:35–36

Spulber N, Sabbaghi A (1998) Economics of Water Resources: From Regulation to Privatization (2nd ed.). Kluwer Academic Publisher, Dordrecht

Stanger G (2000) Water conservation: what can we do? In: Goosen M F A, Shayya W H (eds) Water Management, Purification & Conservation in Arid Climates. Technomics Publ. Co., Lancaster, PA, pp. 1–28

Steduto P (1996) Water use efficiency. In: Pereira L S, Feddes R, Gilley J R, Lesaffre B (eds) Sustainability of Irrigated Agriculture, Kluwer, Dordrecht, pp. 193–209

Steinemann A C (2006) Using climate forecasts for drought management. J Appl Meteo Climat 45:1353–1361

Stewart J I (1988) Response Farming in Rainfed Agriculture. The WHARF Foundation Press, Davis, CA

Svendsen M (ed) (2005) Irrigation and River Basin Management. Options for Governance and Institutions. CABI Publishing, Wallingford

Tadesse T, Wilhite D A, Hayes M J, Harms S K, Goddard S (2005) Discovering associations between climatic and oceanic parameters to monitor drought in Nebraska using data-mining techniques. J Climate 18:1541–1550

Tanji K K, Kielen N C (2002) Agricultural Drainage Water Management in Arid and Semi-arid Areas. FAO Irrigation and Drainage Paper 61, Rome

Tarjuelo J M, de Juan J A (1999) Crop water management. In: van Lier H N, Pereira L S, Steiner F R (eds) CIGR Handbook of Agricultural Engineering, vol. I: Land and Water Engineering, ASAE, St. Joseph, MI, pp. 380–429

Tate E L, Gustard A (2000) Drought definition: a hydrological perspective. In: Vogt J J, Somma F (eds) Drought and Drought Mitigation in Europe. Kluwer, Dordrecht, pp. 23–48

The World Bank (1992) World Development Report 1992: Development and the Environment. Oxford University Press, New York

Thomas D (1998) Desertification and the CCD: issues and links to poverty, natural resources and policies. http://www.panrusa.group.shef.ac.uk/pdfs/WP1.pdf

Thomas T H, Martinson D B (2007) Roofwater Harvesting: A Handbook for Practitioners. IRC International Water and Sanitation Centre, Delft. http://www.irc.nl

Thyer M, Frost A J, Kuczera G (2006) Parameter estimation and model identification for stochastic models of annual hydrological data: Is the observed record long enough? J Hydrol 330:313–328

Tian L, Tang Y P, Wang Y Q (2007) Economic evaluation of seawater desalination for a nuclear heating reactor with multi-effect distillation. Desalination 180:53–61

Tiercelin J R, Vidal A (eds) (2006) Traité d'Irrigation, 2ème edition, Lavoisier, Technique & Documentation, Paris

Tiffen M (1989) Guidelines for Incorporation of Health Safeguards into Irrigation Projects through Intersectorial Cooperation. WHO/FAO/UNEP, Geneva

Topper R, Barkmann P, Bird D, Sares M (2004) Artificial Recharge of Ground Water in Colorado – A Statewide Assessment. Colorado Geological Survey, Denver

Trieb F, Müller-Steinhagen H (2008) Concentrating solar power for seawater desalination in the Middle East and North Africa. Desalination 220:165–183

Trigo R M, Pozo-Vázquez D, Osborn T J, Castro-Díez Y, Gámiz-Fortis S, Esteban-Parra M J (2004) North Atlantic oscillation influence on precipitation, river flow and water resources in the Iberian Peninsula. Int J Climat 24:925–944

Tsakiris G (ed) (2002) Water Resources Management in the Era of Transition (Proc. Int. Conf., Athens), European Water Res. Assoc., Athens

Tyagi N K (1996) Salinity management in irrigated agriculture. In. Pereira L S, Feddes R A, Gilley J R, Lesaffre B (eds) Sustainability of Irrigated Agriculture. Kluwer Academic Publisher, Dordrecht, pp. 345–358

Tyagi N K, Minhas P S (eds) (1998) Agricultural Salinity Management in India. Central Soil Salinity Research Institute, Karnal, India

UN (1975) Ground-water Storage and Artificial Recharge. Natural Resources/Water Series No. 2, Department of Economic and Social Affairs. United Nations, New York

UNCCD (1994) United Nations convention to combat desertification. UN, New York. http://www.unccd.int/convention/text/convention.php

UNCCD (1997) Comics to Combat Desertification. Secretariat of the UN Convention to Combat Desertification, Geneva

UNCED (1992) Managing fragile ecosystems. Combating desertification and drought. Chapter 12 in UN Conference on Environment and Development at Rio de Janeiro. UN, New York

UNDP (1990) Spate Irrigation. Proc. Expert Consultation UNDP and FAO (Aden, Yemen, 1987), UNDP, New York

UNDP (2006) Human Development Report 2006. Beyond Scarcity: Power, Poverty and the Global Water Crisis. Palgrave Macmillan, Houndmills, Basingstoke, Hampshire and New York

UNDP (2007) Human Development Report 2007/2008. Fighting Climate Change: Human Solidarity in a Divided World. Palgrave Macmillan, Houndmills, Basingstoke, Hampshire and New York

UNEP (2001) Seawater Desalination in Mediterranean Countries: Assessment of Environmental Impacts and Proposed Guidelines for the Management of Brine. Mediterranean Action Plan Meeting of the MED POL National Coordinators. (Venice, Italy, 28–31 May 2001), United Nations Environment Programme

UNEP (2002a) Global Environment Outlook GEO 3. Past, Present and Future Perspectives. UNEP, Nairobi and Earthscan Publications, London

UNEP (2002b) International Source Book on Environmentally Sound Technologies for Wastewater and Stormwater Management, UNEP, IETC, Osaka

UNEP (2002c) Rainwater Harvesting and Utilization. UNEP, IETC, Osaka

UNEP (2007) Global Environment Outlook GEO 4. Environment and Development. UNEP, Nairobi

UNESCO (1979) Map of the World Distribution of Arid Regions. MAB Technical Notes 7, UNESCO, Paris

UNESCO (1982) Methods of Hydrological Computations for Water Projects. Studies and Reports in Hydrology n 38, UNESCO, Paris

UNESCO (1987) Casebook of Methods for Computing Hydrological Parameters for Water Projects. Studies and Reports in Hydrology n 48, UNESCO, Paris

UNESCO (1997) Water Resources in the OSS Countries. Evaluation, Use and Management. International hydrologic Programme, UNESCO, Paris

UNESCO (1999) Integrated Drought Management: Lessons from Sub-Saharan Africa (Proc. Int. Conf., Pretoria, South Africa). IHP-V, Technical Documents in Hydrology No. 35, UNESCO

UNESCO (2000) Hydrological Research and Water Resources Management Strategies in Arid and Semi-Arid Zones (Proc. Int Symp., Tashkent, Uzbekistan, Sept. 1995). IHP-V, Technical Documents in Hydrology, No. 32, UNESCO, Paris

UNESCO (2006) Water: a Shared Responsibility. The United Nations World Water Development Report 2, UNESCO, Paris, and Berghahn Books, New York

Unger P W (1994) Tillage Systems for Soil and Water Conservation, FAO Soils Bull. 54, FAO, Rome

Unger P W, Howell T A (1999) Agricultural water conservation – a global perspective. In: Kirkham M B (ed) Water Use in Crop Production. The Haworth Press, New York, pp. 1–36

US National Assessment (2000) Draft Report of the Water Sector of the National Assessment of the Potential Consequences of Climate Variability and Change. http://www.nacc.ugcrp.gov/

USBR (1985) Ground Water Manual. A Guide for the Investigation, Development, and Management of Ground-Water Resources. US Bureau of Reclamation, USDI, US Gov. Print. Office, Washington, DC

USGS (1982) Basic Ground-Water Hydrology. United States Geological Survey, Water Supply Paper 2220. http://pubs.er.usgs.gov/usgspubs/wsp/wsp2220

USGS (1999) Sustainability of Ground-Water Resources. United States Geological Survey, Circular 1186

USGS, 2000. Land Subsidence in the United States. http://water.usgs.gov/ogw/pubs/fs00165/

Vakkilainen P, Varis O (1999) Will Water Be Enough, Will Food Be Enough? IHP-V, Technical Documents in Hydrology, No. 24, UNESCO, Paris

Van den Dool H M, Peng P, Johansson Å, Chelliah M, Shabbar A, Saha S (2006) Seasonal-to-decadal predictability and prediction of North American climate—the Atlantic influence. J Climate 19:6005–6024

van der Molen W H, Martínez Beltrán J, Ochs W J (2007) Guidelines and computer programs for the planning and design of land drainage systems. FAO Irrigation and Drainage Paper 62, FAO, Rome

van Lanen H A J (ed) (1998) Monitoring for Groundwater Management in Semi- Arid Regions. Studies and Reports in Hydrology 57, UNESCO, Paris

van Lier H N, Pereira L S, Steiner F R (eds) (1999) CIGR Handbook of Agricultural Engineering, vol. I: Land and Water Engineering, ASAE, St. Joseph, MI

Verdon D C, Wyatt A M, Kiem A S, Franks S W (2004) Multidecadal variability of rainfall and streamflow: Eastern Australia. Water Resour Res. 40. doi:10.1029/2004WR003234

Versteeg N, Tolboom J (2003) Water Conservation: Urban Utilities. Technical Note F.1, The World Bank, Washington, DC

Vicente-Serrano S M, González-Hidalgo J C, De Luis M, Raventós J (2004) Drought patterns in the Mediterranean area: The Valencia region (Eastern Spain). Climate Res 26:5–15

Vlachos E, James L D (1983) Drought Impacts. In: Yevjevich V, Cunha L V, Vlachos E (eds) Coping With Droughts. Water Resources Publications, Littleton, CO, pp. 44–73

Vogt J V, Somma F (eds) (2000) Drought and Drought Mitigation in Europe. Kluwer, Dordrecht

Vrba J, Lipponen A (eds) (2007) Groundwater Resources Sustainability Indicators. IHP-VI Series on Groundwater No. 14, UNESCO, Paris

Waggoner P E (1994) How Much Land Can Ten Billion People Spare for Nature? Council for Agricultural Science and Technology, The Rockfeller University, New York

WCED (1987) Our Common Future. World Commission for Environment and Development, Oxford University Press, New York

Wedgbrow C S, Wilby R L, Fox H R, O'Hare G (2002) Prospects for seasonal forecasting of summer drought and low river flow anomalies in England and Wales. Int J Climat 22:219–236

Westcot D W (1997) Quality Control of Wastewater for Irrigated Agriculture. FAO Water Report 10, FAO, Rome

Wheater H, Al-Weshah R A (eds) (2002) Hydrology of Wadi Systems. IHP-V, Technical Documents in Hydrology No. 55, UNESCO, Paris

WHO (2003) Emerging Issues in Water and Infectious Disease. World Health Organiz., Geneva

WHO (2004) Water, Sanitation and Hygiene Links to Health. Facts and Figures World Health Organiz., Geneva. http://www.who.int/water_sanitation_health/publications/facts2004/en

WHO (2006a) Guidelines for Drinking-water Quality, incorporating First Addendum. Vol. 1, Recommendations. 3rd ed. (Electronic version for the Web). World Health Organiz., Geneva. http://www.who.int/water_sanitation_health/dwq/gdwq3rev/en/

WHO (2006b) Meeting the MDG Drinking Water and Sanitation Target: the Urban and Rural Challenge of the Decade. WHO, Geneva. http://www.wssinfo.org/en/40_mdg2006.html

WHO (2006c) WHO Guidelines for the Safe Use of Wastewater, Excreta and Greywater. Volume I: Policy and Regulatory Aspects. World Health Organization, Geneva

WHO (2006d) WHO Guidelines for the Safe Use of Wastewater, Excreta and Greywater. Volume II: Wastewater Use in Agriculture. World Health Organization, Geneva

WHO (2006e) WHO Guidelines for the Safe Use of Wastewater, Excreta and Greywater. Volume III: Wastewater and Excreta Use in Aquaculture. World Health Organization, Geneva

WHO (2006f) WHO Guidelines for the Safe Use of Wastewater, Excreta and Greywater. Volume IV: Excreta and Greywater Use in Agriculture. World Health Organization, Geneva

Wilhite D A (ed) (2005a) Drought and Water Crises. Science, Technology, and Management. Taylor and Francis, Boca Raton

Wilhite D A (2005b) Drought policy and preparedness: the Australian experience in an international context. In: Botterill L C, Wilhite D A (eds) From Disaster Response to Risk Management: Australia's National Drought Policy. Springer, Dordrecht, pp. 157–183

Wilhite D A, Buchanan-Smith M (2005) Drought as hazard: Understanding the natural and social context. In: Wilhite D A (ed) Drought and Water Crises. Science, Technology, and Management. Taylor and Francis, Boca Raton. pp. 3–29

Wilhite D A, Glantz M H (1987) Understanding the drought phenomenon: The role of definitions. In: Wilhite D A, Easterling W E, Wood D A (eds) Planning for Drought. Vestview Press, Boulder, CO, pp. 11–27

Wilhite D A, Easterling W E, Wood D A (eds) (1987) Planning for Drought. Toward a Reduction of Societal Vulnerability. Westview Press, Boulder and London

Wilhite D A, Sivakumar M V K, Wood D A (eds) (2000) Early Warning Systems for Drought Preparedness and Drought Management. (Proc. Expert Group Meeting, Lisbon). World Meteor. Organiz., Geneva. http://drought.unl.edu/monitor/EWS/EWS_WMO.html

Wilkinson R E (ed) (2000) Plant - Environment Interactions. Marcel Dekker, New York

WMO (1994) Guide to Hydrological Practices – Data Acquisition and Processing, Analysis, Forecasting and other Applications. World Meteorological Organization, Geneva

WMO (1996) Water Resource Management and Desertification: Problems and Challenges. WMO, Geneva

WMO (2005) Climate and Land Degradation. WMO Publ. No. 989, Geneva

WMO (2006) Drought Monitoring and Early Warning: Concepts, Progress and Future Challenges. WMO Publ. No. 1006, Geneva

Wootton T P et al. (eds) (1996) Hydrology Handbook, ASCE Manuals and Reports on Engineering Practice No. 28, ASCE, New York

World Bank (2006) Water Management in Agriculture: Ten years of World Bank assistance, 1994–2004. The World Bank, Washington DC

Yassoglou N J (1998) Semi-natural environments and processes. Soil: Mediterranean soil types in a desertification context. In: Mairota P, Thornes J B, Geeson N (eds) Atlas of Mediterranean Environments in Europe: The Desertification Context from MEDALUS. John Wiley & Sons, Chichester

Yevjevich V, Cunha L V, Vlachos E (eds) (1983) Coping with Droughts, Water Resources Publications, Littleton, Colorado

Yevjevich V (2001) Water diversions and interbasin transfers. Water Internat 26:342–348

Zairi A, El Amami H, Slatni A, Pereira L S, Rodrigues P N, Machado T (2003) Coping with drought: deficit irrigation strategies for cereals and field horticultural crops in Central Tunisia. In: Rossi G, Cancelliere A, Pereira L S, Oweis T, Shatanawi M, Zairi A (eds) Tools for Drought Mitigation in Mediterranean Regions. Kluwer, Dordrecht, pp. 181–201

Zaporozec A, Miller J C (2000) Ground-Water Pollution. UNESCO IHP, Paris

Zardari N H, Cordery I (2007) The use of a multicriteria method in irrigation water allocations. Agric J 2:236–241

Index